Volume 17

CRM MONOGRAPH SERIES

Centre de Recherches Mathématiques
Université de Montréal

Acyclic Models

Michael Barr
Peter Redpath Emeritus Professor
of Pure Mathematics

American Mathematical Society
Providence, Rhode Island USA

The production of this volume was supported in part by the Fonds pour la Formation de Chercheurs et l'Aide à la Recherche (Fonds FCAR) and the Natural Sciences and Engineering Research Council of Canada (NSERC).

2000 *Mathematics Subject Classification.* Primary 18G60.

Library of Congress Cataloging-in-Publication Data

Barr, Michael.
Acyclic models / Michael Barr.
p. cm. — (CRM monograph series, ISSN 1065-8599 ; v. 17)
Includes bibliographical references and index.
ISBN 0-8218-2877-0 (alk. paper)
1. Acyclic models. I. Title. II. Series.

QA169.B344 2002
512′.55—dc21 2002021460

∞ The paper used in this book is acid-free and falls within the guidelines
established to ensure permanence and durability.
This volume was submitted to the American Mathematical Society
in camera ready form by the Centre de Recherches Mathématiques.
Visit the AMS home page at URL: http://www.ams.org/

10 9 8 7 6 5 4 3 2 1   07 06 05 04 03 02

For
Zachary, Madeline, Keenan, and Noah
and
all my other grandchildren yet unborn

Contents

Preface

During the late 1940s and the early 1950s cohomology theories had been defined in three different areas of abstract algebra. These were the theories of Hochschild [1945, 1946] for associative algebras, Eilenberg & Mac Lane [1947] for groups, and Eilenberg & Chevalley [1948] for Lie algebras. In part, these theories grew out of the fact that Eilenberg and Mac Lane had noticed, as early as 1939 or 1940, that certain computations involving extensions of groups greatly resembled certain computations in the cohomology theory of topological spaces.

The three definitions were similar but not identical and had a certain *ad hoc* aspect. Then Cartan and Eilenberg showed in their influential book [1956] that there was a uniform definition of homology and cohomology that united the three examples. Here is a brief description of their approach. Let $\mathscr{C}$ stand for one of the three categories groups, associative algebras, or Lie algebras and let Assoc be the fixed category of associative algebras. Then they defined a functor $\text{env}\colon \mathscr{C} \longrightarrow \mathsf{Assoc}$ (for enveloping algebra) with the property that for any object C of $\mathscr{C}$, a left module for the associative algebra $\text{env}(\mathscr{C})$ was the same thing as a coefficient module for the usual cohomology of $\mathscr{C}$. In the case of groups the functor assigned to each group its group ring. For an associative algebra A, the enveloping algebra is $A \otimes A^{\text{op}}$ and for a Lie algebras it was its usual enveloping algebra, which will be described in 6.5. In each case, the crucial fact was that the cohomology vanished when the coefficient module was injective. It became clear somewhat later, however, that the Cartan-Eilenberg approach was limited to those cohomology theories that vanished on injective modules, which was not generally the case.

Around 1960 my thesis supervisor, the late David K. Harrison, produced a cohomology theory for commutative algebras [1962]. Somewhat surprisingly, I showed by simple example that the cohomology did not necessarily vanish with injective coefficients, so that the Cartan–Eilenberg approach could not work in that case. The counter-example appeared as Barr [1968a]. (As an amusing sidelight, that note was actually written, submitted, refereed, and accepted for publication by Murray Gerstenhaber, then editor of the Bulletin of the AMS who thought, against my judgment, that it should be published.)

In 1962 I arrived at Columbia University as a newly minted Ph.D. There I found that Eilenberg had gathered round himself a remarkable collection of graduate students and young researchers interested in homological algebra, algebraic topology, and category theory. There may have been others, but the ones I recall are Harry Appelgate, Jon Beck, Peter Freyd Joe Johnson, Bill Lawvere, Fred Linton, Barry Mitchell, George Rinehart, and Myles Tierney. Unfortunately, Peter Freyd and Bill Lawvere were no longer in residence, but it was still a remarkable collection.

The Cartan-Eilenberg to cohomology can be characterized as cohomology computed via an injective resolution of the coefficient variable. Rinehart and I were interested in trying to define cohomology by resolving the algebra variable instead of the coefficient module. Our attempts Barr [1965a,b] and Barr & Rinehart [1966] were not particularly successful, although they did point to the central role of the derivations functor. In retrospect, one can say that the basic reason is that these categories of algebras are not abelian and that simplicial objects replace chain complexes in a non-abelian setting. I was not at that time at all aware of simplicial objects. That I learned from Jon Beck, who came to this subject from topology.

During the time I was at Columbia, Beck was writing his thesis (it was not presented until 1967, but there was a complete draft by 1964) defining and studying the essential properties of cotriple cohomology using a simplicial resolution that came from a cotriple. There were four (including the cohomology of commutative algebras) theories to compare it to, as described above. In all cases it was clear that the cotriple H^1 was equivalent to the Cartan–Eilenberg, respectively Harrison, H^2 since they both classified "singular" extensions. (There is always a dimension shift by one, which I will cease mentioning.) Beck and I spent much of the calendar year 1964—first at Columbia, then at the University of Illinois in Urbana—attempting to show that the cotriple cohomology was the same as the older ones in all degrees. This task seemed, at the time, hopeless. The cochain complexes that resulted from the theories were utterly different and we just had no handle on the question.

Then over the Christmas vacation, 1964, Beck went to New York and visited Appelgate, who was writing a thesis on acyclic models. He suggested that we try to use this technique on our problem. His version was based on using Kan extensions to induce a cotriple on a functor category. When Beck returned to Urbana, he told me about this. We quickly determined that in our case, we already had a cotriple, the one we were using to define the cohomology, and did not need to go into a functor category. Within a matter of days, we had worked out the simple version of acyclic models that we needed and verified the hypotheses for the cases of groups and associative algebras Barr & Beck [1966]. The use of acyclic models had turned a seeming impossibility into a near triviality. Some years later, I was able to verify the hypotheses for Harrison's commutative cohomology, but only for algebras over a field of characteristic 0. In the meantime, Quillen produced an example showing that the Harrison cohomology over a field of finite characteristic was not a cotriple cohomology. As for Lie algebras, this was not actually dealt with until Barr [1996a] when I examined the Cartan–Eilenberg theory from an abstract point of view.

Somewhat after the Barr-Beck developments, Michel André [1967, 1974] observed that there was a simple version of acyclic models based on an easy spectral sequence argument. This was weaker than the original version in several ways, including that its conclusion was a homology isomorphism rather than a homotopy equivalence. An additional flaw was that it did not conclude that the homology isomorphism was natural, but showed only that, for each object, the cohomology groups of that object were isomorphic. Still, it was easy to apply and André made great use of it in his study of cohomology of commutative rings (the definition he used was equivalent to the cotriple cohomology, not to the Harrison cohomology).

Partly to summarize in one place of all this development, partly because a powerful technique seemed in some danger of disappearing without a trace, I decided, sometime around 1990, to try to write a book on acyclic models. I produced some notes and then offered a course on the subject in the early 1990s. Among the registered students there was one, named Rob Milson, who was writing a thesis on differential geometry and applied mathematics. Not only did he register for the course, he attended all the lectures and studied the material carefully. Then he came up with a series of questions. My attempts to answer these questions led to a wholly different way of looking at acyclic models.

One question that Milson asked concerned the naturality of André's homology version of acyclic models. I thought that this could be settled by showing that André's isomorphism was induced by a natural transformation between the chain complexes in the category of additive relations that turned out to be a function and an isomorphism when you pass to homology. I am reasonably confident that this would have worked. However, in order to explore it, I began by looking at the following diagram (which I take from page 102)

$$\begin{array}{ccccc}
K_{-1}G^{\bullet+1} & \xleftarrow{\alpha G^{\bullet+1}} & K_{\bullet}G^{\bullet+1} & \xrightarrow{K_{\bullet}\epsilon} & K_{\bullet} \\
{\scriptstyle f_{-1}G^{\bullet+1}}\downarrow & & & & \\
L_{-1}G^{\bullet+1} & \xleftarrow[\beta G^{\bullet+1}]{} & L_{\bullet}G^{\bullet+1} & \xrightarrow[L_{\bullet}\epsilon]{} & L_{\bullet}
\end{array}$$

in which K and L are chain complex functors and G is the functor part of a cotriple. The goal is to construct a natural transformation from the homology of K to that of L. The meaning of the rest of this diagram, in particular the meaning of $G^{\bullet+1}$ will be explained in chapter 5. Three of the five arrows are in the appropriate direction. I was struck immediately by the fact that the standard acyclic models theorems have two principal hypotheses. In the homology version, one of these principal hypotheses is that $K\epsilon$ induces an isomorphism in homology and the other one is that $\beta G^{\bullet+1}$ does. In the homotopy version, the two principal hypotheses imply immediately that $K\epsilon$ and $\beta G^{\bullet+1}$ have homotopy inverses. Thus if you formally invert the maps that are homology isomorphisms, respectively homotopy equivalences, the required transformation $K \longrightarrow L$ appears immediately. There is a well-known theory, developed in [Gabriel & Zisman, 1967], of inversion of a class of arrows in a category. Using this theory allows us to define the transformation, show that it is natural, and derive some of its properties, not only for homology and homotopy, but also for possible classes of arrows intermediate to those. The most interesting one is that of weak homotopy: the arrow $K \longrightarrow L$ is natural and has, for each object, a homotopy inverse, which is not necessarily natural. This weak homotopy version, which answers another question raised by Milson, applies in the following situation. Suppose M is a C^p manifold (of some dimension n) and q is an integer with $0 \leq q \leq p$. Let $C_i^q(M)$ denote the subgroup of $C_i(M) = C_i^0 M$ that is generated by the singular i-simplexes that are q times differentiable. Is the inclusion $C_i^q(M) \longrightarrow C_i(M)$ a homotopy equivalence? The answer, as best we know it, is that it is for each M but we know of no homotopy inverse that is natural as a functor of M. Before the discovery of Theorem 5.3.1, I had attempted to prove this by a direct computation, but had got nowhere. The acyclic models proof is sufficiently constructive as to give, at least in principle, a direct computation, but I have not tried it. The main results appeared in Barr [1996b].

This example illustrates another fact. As we will see in 5.3.1, not only does f exist in the fraction category, but it is the unique extension of f_{-1}. Thus, if we already have $f\colon K \longrightarrow L$ such that f_{-1} is an isomorphism, then we know that in the fraction category f induces the homology isomorphism or the (weak) homotopy equivalence.

The upshot of this was that I had to throw away the preliminary notes and redo the book from the beginning; the book you see before you is the result.

This book could be used as a text for a somewhat idiosyncratic course that serves as an introduction to both homological algebra and algebraic topology. The centerpiece of the book is the main theorem on acyclic models that was discovered only in 1993, as just described. Although various forms of acyclic models have long been known (going back at least to the Eilenberg–Zilber [1953] theorem), they were mostly in the form of a technique, rather than an explicit theorem.

Aside from the acyclic models theorem itself, this book includes the mathematics necessary to understand and apply the basic theorem as well as some of what is needed to understand the examples. The only prerequisite is familiarity with basic algebra and topology. In a very few places, specific results not found in basic courses are used. An example is the Poincaré–Witt theorem on the structure of the enveloping algebra associated to a Lie algebra in the discussion of the Lie algebra cohomology. Aside from that and one or two other places the book is almost entirely self contained.

Chapter 1 is a general introduction to category theory. Chapter 2 is an introduction to abelian categories and also to homological algebra. Chain and cochain complexes are defined as well as Ext and Tor. Chapter 3 introduces the homology of chain complexes and also discusses simplicial objects and the associated chain complexes. Chapter 4 is about that part of the theory of triples (= monads) that is needed for acyclic models. Chapter 5 proves the main acyclic models theorem and draws some conclusions, including the older versions of the theorem. The remaining chapters give applications of the theory. Chapter 6 discusses the homological algebra of Cartan & Eilenberg [1956] and uses acyclic models to give general criteria for the cohomology theories described there to be equivalent to cotriple cohomologies. Then these criteria are used to show that the various theories there are, in fact, cotriple cohomology theories. In Chapter 7 we show that other cohomology theories in algebra, notably the characteristic 0 cohomology theory for commutative algebras, which does not fit the Cartan–Eilenberg pattern, is also a cotriple cohomology theory. Finally, in Chapter 8 we give applications to topology, including proofs of the equivalence of singular and simplicial homology on triangulable spaces, a proof of the equivalence of oriented and ordered chain complexes, a proof of the Mayer–Vietoris theorem, and a sketch of an acyclic models proof of the de Rham theorem. Notably absent is any use of simplicial approximation (although subdivision is used).

I would like to thank an anonymous referee for catching a number of embarrassing errors. I did not follow his (or her; I will adhere to the former practive of using these pronouns androgenously) advice in all matters, however. He finds the last chapter on applications of algebraic topology without interest. I think that it is at least moderately interesting that one can develop the theory without simplicial approximation and that one can go quite far without spectral sequences.

Chapter 1 consists largely of Chapter 1 of Barr & Wells [1984] and Chapter 4 is mainly part of Chapter 4 of the same book. I would like to thank Charles Wells for permission to use this material.

Not every part of Chapter 1 is actually required for the rest of the book. I have kept them so that the book would also give a self-contained introduction to categories. The following are the sections and parts of sections of Chapter 1 that are used in the rest of the book: 1, 2.1–2.5, 3, 4.1, 5.1–5.2, 6.1–6.3 7.1–7.3, 8, 9.1–9.3, 10, 11.

CHAPTER 1

Categories

1. Introduction

The conceptual basis for the acyclic models theorem that is the main topic of this book is the notion of category. The language of categories is a convenience in many areas of mathematics, but in the understanding of the role of acyclic models, it is a necessity. The results cannot be stated, let alone proved, without reference to categories. In this chapter, we give a brief, but fairly complete, introduction to the subject.

2. Definition of category

2.1. A **category** $\mathscr{C}$ consists of two collections, $\mathrm{Ob}(\mathscr{C})$, whose elements are the **objects** of $\mathscr{C}$, and $\mathrm{Ar}(\mathscr{C})$, the **arrows** (or **morphisms** or **maps**) of $\mathscr{C}$. To each arrow is assigned a pair of objects, called the **source** (or **domain**) and the **target** (or **codomain**) of the arrow. The notation $f: A \longrightarrow B$ means that f as an arrow with source A and target B. If $f: A \longrightarrow B$ and $g: B \longrightarrow C$ are two arrows, there is an arrow $g \circ f: A \longrightarrow C$ called the **composite** of g and f. The composite is not defined otherwise. We often write gf instead of $g \circ f$ when there is no danger of confusion. For each object A there is an arrow id_A (often written 1_A or just 1, depending on the context), called the **identity** of A, whose source and target are both A. These data are subject to the following axioms:

(1) for $f: A \longrightarrow B$,

$$f \circ \mathrm{id}_A = \mathrm{id}_B \circ f = f$$

(2) for $f: A \longrightarrow B$, $g: B \longrightarrow C$, $h: C \longrightarrow D$,

$$h \circ (g \circ f) = (h \circ g) \circ f$$

A category consists of two "collections", the one of sets and the one of arrows. These collections are not assumed to be sets and in many interesting cases they are not, as will be seen. When the collection of arrows is a set then the category is said to be **small**. It follows in that case that the collection of objects is also a set since there is one-one correspondence between the objects and the identity arrows.

While we do not suppose in general that the arrows form a set, we do usually suppose (and will, unless it is explicitly mentioned to the contrary) that when we fix two objects A and B of $\mathscr{C}$, that the collection of arrows with source A and target B is a set. This set is denoted $\mathrm{Hom}_{\mathscr{C}}(A, B)$. We will omit the subscript denoting the category whenever we can get away with it. A set of the form $\mathrm{Hom}(A, B)$ is called a **homset**. Categories that satisfy this condition are said to be **locally small**.

Many familiar examples of categories will occur immediately to the reader, such as the category Set of sets and set functions, the category Grp of groups and homomorphisms, and the category Top of topological spaces and continuous maps.

In each of these cases, the composition operation on arrows is the usual composition of functions.

A more interesting example is the category whose objects are topological spaces and whose arrows are homotopy classes of continuous maps. Because homotopy is compatible with composition, homotopy classes of continuous functions behave like functions (they have sources and targets, they compose, etc.) but are not functions. This category is usually known as the category of homotopy types.

All but the last example are of categories whose objects are sets with mathematical structure and the morphisms are functions which preserve the structure. Many mathematical structures are themselves categories. For example, one can consider any group G as a category with exactly one object; its arrows are the elements of G regarded as having the single object as both source and target. Composition is the group multiplication, and the group identity is the identity arrow. This construction works for monoids as well. In fact, a monoid can be defined as a category with exactly one object.

A poset (partially ordered set) can also be regarded as a category: its objects are its elements, and there is exactly one arrow from an element x to an element y if and only if $x \leq y$; otherwise there are no arrows from x to y. Composition is forced by transitivity and identity arrows by reflexivity. Thus a category can be thought of as a generalized poset. This perception is important, since many of the fundamental concepts of category theory specialize to nontrivial and often well-known concepts for posets (the reader is urged to fill in the details in each case).

In the above examples, we have described categories by specifying both their objects and their arrows. Informally, it is very common to name the objects only; the reader is supposed to supply the arrows based on his general knowledge. If there is any doubt, it is, of course, necessary to describe the arrows as well. Sometimes there are two or more categories in general use with the same objects but different arrows. For example, the following three categories all have the same objects: complete sup-semilattices, complete inf-semilattices, complete lattices. Further variations can be created according as the arrows are required to preserve the top (empty inf) or bottom (empty sup) or both.

2.2. Some constructions for categories. A **subcategory** $\mathscr{D}$ of a category $\mathscr{C}$ is a pair of subsets D_O and D_A of the objects and arrows of $\mathscr{C}$ respectively, with the following properties.

(1) If $f \in D_A$ then the source and target of f are in D_O.
(2) If $C \in D_O$, then $\mathrm{id}_C \in D_A$.
(3) If f, $g \in D_A$ are a composable pair of arrows then $g \circ f \in D_A$.

The subcategory is **full** if for any $C, D \in D_O$, if $f{:}\,C \longrightarrow D$ in $\mathscr{C}$, then $f \in D_A$. For example, the category of Abelian groups is a full subcategory of the category of groups (every homomorphism of groups between Abelian groups is a homomorphism of Abelian groups), whereas the category of monoids (semigroups with identity element) is a subcategory, but not a full subcategory, of the category of semigroups (a semigroup homomorphism need not preserve 1).

One also constructs the **product** $\mathscr{C} \times \mathscr{D}$ of two categories $\mathscr{C}$ and $\mathscr{D}$ in the obvious way: the objects of $\mathscr{C} \times \mathscr{D}$ are pairs (A, B) with A an object of $\mathscr{C}$ and B

an object of $\mathscr{D}$. An arrow

$$(f, g): (A, B) \longrightarrow (A', B')$$

has $f: A \longrightarrow A'$ in $\mathscr{C}$ and $g: B \longrightarrow B'$ in $\mathscr{D}$. Composition is coordinatewise.

To define the next concept, we need the idea of commutative diagram. A diagram is said to commute if any two paths between the same nodes compose to give the same morphism. The formal definition of diagram and commutative diagram is given in 8.1 below.

If A is any object of a category $\mathscr{C}$, the **slice category** $\mathscr{C}/A$ of objects of $\mathscr{C}$ **over** A has as objects all arrows of $\mathscr{C}$ with target A. An arrow of $\mathscr{C}/A$ from $f: B \longrightarrow A$ to $g: C \longrightarrow A$ is an arrow $h: B \longrightarrow C$ making the following diagram commute.

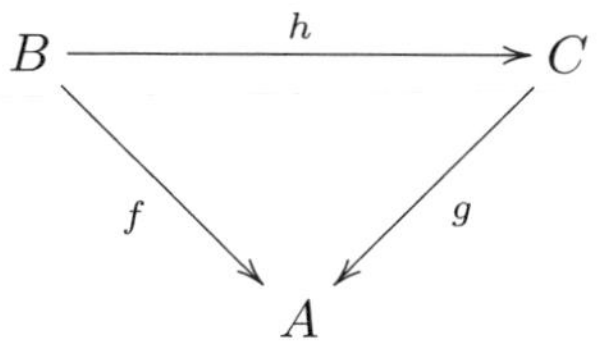

In this case, one sometimes writes $h: f \longrightarrow g$ *over* A.

It is useful to think of an object of Set/A as an A-indexed family of disjoint sets (the inverse images of the elements of A). The commutativity of the above diagram means that the function h is consistent with the decomposition of B and C into disjoint sets.

2.3. Definitions without using elements. The introduction of categories as a part of the language of mathematics has made possible a fundamental, intrinsically categorical technique: the element-free definition of mathematical properties by means of commutative diagrams, limits and adjoints. (Limits and adjoints are defined later in this chapter.) By the use of this technique, category theory has made mathematically precise the unity of a variety of concepts in different branches of mathematics, such as the many product constructions which occur all over mathematics (described in Section 8) or the ubiquitous concept of isomorphism, discussed below. Besides explicating the unity of concepts, categorical techniques for defining concepts without mentioning elements have enabled mathematicians to provide a useful axiomatic basis for algebraic topology, homological algebra and other theories.

Despite the possibility of giving element-free definitions of these constructions, it remains intuitively helpful to think of them as being defined with elements. Fortunately, this can be done: In Section 5, we introduce a more general notion of element of an object in a category (more general even when the category is Set) which in many circumstances makes categorical definitions resemble familiar definitions involving elements of sets, and which also provides an explication of the old notion of variable quantity.

2.4. Isomorphisms and terminal objects. The notion of isomorphism can be given an element-free definition for any category: An arrow $f: A \longrightarrow B$ in a category is an **isomorphism** if it has an **inverse**, namely an arrow $g: B \longrightarrow A$ for which $f \circ g = \mathrm{id}_B$ and $g \circ f = \mathrm{id}_A$. In other words, both triangles of the following diagram must commute:

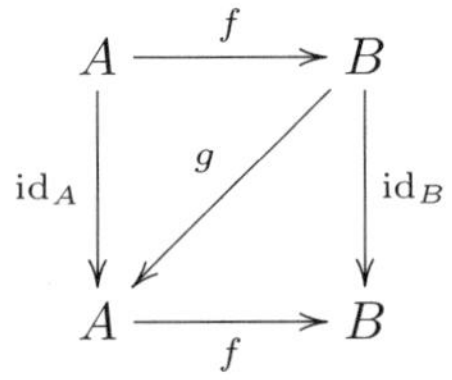

In a group regarded as a category, every arrow is invertible, whereas in a poset regarded as a category, the only invertible arrows are the identity arrows (which are invertible in any category).

It is easy to check that an isomorphism in Grp is what is usually called an isomorphism (commonly defined as a bijective homomorphism, but some newer texts give the definition above). An isomorphism in Set is a bijective function, and an isomorphism in Top is a homeomorphism.

Singleton sets in Set can be characterized without mentioning elements, too. A **terminal object** in a category $\mathscr{C}$ is an object T with the property that for every object A of $\mathscr{C}$ there is exactly one arrow from A to T. It is easy to see that terminal objects in Set, Top, and Grp are all one element sets with the only possible structure in the case of the last two categories.

2.5. Duality. If $\mathscr{C}$ is a category, then we define $\mathscr{C}^{\mathrm{op}}$ to be the category with the same objects and arrows as $\mathscr{C}$, but an arrow $f\colon A \longrightarrow B$ in $\mathscr{C}$ is regarded as an arrow from B to A in $\mathscr{C}^{\mathrm{op}}$. In other words, for all objects A and B of $\mathscr{C}$,

$$\mathrm{Hom}_{\mathscr{C}}(A,B) = \mathrm{Hom}_{\mathscr{C}^{\mathrm{op}}}(B,A)$$

If $f\colon A \longrightarrow B$ and $g\colon B \longrightarrow C$ in $\mathscr{C}$, then the composite $f \circ g$ in $\mathscr{C}^{\mathrm{op}}$ is by definition the composite $g \circ f$ in $\mathscr{C}$. The category $\mathscr{C}^{\mathrm{op}}$ is called the **opposite category** of $\mathscr{C}$.

If P is a property that objects or arrows in a category may have, then the **dual** of P is the property of having P in the opposite category. As an example, consider the property of being a terminal object. If an object A of a category $\mathscr{C}$ is a terminal object in $\mathscr{C}^{\mathrm{op}}$, then $\mathrm{Hom}_{\mathscr{C}}(B,A)$ has exactly one arrow for every object B of $\mathscr{C}$. Thus the dual property of being a terminal object is the property: $\mathrm{Hom}(A,B)$ has exactly one arrow for each object B. An object A with this property is called an **initial object**. In Set and Top, the empty set is the initial object (see "Fine points" below). In Grp, on the other hand, the one-element group is both an initial and a terminal object.

Clearly if property P is dual to property Q then property Q is dual to property P. Thus being an initial object and being a terminal object are dual properties. Observe that being an isomorphism is a self-dual property.

Constructions may also have duals. For example, the dual to the category of objects over A is the category of objects **under** A. An object is an arrow *from* A and an arrow from the object $f\colon A \longrightarrow B$ to the object $g\colon A \longrightarrow C$ is an arrow h from B to C for which $h \circ f = g$.

Often a property and its dual each have their own names; when they do not (and sometimes when they do) the dual property is named by prefixing "co-". For example, one could, and some sources do, call an initial object "coterminal", or a terminal object "coinitial".

2.6. Definition of category by commutative diagrams. The notion of category itself can be defined in an element-free way. We describe the idea behind this alternate definition here, but some of the sets we construct are defined in terms of elements. In Section 7, we show how to define these sets without mentioning elements (by pullback diagrams).

Before giving the definition, we mention several notational conventions that will recur throughout the book.

(1) If X and Y are sets, $p_1: X \times Y \longrightarrow X$ and $p_2: X \times Y \longrightarrow Y$ are the coordinate projections.

(2) If X, Y and Z are sets and $f: X \longrightarrow Y$, $g: X \longrightarrow Z$ are functions,

$$(f, g): X \longrightarrow Y \times Z$$

is the function whose value at $a \in X$ is $(f(a), g(a))$.

(3) If X, Y, Z, and W are sets and $f: X \longrightarrow Z$, $g: Y \longrightarrow W$ are functions, then

$$f \times g: X \times Y \longrightarrow Z \times W$$

is the function whose value at (a, b) is $(f(a), g(b))$. This notation is also used for maps defined on subsets of product sets (as in 4 below).

A category consists of two sets A and O and four functions $d^0, d^1: A \longrightarrow O$, $u: O \longrightarrow A$ and $m: P \longrightarrow A$, where P is the set

$$\{(f, g) \mid d^0(f) = d^1(g)\}$$

of composable pairs of arrows for which the following Diagrams 1–4 commute. For example, the right diagram of (2) below says that $d^1 \circ p_1 = d^1 \circ m$. The interpretation of these sets and functions is: A is the set of arrows, O is the set of objects, d^0 takes an arrow to its domain, d^1 takes an arrow to its codomain, u takes an object to the identity arrow on that object, and m takes a composable pair of arrows to their composite: $m(f, g) = fg$.

(1)

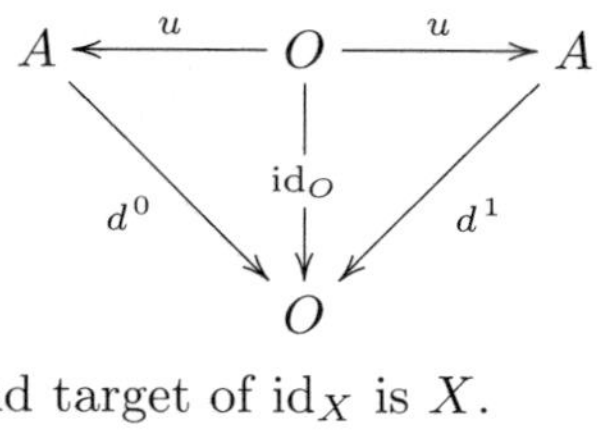

This says that the source and target of id_X is X.

(2)

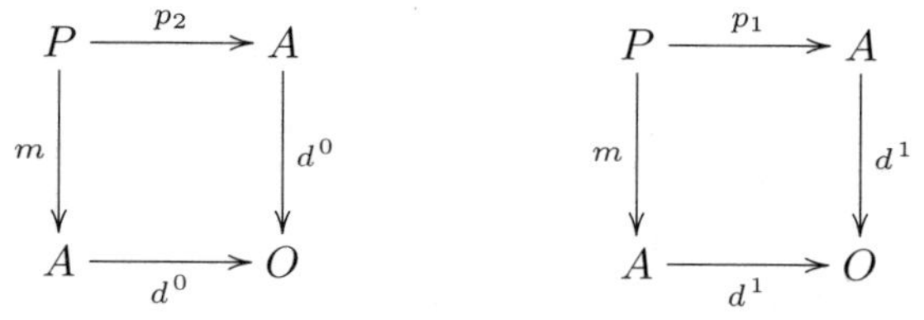

This says that the source of $f \circ g$ is that of g and its target is that of f.

(3)

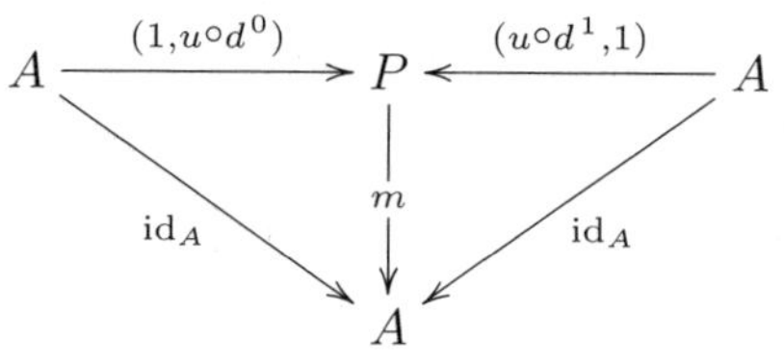

This characterizes the left and right identity laws.

In the next diagram, Q is the set of composable triples of arrows:

$$Q = \{(f, g, h) \mid d^1(h) = d^0(g) \text{ and } d^1(g) = d^0(f)\}$$

(4)
$$\begin{array}{ccc} Q & \xrightarrow{1\times m} & P \\ {\scriptstyle m\times 1}\downarrow & & \downarrow{\scriptstyle m} \\ P & \xrightarrow[m]{} & A \end{array}$$

This is associativity of composition.

It is straightforward to check that this definition is equivalent to the first one. We will treat diagrams more formally in Section 8.

The diagrams just given actually describe geometric objects, namely the classifying space of the category. Indeed, the functions between O, A, P and Q generated by u, d^0, d^1, m and the coordinate maps form a simplicial set truncated in dimension three. Simplicial sets are defined in 3.3. But the reader needs no knowledge of simplicial sets at this point.

2.7. Fine points. Note that a category may be empty, that is have no objects and (of course) no arrows. Observe that a subcategory of a monoid regarded as a category may be empty; if it is not empty, then it is a submonoid. This should cause no more difficulty than the fact that a submonoid of a group may not be a subgroup. The basic reason is that a monoid must have exactly one object, while a subcategory need not have any.

It is important to observe that in categories such as Set, Grp and Top in which the arrows are actually functions, the definition of category requires that the function have a uniquely specified domain and codomain, so that for example in Top the continuous function from the set $\mathbf{R}$ of real numbers to the set $\mathbf{R}^+$ of nonnegative real numbers which takes a number to its square is different from the function from $\mathbf{R}$ to $\mathbf{R}$ which does the same thing, and both of these are different from the squaring function from $\mathbf{R}^+$ to $\mathbf{R}^+$.

A definition of "function" in Set which fits this requirement is this: A **function** is an ordered triple (A, G, B) where A and B are sets and G is a subset of $A \times B$ with the property that for each $x \in A$ there is exactly one $y \in B$ such that $(x, y) \in G$. This is equivalent to saying that the composite

$$G \hookrightarrow A \times B \longrightarrow A$$

is an isomorphism (the second function is projection on the first coordinate). Then the domain of the function is the set A and the codomain is B. As a consequence of this definition, A is empty if and only if G is empty, but B may or may not be empty. Thus there is exactly one function, namely $(\emptyset, \emptyset, B)$, from the empty set to each set B, so that the empty set is the initial object in Set, as claimed previously. (Note also that if (A, G, B) is a function then G uniquely determines A but not B. This asymmetry is reversed in the next paragraph.)

An equivalent definition of function is a triple (A, G^*, B) where G^* is the quotient of the disjoint union $A + B$ by an equivalence relation for which each element of B is contained in exactly one equivalence class. In other words, the composite

$$B \longrightarrow A + B \twoheadrightarrow G^*$$

is an isomorphism, where the first arrow is the inclusion into the sum and the second is the quotient mapping. This notion actually corresponds to the intuitive picture of function frequently drawn for elementary calculus students which illustrates the squaring function from $\{-2,-1,0,1,2\}$ to $\{0,1,2,3,4\}$ this way:

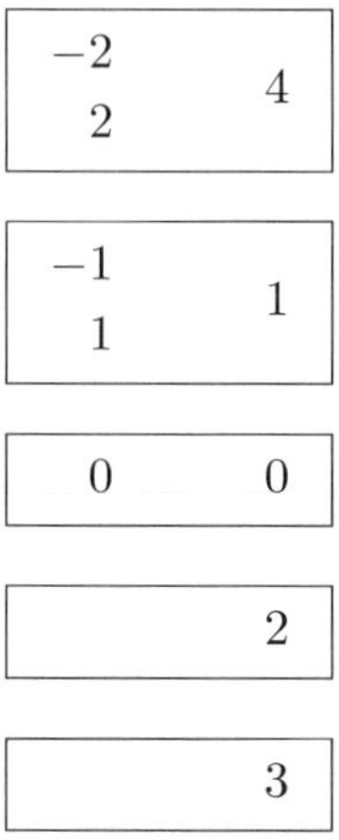

The set G is called the **graph** and G^* the **cograph** of the function. We will see in Section 1.8 that the graph and cograph are dual to each other.

2.8. Exercises.

1. Show that the following definition of category which is sometimes used is equivalent to the definition given in this section: A **category** is a set with a partially defined binary operation denoted "$\circ$" with the following properties:

(a) the following statements are equivalent:

(i) $f \circ g$ and $g \circ h$ are both defined;
(ii) $f \circ (g \circ h)$ is defined;
(iii) $(f \circ g) \circ h$ is defined;

(b) if $(f \circ g) \circ h$ is defined, then $(f \circ g) \circ h = f \circ (g \circ h)$;

(c) for any f, there are elements e and e' for which $e \circ f$ is defined and equal to f and $f \circ e'$ is defined and equal to f.

2. Verify that the following constructions produce categories.

(a) For any category $\mathscr{C}$, the **arrow category** $\mathrm{Ar}(\mathscr{C})$ of arrows of $\mathscr{C}$ has as objects the arrows of $\mathscr{C}$, and an arrow from $f\colon A \longrightarrow B$ to $g\colon A' \longrightarrow B'$ is a pair of arrows $h\colon A \longrightarrow A'$ and $k\colon B \longrightarrow B'$ making the following diagram commute:

$$\begin{array}{ccc} A & \xrightarrow{\;h\;} & A' \\ {\scriptstyle f}\downarrow & & \downarrow{\scriptstyle g} \\ B & \xrightarrow[\;k\;]{} & B' \end{array}$$

(b) The **twisted arrow category of** $\mathscr{C}$ is defined the same way as the arrow category except that the direction of k is reversed.

3. **(a)** Show that $h\colon f \longrightarrow g$ is an isomorphism in the category of objects of $\mathscr{C}$ over A if and only if h is an isomorphism of $\mathscr{C}$.

(b) Give an example of objects A, B and C in a category $\mathscr{C}$ and arrows $f: B \longrightarrow A$ and $g: C \longrightarrow A$ such that B and C are isomorphic in $\mathscr{C}$ but f and g are not isomorphic in $\mathscr{C}/A$.

4. Describe the isomorphisms, initial objects, and terminal objects (if they exist) in each of the categories in Exercise 2.

5. Describe the initial and terminal objects, if they exist, in a poset regarded as a category.

6. Show that any two terminal objects in a category are isomorphic by a unique isomorphism.

7. **(a)** Prove that for any category $\mathscr{C}$ and any arrows f and g of $\mathscr{C}$ such that the target of g is isomorphic to the source of f, there is an arrow f' which (i) is isomorphic to f in $\mathrm{Ar}(\mathscr{C})$ and (ii) has source the same as the target of g. ($\mathrm{Ar}(\mathscr{C})$ is defined in Exercise 2 above.)

(b) Use the fact given in (a) to describe a suitable definition of domain, codomain and composition for a category with one object chosen for each isomorphism class of objects of $\mathscr{C}$ and one arrow from each isomorphism class of objects of $\mathrm{Ar}(\mathscr{C})$. Such a category is called a **skeleton** of $\mathscr{C}$.

8. A category is **connected** if it is possible to go from any object to any other object of the category along a path of "composable" *forward or backward* arrows. Make this definition precise and prove that every category is a union of disjoint connected subcategories in a unique way.

9. A **preorder** is a set with a reflexive, transitive relation defined on it. Explain how to regard a preorder as a category with at most one arrow from any object A to any object B.

10. **(a)** Describe the opposite of a group regarded as a category. Show that it is isomorphic to, but not necessarily the same as, the original group.

(b) Do the same for a monoid, but show that the opposite need not be isomorphic to the original monoid.

(c) Do the same as (b) for posets.

11. An **arrow congruence** on a category $\mathscr{C}$ is an equivalence relation E on the arrows for which

(i) fEf' implies that f and f' have the same domain and codomain.
(ii) If fEf' and gEg' and $f \circ g$ is defined, then $(f \circ g)E(f' \circ g')$.

There are more general congruences in which objects are identified. These are considerably more complicated since new composites are formed when the target of one arrow gets identified with the source of another.

(a) Show that any relation R on the arrows of $\mathscr{C}$ generates a unique congruence on $\mathscr{C}$.

(b) Given a congruence E on $\mathscr{C}$, define the **quotient category** $\mathscr{C}/E$ in the obvious way (same objects as $\mathscr{C}$) and show that it is a category. This notation conflicts with the slice notation, but context should make it clear. In any case, quotient categories are not formed very often.

(Thus any set of diagrams in $\mathscr{C}$ generate a congruence E on $\mathscr{C}$ with the property that $\mathscr{C}/E$ is the largest quotient in which the diagrams commute.)

12. Show that in a category with an initial object 0 and a terminal object 1, $0 \cong 1$ if and only if there is a map $1 \longrightarrow 0$.

3. Functors

3.1. Like every other kind of mathematical structured object, categories come equipped with a notion of morphism. It is natural to define a morphism of categories to be a map which takes objects to objects, arrows to arrows, and preserves source, target, identities and composition.

If $\mathscr{C}$ and $\mathscr{D}$ are categories, a **functor** $F: \mathscr{C} \longrightarrow \mathscr{D}$ is a map for which

(1) if $f: A \longrightarrow B$ is an arrow of $\mathscr{C}$, then $Ff: FA \longrightarrow FB$ is an arrow of $\mathscr{D}$;
(2) $F(\mathrm{id}_A) = \mathrm{id}_{FA}$; and
(3) if $g: B \longrightarrow C$, then $F(g \circ f) = Fg \circ Ff$.

If $F: \mathscr{C} \longrightarrow \mathscr{D}$ is a functor, then $F^{\mathrm{op}}: \mathscr{C}^{\mathrm{op}} \longrightarrow \mathscr{D}^{\mathrm{op}}$ is the functor which does the same thing as F to objects and arrows.

A functor $F: \mathscr{C}^{\mathrm{op}} \longrightarrow \mathscr{D}$ is called a **contravariant functor** from $\mathscr{C}$ to $\mathscr{D}$. In this case, F^{op} goes from $\mathscr{C}$ to $\mathscr{D}^{\mathrm{op}}$. For emphasis, a functor from $\mathscr{C}$ to $\mathscr{D}$ is occasionally called a **covariant functor**.

$F: \mathscr{C} \longrightarrow \mathscr{D}$ is **faithful** if it is injective when restricted to each homset, and it is **full** if it is surjective on each homset, i.e., if for every pair of objects A and B, every arrow in $\mathrm{Hom}(FA, FB)$ is F of some arrow in $\mathrm{Hom}(A, B)$. Some sources use the phrase "fully faithful" to describe a functor which is full and faithful.

F **preserves** a property P that an arrow may have if $F(f)$ has property P whenever f has. It **reflects** property P if f has the property whenever $F(f)$ has. For example, any functor must preserve isomorphisms (Exercise 1), but a functor need not reflect them.

Here are some examples of functors:

(1) For any category $\mathscr{C}$, there is an identity functor $\mathrm{id}_{\mathscr{C}}: \mathscr{C} \longrightarrow \mathscr{C}$.
(2) The categories Grp and Top are typical of many categories considered in mathematics in that their objects are sets with some sort of structure on them and their arrows are functions which preserve that structure. For any such category $\mathscr{C}$, there is an **underlying set functor** $U: \mathscr{C} \longrightarrow \mathsf{Set}$ which assigns to each object its set of elements and to each arrow the function associated to it. Such a functor is also called a **forgetful functor**, the idea being that it forgets the structure on the set. Such functors are always faithful and rarely full.
(3) Many other mathematical constructions, such as the double dual functor on vector spaces, the commutator subgroup of a group or the fundamental group of a path connected space, are the object maps of functors (in the latter case the domain is the category of pointed topological spaces and base-point-preserving maps). There are, on the other hand, some canonical constructions which do not extend to maps. Examples include the center of a group or ring, and groups of automorphisms quite generally. See Exercise 8 and Exercise 9.
(4) For any set A, let FA denote the free group generated by A. The defining property of free groups allows you to conclude that if $f: A \longrightarrow B$ is any function, there is a unique homomorphism $Ff: FA \longrightarrow FB$ with the property that $Ff \circ i = j \circ f$, where $i: A \longrightarrow FA$ and $j: B \longrightarrow FB$ are the inclusions. It is an easy exercise to see that this makes F a functor

from Set to Grp. Analogous functors can be defined for the category of monoids, the category of Abelian groups, and the category of R-modules for any ring R.

(5) For a category $\mathscr{C}$, $\mathrm{Hom}_{\mathscr{C}} = \mathrm{Hom}$ is a functor in each variable separately, as follows: For fixed object A, $\mathrm{Hom}(A, f) \colon \mathrm{Hom}(A, B) \longrightarrow \mathrm{Hom}(A, C)$ is defined for each arrow $f \colon B \longrightarrow C$ by requiring that $\mathrm{Hom}(A, f)(g) = f \circ g$ for $g \in \mathrm{Hom}(A, B)$; this makes $\mathrm{Hom}(A, -) \colon \mathscr{C} \longrightarrow \mathsf{Set}$ a functor. Similarly, for a fixed object B, $\mathrm{Hom}(-, B)$ is a functor from $\mathscr{C}^{\mathrm{op}}$ to Set; $\mathrm{Hom}(h, B)$ is composition with h on the right instead of on the left. $\mathrm{Hom}(A, -)$ and $\mathrm{Hom}(-, B)$ are the **covariant** and **contravariant hom functors**, respectively. $\mathrm{Hom}(-, -)$ is also a Set-valued functor, with domain $\mathscr{C}^{\mathrm{op}} \times \mathscr{C}$. A familiar example of a contravariant hom functor is the functor which takes a vector space to the underlying set of its dual.

(6) The **powerset** (set of subsets) of a set is the object map of an important contravariant functor $\mathbf{P}$ from Set to Set which plays a central role in this book. The map from $\mathbf{P}B$ to $\mathbf{P}A$ induced by a function $f \colon A \longrightarrow B$ is the inverse image map; precisely, if $B_0 \in \mathbf{P}B$, i.e. $B_0 \subseteq B$, then

$$\mathbf{P}f(B_0) = \{x \in A \mid f(x) \in B_0\}$$

The object function $\mathbf{P}$ can also be made into a covariant functor, in at least two different ways (Exercise 6).

(7) If G and H are groups considered as categories with a single object, then a functor from G to H is exactly a group homomorphism.

(8) If P and Q are posets, a functor from P to Q is exactly a nondecreasing map. A contravariant functor is a nonincreasing map.

3.2. Isomorphism and equivalence of categories. The composite of functors is a functor, so the collection of categories and functors is itself a category, denoted Cat. If $\mathscr{C}$ and $\mathscr{D}$ are categories and $F \colon \mathscr{C} \longrightarrow \mathscr{D}$ is a functor which has an inverse $G \colon \mathscr{D} \longrightarrow \mathscr{C}$, so that it is an isomorphism in the category of categories, then naturally $\mathscr{C}$ and $\mathscr{D}$ are said to be **isomorphic**.

However, the notion of isomorphism does not capture the most useful sense in which two categories can be said to be essentially the same; that is the notion of equivalence. A functor $F \colon \mathscr{C} \longrightarrow \mathscr{D}$ is said to be an **equivalence** if it is full and faithful and has the property that for any object B of $\mathscr{D}$ there is an object A of $\mathscr{C}$ for which $F(A)$ is isomorphic to B. The definition appears asymmetrical but in fact given the axiom of choice if there is an equivalence from $\mathscr{C}$ to $\mathscr{D}$ then there is an equivalence from $\mathscr{D}$ to $\mathscr{C}$ (Exercise 11).

The notion of equivalence captures the perception that, for example, for most purposes you are not changing group theory if you want to work in a category of groups which contains only a countable number (or finite, or whatever) of copies of each isomorphism type of groups and all the homomorphisms between them.

Statements in Section 2 like, "A group may be regarded as a category with one object in which all arrows are isomorphisms" can be made precise using the notion of equivalence: The category of groups and homomorphisms is equivalent to the category of categories with exactly one object in which each arrow is an isomorphism, and all functors between them. Any isomorphism between these categories would seem to require an axiom of choice for proper classes.

3.3. Comma categories. Let $\mathscr{A}$, $\mathscr{C}$ and $\mathscr{D}$ be categories and $F: \mathscr{C} \longrightarrow \mathscr{A}$, $G: \mathscr{D} \longrightarrow \mathscr{A}$ be functors. From these ingredients we construct the **comma category** (F, G) which is a generalization of the slice $\mathscr{A}/A$ of a category over an object discussed in Section 2. The objects of (F, G) are triples (C, f, D) with $f: FC \longrightarrow GD$ an arrow of $\mathscr{A}$ and C, D objects of $\mathscr{C}$ and $\mathscr{D}$ respectively. An arrow $(h, k): (C, f, D) \longrightarrow (C', f', D')$ consists of $h: C \longrightarrow C'$ and $k: D \longrightarrow D'$ making

$$\begin{array}{ccc} FC & \xrightarrow{Fh} & FC' \\ {\scriptstyle f}\Big\downarrow & & \Big\downarrow{\scriptstyle f'} \\ GD & \xrightarrow[Gk]{} & GD' \end{array}$$

commute. It is easy to verify that coordinatewise composition makes (F, G) a category.

When A is an object of $\mathscr{A}$, we can consider it as a functor $A: 1 \longrightarrow \mathscr{A}$. Then the comma category $(\mathrm{Id}_{\mathscr{A}}, A)$ is just the slice $\mathscr{A}/A$ defined in Section 2. The category of arrows under an object is similarly a comma category.

Each comma category (F, G) is equipped with two **projections** $p_1: (F, G) \longrightarrow \mathscr{C}$ projecting objects and arrows onto their first coordinates, and $p_2: (F, G) \longrightarrow \mathscr{D}$ projecting objects onto their third coordinates and arrows onto their second.

3.4. Exercises.

1. Show that functors preserve isomorphisms, but do not necessarily reflect them.

2. Use the concept of arrow category to describe a functor which takes a group homomorphism to its kernel.

3. Show that the following define functors:

(a) the projection map from a product $\mathscr{C} \times \mathscr{D}$ of categories to one of them;

(b) for $\mathscr{C}$ a category and an object A of $\mathscr{C}$, the constant map from a category $\mathscr{B}$ to $\mathscr{C}$ which takes every object to A and every arrow to id_A;

(c) the forgetful functor from the category $\mathscr{C}/A$ of objects over A to $\mathscr{C}$ which takes an object $B \longrightarrow A$ to B and an arrow $h: B \longrightarrow C$ over A to itself.

4. Show that the functor P of Example 6 is faithful but not full and reflects isomorphisms.

5. Give examples showing that functors need not preserve or reflect initial or terminal objects.

6. Show that the map which takes a set to its powerset is the object map of at least two covariant functors from Set to Set: If $f: A \longrightarrow B$, one functor takes a subset A_0 of A to its image $f_!(A_0) = f(A_0)$, and the other takes A_0 to the set

$$f_*(A_0) = \{y \in B \mid \text{if } f(x) = y \text{ then } x \in A_0\} = \{y \in B \mid f^{-1}(y) \subseteq A_0\}$$

Show that $f^{-1}(B) \subseteq A$ if and only if $B \subseteq f_*(A)$ and that $A \subseteq f^{-1}(B)$ if and only if $f_!(A) \subseteq B$.

7. Show that the definition given in Example 4 makes the free group construction F a functor.

8. Show that there is no functor from Grp to Grp which takes each group to its center. (Hint: Consider the group G consisting of all pairs (a, b) where a is any integer and b is 0 or 1, with multiplication

$$(a, b)(c, d) = (a + (-1)^b c, b + d)$$

the addition in the second coordinate being (mod 2).)

9. Show that there is no functor from Grp to Grp which takes each group to its automorphism group.

10. Show that every category is equivalent to its skeleton (see Exercise 7 of Section 2).

11. Show that equivalence is an equivalence relation on any set of categories. (This exercise is easier to do after you do Exercise 7 of Section 4).

12. **(a)** Make the statement "a preordered set can be regarded as a category in which there is no more than one arrow between any two objects" precise by defining a subcategory of the category of categories and functors that the category of preordered sets and order-preserving maps is equivalent to (see Exercise 9 of Section 2).

(b) Show that, when regarded as a category, every preordered set is equivalent to a poset.

13. An **atom** in a Boolean algebra is an element greater than 0 but with no elements between it and 0. A Boolean algebra is **atomic** if every element x of the algebra is the join of all the atoms smaller than x. A Boolean algebra is **complete** if every subset has an infimum and a supremum. A **CABA** is a complete atomic Boolean algebra. A CABA homomorphism is a Boolean algebra homomorphism between CABA's which preserves all infs and sups (not just finite ones, which any Boolean algebra homomorphism would do). Show that the opposite of the category of sets is equivalent to the category of CABA's and CABA homomorphisms.

14. An **upper semilattice** is a partially ordered set in which each finite subset (including the empty set) of elements has a least upper bound. Show that the category of upper semilattices and functions which preserve the least upper bound of any finite subset (and hence preserve the ordering) is equivalent to the category of commutative monoids in which every element is idempotent and monoid homomorphisms.

15. Show that the arrow and twisted arrow categories of Exercise 2 of Section 2 are comma categories.

16. Show that the category Set of sets nor the category Ab of abelian groups is equivalent to its opposite category. (Hint: Find a property of the category for which the dual property is not satisfied.)

4. Natural transformations

4.1. In topology, a homotopy from $f: A \longrightarrow B$ to $g: A \longrightarrow B$ is given by a path in B from fx to gx for each element $x \in A$ such that the paths fit together continuously. A natural transformation is analogously a deformation of one *functor* to another.

If $F: \mathscr{C} \longrightarrow \mathscr{D}$ and $G: \mathscr{C} \longrightarrow \mathscr{D}$ are two functors, $\lambda: F \longrightarrow G$ is a **natural transformation** from F to G if λ is a collection of arrows $\lambda C: FC \longrightarrow GC$, one for each object C of $\mathscr{C}$, such that for each arrow $g: C \longrightarrow C'$ of $\mathscr{C}$ the following diagram commutes:

$$\begin{array}{ccc} FC & \xrightarrow{\lambda C} & GC \\ {\scriptstyle Fg}\downarrow & & \downarrow{\scriptstyle Gg} \\ FC' & \xrightarrow[\lambda C']{} & GC' \end{array}$$

The arrows λC are the **components** of λ.

The natural transformation λ is a **natural equivalence** if each component of λ is an isomorphism in $\mathscr{D}$.

The natural map of a vector space to its double dual is a natural transformation from the identity functor on the category of vector spaces and linear maps to the double dual functor. When restricted to finite dimensional vector spaces, it is a natural equivalence. As another example, let $n > 1$ be a positive integer and let GL_n denote the functor from the category of commutative rings with unity to the category of groups which takes a ring to the group of invertible $n \times n$ matrices with entries from the ring, and let Un denote the group of units functor (which is actually GL_1). Then the determinant map is a natural transformation from GL_n to Un. The Hurewicz transformation from the fundamental group of a topological space to its first homology group is also a natural transformation of functors.

4.2. Functor categories. Let $\mathscr{C}$ and $\mathscr{D}$ be categories with $\mathscr{C}$ small. The collection $\mathrm{Func}(\mathscr{C}, \mathscr{D})$ of functors from $\mathscr{C}$ to $\mathscr{D}$ is category with natural transformations as arrows. If F and G are functors, a natural transformation λ requires, for each object C of $\mathscr{C}$, an element of $\mathrm{Hom}_{\mathscr{D}}(FC, GC)$, subject to the naturality conditions. If $\mathscr{C}$ is small, there is no more than a set of such natural transformations $F \longrightarrow G$ and so this collection is a set. If $\lambda: F \longrightarrow G$ and $\mu: G \longrightarrow H$ are natural transformations, their composite $\mu \circ \lambda$ is defined by requiring its component at C to be $\mu C \circ \lambda C$. Of course, $\mathrm{Func}(\mathscr{C}, \mathscr{D})$ is just $\mathrm{Hom}_{\mathsf{Cat}}(\mathscr{C}, \mathscr{D})$, and so is already a functor in each variable to Set. It is easy to check that for any $F: \mathscr{D} \longrightarrow \mathscr{E}$,

$$\mathrm{Func}(\mathscr{C}, F): \mathrm{Func}(\mathscr{C}, \mathscr{D}) \longrightarrow \mathrm{Func}(\mathscr{C}, \mathscr{E})$$

is actually a functor and not only a Set-function, and similarly for $\mathrm{Func}(F, \mathscr{C})$, so that in each variable Func is actually a Cat-valued functor.

We denote the hom functor in $\mathrm{Func}(\mathscr{C}, \mathscr{D})$ by $\mathbf{Nat}(F, G)$ for functors F, $G: \mathscr{C} \longrightarrow \mathscr{D}$. A category of the form $\mathrm{Func}(\mathscr{C}, \mathscr{D})$ is called a **functor category** and is frequently denoted $\mathscr{D}^{\mathscr{C}}$ especially in the later chapters on sheaves.

4.3. Notation for natural transformations. Suppose there are categories and functors as shown in this diagram:

$$\mathscr{B} \xrightarrow{H} \mathscr{C} \underset{G}{\overset{F}{\rightrightarrows}} \mathscr{D} \xrightarrow{K} \mathscr{E} \qquad (\Downarrow \lambda \text{ between } F \text{ and } G)$$

Note that in diagrams, we often denote a natural transformation by a double arrow: $\lambda: F \Rightarrow G$.

Suppose $\lambda: F \longrightarrow G$ is a natural transformation. Then λ induces two natural transformations $K\lambda: KF \longrightarrow KG$ and $\lambda H: FH \longrightarrow GH$. The component of $K\lambda$ at an object C of $\mathscr{C}$ is

$$K(\lambda C): KFC \longrightarrow KGC$$

Then $K\lambda$ is a natural transformation simply because K, like any functor, takes commutative diagrams to commutative diagrams. The component of λH at an object B of $\mathscr{B}$ is the component of λ at HB. λH is a natural transformation because H is defined on morphisms.

We should point out that although the notations $K\lambda$ and λH look formally dual, they are quite different in meaning. The first is the result of applying a functor to a value of a natural transformation (which is a morphism in the codomain category) while the second is the result of taking the component of a natural transformation at a value of a functor. Nonetheless, the formal properties of the two quite different operations are the same. This is why we use the parallel notation when many other writers use distinct notation. (Compare the use of $\langle f, v\rangle$ for $f(v)$ by many analysts.) Thus advances mathematics.

Exercise 6 below states a number of identities which hold for natural transformations. Some of them are used later in the book, particularly in triple theory.

4.4. Exercises.

1. Show how to describe a natural transformation as a functor from an arrow category to a functor category.

2. What is a natural transformation from one group homomorphism to another?

3. Let $R: \mathscr{C} \longrightarrow \mathscr{D}$ be a functor. Show that $f \mapsto Rf$ is a natural transformation $\mathrm{Hom}_{\mathscr{C}}(C, -) \longrightarrow \mathrm{Hom}_D(RC, R(-))$ for any object C of $\mathscr{C}$.

4. (a) Show that the inclusion of a set A into the free group FA generated by A determines a natural transformation from the identity functor on Set to the functor UF where U is the underlying set functor.

(b) Find a natural transformation from $FU: \mathsf{Grp} \longrightarrow \mathsf{Grp}$ to the identity functor on Grp which takes a one letter word of FUG to itself. Show that there is only one such.

5. In Section 3, we mentioned three ways of defining the powerset as a functor. (See Exercise 6.) For which of these definitions do the maps which take each element x of a set A to the set $\{x\}$ (the "singleton" maps) form a natural transformation from the identity functor to the powerset functor?

6. Let categories and functors be given as in the following diagram.

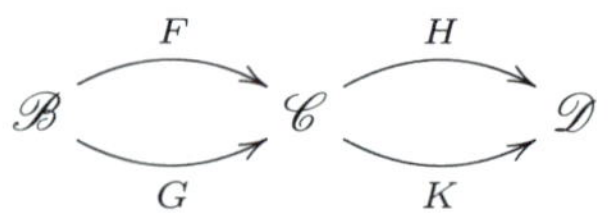

Suppose $\kappa: F \longrightarrow G$ and $\mu: H \longrightarrow K$ are natural transformations.

(a) Show that this diagram commutes:

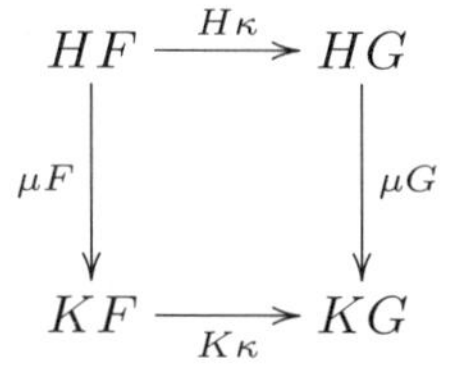

(b) Define $\mu\kappa$ by requiring that its component at B be $\mu GB \circ H\kappa B$, which by (a) is $K\kappa B \circ \mu FB$. Show that $\mu\kappa$ is a natural transformation from $H \circ F$ to $K \circ G$. This defines a composition operation, called **application**, on natural transformations. Although it has the syntax of a composition law, as we will see below, semantically it is the result of *applying* μ to κ. In many, especially older works, it is denoted $\mu * \kappa$, and these books often use juxtaposition to denote composition.

(c) Show that $H\kappa$ and μG have the same interpretation whether thought of as instances of application of a functor to a natural transformation, resp. evaluation of a natural transformation at a functor, or as examples of an application operation where the name of a functor is used to stand for the identity natural transformation. (This exercise may well take longer to understand than to do.)

(d) Show that application as defined above is associative in the sense that if $(\mu\kappa)\beta$ is defined, then so is $\mu(\kappa\beta)$ and they are equal.

(e) Show that the following rules hold, where $\circ$ denotes the composition of natural transformations defined earlier in this chapter. These are called **Godement's rules**. In each case, the meaning of the rule is that if one side is defined, then so is the other and they are equal. They all refer to this diagram, and the name of a functor is used to denote the identity natural transformation from that functor to itself. The other natural transformations are $\kappa\colon F_1 \longrightarrow F_2$, $\lambda\colon F_2 \longrightarrow F_3$, $\mu\colon G_1 \longrightarrow G_2$, and $\nu\colon G_2 \longrightarrow G_3$.

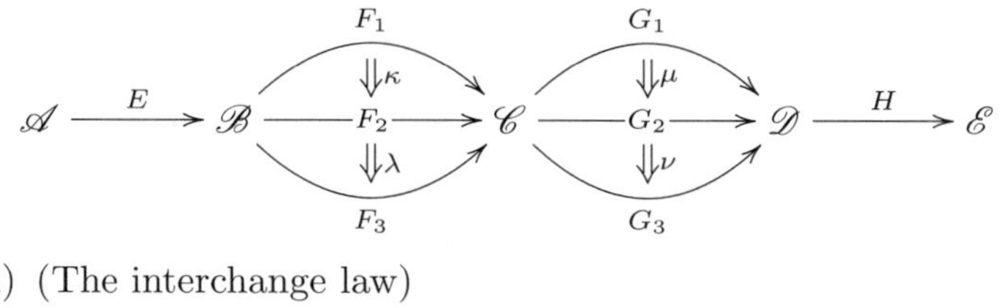

(i) (The interchange law)

$$(\nu \circ \mu)(\lambda \circ \kappa) = (\nu\lambda) \circ (\mu\kappa)$$

(ii) $(H \circ G_1)\kappa = H(G_1\kappa)$.
(iii) $\mu(F_1 \circ E) = (\mu F_1)E$.
(iv) $G_1(\lambda \circ \kappa)E = (G_1\lambda E) \circ (G_1\kappa E)$.
(v) $(\mu F_2) \circ (G_1\kappa) = (G_2\kappa) \circ (\mu F_1)$.

7. Show that two categories $\mathscr{C}$ and $\mathscr{D}$ are equivalent if and only if there are functors $F\colon \mathscr{C} \longrightarrow \mathscr{D}$ and $G\colon \mathscr{D} \longrightarrow \mathscr{C}$ such that $G \circ F$ is naturally equivalent to $\mathrm{id}_{\mathscr{C}}$ and $F \circ G$ is naturally equivalent to $\mathrm{id}_{\mathscr{D}}$.

5. Elements and subobjects

5.1. Elements. One of the important perceptions of category theory is that an arrow $x\colon T \longrightarrow A$ in a category can be regarded as an **element of A defined**

over T. The idea is that x is a *variable* element of A, meaning about the same thing as the word "quantity" in such sentences as, "The quantity x^2 is nonnegative", found in older calculus books.

One must not get carried away by this idea and introduce elements everywhere. One of the main benefits of category theory is you do not have to do things in terms of elements unless it is advantageous to. In 4.2 is a construction that is almost impossible to understand in terms of elements, but is very easy with the correct conceptual framework. On the other hand, we will see many examples later in which the use of elements leads to a substantial simplification. The point is not to allow a tool to become a straitjacket.

When $x: T \longrightarrow A$ is thought of as an element of A defined on T, we say that T is the **domain of variation** of the element x. It is often useful to think of x as an element of A defined in terms of a parameter in T. A related point of view is that x is a set of elements of A indexed by T. By the way, this is distinct from the idea that x is a family of disjoint subsets of T indexed by A, as mentioned in 2.2.

The notation "$x \in^T A$" is a useful quick way of saying that x is an element of A defined on T. This notation will be extended when we consider subobjects later in this section.

If $x \in^T A$ and $f: A \longrightarrow B$, then $f \circ x \in^T B$; thus morphisms can be regarded as functions taking elements to elements. The Yoneda Lemma, Theorem 2 of the next section, says (among other things) that any function which takes elements to elements in a coherent way in a sense that will be defined precisely "is" a morphism of the category. Because of this, we will write $f(x)$ for $f \circ x$ when it is helpful to think of x as a generalized element.

Note that every object A has at least one element id_A, its **generic element**.

If A is an object of a category $\mathscr{C}$ and $F: \mathscr{C} \longrightarrow \mathscr{D}$ is a functor, then F takes any element of A to an element of FA in such a way that (i) generic elements are taken to generic elements, and (ii) the action of F on elements commutes with change of the domain of variation of the element. (If you spell those two conditions out, they are essentially the definition of functor.)

Isomorphisms can be described in terms of elements, too: An arrow $f: A \longrightarrow B$ is an isomorphism if and only if f (thought of as a function) is a bijection between the elements of A defined on T and the elements of B defined on T for all objects T of $\mathscr{C}$. (To get the inverse, apply this fact to the element $\mathrm{id}_A: A \longrightarrow A$.) And a terminal object is a singleton in a very strong sense—for any domain of variation it has exactly one element.

In the rest of this section we will develop the idea of element further and use it to define subobjects, which correspond to subsets of a set.

5.2. Monomorphisms and epimorphisms. An arrow $f: A \longrightarrow B$ is a **monomorphism** (or just a "mono", adjective "monic"), if f (i.e., $\mathrm{Hom}(T, f)$) is injective (one to one) on elements defined on each object T —in other words, for every pair x, y of elements of A defined on T, $f(x) = f(y)$ implies $x = y$.

In terms of composition, this says that f is left cancelable, i.e, if $f \circ x = f \circ y$, then $x = y$. This has a dual concept: The arrow f is an **epimorphism** ("an epi", "epic") if it is *right* cancelable. This is true if and only if the contravariant functor $\mathrm{Hom}(f, T)$ is injective (not surjective!) for every object T. Note that surjectivity is not readily described in terms of generalized elements.

In Set, every monic is injective and every epic is surjective (onto). The same is true of Grp, but the fact that epis are surjective in Grp is moderately hard to prove (Exercise 2). On the other hand, any dense map, surjective or not, is epi in the category of Hausdorff spaces and continuous maps.

An arrow $f: A \longrightarrow B$ which is "surjective on elements", in other words for which $\mathrm{Hom}(T, f)$ is surjective for every object T, is necessarily an epimorphism and is called a **split epimorphism**. An equivalent definition is that there is an arrow $g: B \longrightarrow A$ which is a right inverse to f, so that $f \circ g = \mathrm{id}_B$. The Axiom of Choice is equivalent to the statement that every epi in Set is split. In general, in categories of sets with structure and structure preserving functions, split epis are surjective and (as already pointed out) surjective maps are epic (see Exercise 6), but the converses often do not hold. We have already mentioned Hausdorff spaces as a category in which there are nonsurjective epimorphisms; another example is the embedding of the ring of integers in the field of rational numbers in the category of rings and ring homomorphisms. As for the other converse, in the category of groups the (unique) surjective homomorphism from the cyclic group of order 4 to the cyclic group of order 2 is an epimorphism which is not split.

An arrow with a *left* inverse is necessarily a monomorphism and is called a **split monomorphism**. Split monos in Top are called retractions; in fact the word "retraction" is sometimes used to denote a split mono in any category.

The property of being a split mono or split epi is necessarily preserved by any functor. The property of being monic or epic is certainly not in general preserved by any functor. Indeed, if Ff is epi for every functor F, then f is necessarily a split epi. (Exercise 5.)

Notation: In diagrams, we usually draw an arrow with an arrowhead at its tail:

$$\rightarrowtail$$

to indicate that it is a monomorphism. The usual dual notation for an epimorphism is

$$\twoheadrightarrow$$

However in this book we reserve that latter notation for *regular* epimorphisms to be defined in 8.8.

5.3. Subobjects. We now define the notion of subobject of an object in a category; this idea partly captures and partly generalizes the concept of "subset", "subspace", and so on, familiar in many branches of mathematics.

If $i: A_0 \longrightarrow A$ is a monomorphism and $a: T \longrightarrow A$, we say a **factors through** i (or factors through A_0 if it is clear which monomorphism i is meant) if there is an arrow j for which

(5)

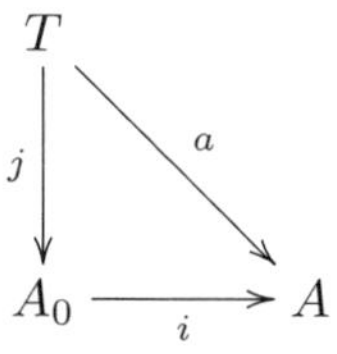

commutes. In this situation we extend the element point of view and say that the element a of A is an element of A_0 (or of i if necessary). This is written "$a \in_A^T A_0$". The subscript A is often omitted if the context makes it clear.

5.4. Lemma. *Let* $i\colon A_0 \longrightarrow A$ *and* $i'\colon A_0' \longrightarrow A$ *be monomorphisms in a category* $\mathscr{C}$. *Then* A_0 *and* A_0' *have the same elements of* A *if and only if they are isomorphic in the category* $\mathscr{C}/A$ *of objects over* A, *in other words if and only if there is an isomorphism* $j\colon A \longrightarrow A'$ *for which*

(6)
$$\begin{array}{ccc} A_0 & & \\ {\scriptstyle j}\downarrow & \searrow^{i} & \\ A_0' & \xrightarrow[i']{} & A \end{array}$$

commutes.

Proof. Suppose A_0 and A_0' have the same elements of A. Since $i \in_A^{A_0} A_0$, it factors through A_0', so there is an arrow $j\colon A_0 \longrightarrow A_0'$ such that (2) commutes. Interchanging A_0 and A_0' we get $k\colon A_0' \longrightarrow A_0$ such that $i \circ k = i'$. Using the fact that i and i' are monic, it is easy to see that j and k must be inverses to each other, so they are isomorphisms.

Conversely, if j is an isomorphism making (2) commute and $a \in_A^T A_0$, so that $a = i \circ u$ for some $u\colon T \longrightarrow A_0$, then $a = i' \circ j \circ u$ so that $a \in_A^T A_0'$. A similar argument interchanging A_0 and A_0' shows that A_0 and A_0' have the same elements of A. □

Two monomorphisms are said to be **equivalent** if they have the same elements. A **subobject** of A is an equivalence class of monomorphisms into A. We will frequently refer to a subobject by naming one of its members, as in "Let $A_0 \rightarrowtail A$ be a subobject of A".

In Set, each subobject of a set A contains exactly one inclusion of a subset into A, and the subobject consists of those injective maps into A which has that subset as image. Thus "subobject" captures the notion of "subset" in Set exactly.

Any map from a terminal object in a category is a monomorphism and so determines a subobject of its target. Because any two terminal objects are isomorphic *by a unique isomorphism* (Exercise 6 of Section 2), that subobject contains exactly one map based on each terminal object. We will henceforth assume that in any category we deal with, we have picked a particular terminal object (if it has one) as the canonical one and call it "*the* terminal object".

5.5. Exercises.

1. Describe initial objects using the terminology of elements, and using the terminology of indexed families of subsets.

(a) Show that in Set, a function if injective if and only if it is a monomorphism and surjective if and only if it is an epimorphism.

(b) Show that every epimorphism in Set is split. (This is the Axiom of Choice).

(c) Show that in the category of Abelian groups and group homomorphisms, a homomorphism is injective if and only if it is a monomorphism and surjective if and only if it is an epimorphism.

(d) Show that neither monos nor epis are necessarily split in the category of Abelian groups.

2. Show that in Grp, every homomorphism is injective if and only if it is a monomorphism and surjective if and only if it is an epimorphism. (If you get stuck trying to show that an epimorphism in Grp is surjective, see the hint on page 21 of Mac Lane [1971].)

3. Show that all epimorphisms are surjective in Top, but not in the category of all Hausdorff spaces and continuous maps.

4. Show that the embedding of an integral domain (assumed commutative with unity) into its field of quotients is an epimorphism in the category of commutative rings and ring homomorphisms. When is it a split epimorphism?

(a) Show that the following two statements about an arrow $f: A \longrightarrow B$ in a category $\mathscr{C}$ are equivalent:

(b) $\text{Hom}(T, f)$ is surjective for every object T of $\mathscr{C}$.

(c) There is an arrow $g: B \longrightarrow A$ such that $f \circ g = \text{id}_B$.

(d) Show that any arrow satisfying the conditions of (a) is an epimorphism.

5. Show that if Ff is epi for every functor F, then f is a split epi.

6. Let $U: \mathscr{C} \longrightarrow \mathsf{Set}$ be a faithful functor and f an arrow of $\mathscr{C}$. (Note that the functors we have called "forgetful"—we have not defined that word formally—are obviously faithful.) Prove:

(a) If Uf is surjective then f is an epimorphism.

(b) If f is a split epimorphism then Uf is surjective.

(c) If Uf is injective then f is a monomorphism.

(d) If f is a split monomorphism, then Uf is injective.

7. A **subfunctor** of a functor $F: \mathscr{C} \longrightarrow \mathsf{Set}$ is a functor G with the properties

(a) $GA \subseteq FA$ for every object A of $\mathscr{C}$.

(b) If $f: A \longrightarrow B$, then $Gf(GA) \subseteq GB$.

Show that the subfunctors of a functor are the "same" as subobjects of the functor in the category $\text{Func}(\mathscr{C}, \mathsf{Set})$.

6. The Yoneda Lemma

6.1. Elements of a functor. A functor $F: \mathscr{C} \longrightarrow \mathsf{Set}$ is an object in the functor category $\text{Func}(\mathscr{C}, \mathsf{Set})$: an "**element**" of F is therefore a natural transformation into F. The Yoneda Lemma, Lemma 1 below, says in effect that the elements of a Set-valued functor F defined (in the sense of Section 5) on the hom functor $\text{Hom}(A, -)$ for some object A of $\mathscr{C}$ are essentially the same as the (ordinary) elements of the set FA. To state this properly requires a bit of machinery.

If $f: A \longrightarrow B$ in $\mathscr{C}$, then f induces a natural transformation from $\text{Hom}(B, -)$ to $\text{Hom}(A, -)$ by composition: the component of this natural transformation at an object C of $\mathscr{C}$ takes an arrow $h: B \longrightarrow C$ to $h \circ f: A \longrightarrow C$. This construction defines a contravariant functor from $\mathscr{C}$ to $\text{Func}(\mathscr{C}, \mathsf{Set})$ called the **Yoneda map**. It is straightforward and very much worthwhile to check that this construction really does give a natural transformation for each arrow f and that the resulting Yoneda map really is a functor.

Because $\mathbf{Nat}(-, -)$ is contravariant in the first variable (it is a special case of Hom), the map which takes an object B of $\mathscr{C}$ and a functor $F: \mathscr{C} \longrightarrow \mathsf{Set}$ to $\mathbf{Nat}(\text{Hom}(B, -), F)$ is a functor from $\mathscr{C} \times \text{Func}(\mathscr{C}, \mathsf{Set})$ to Set. Another such functor is the evaluation functor which takes (B, F) to FB, and (g, λ), where $g: B \longrightarrow A \in \mathscr{C}$ and $\lambda: F \longrightarrow G$ is a natural transformation, to $Gg \circ \lambda B$. Remarkably, these two functors are naturally isomorphic; it is in this sense that the elements of F defined on $\mathrm{Hom}(B, -)$ are the ordinary elements of FB.

6.2. Lemma (Yoneda). *The map* $\phi: \mathbf{Nat}(\mathrm{Hom}(B, -), F) \longrightarrow FB$ *defined by* $\phi(\lambda) = \lambda B(\mathrm{id}_B)$ *is a natural isomorphism of the functors defined in the preceding paragraph.*

Proof. The inverse of ϕ takes an element u of FB to the natural transformation λ defined by requiring that $\lambda A(g) = Fg(u)$ for $g \in \mathrm{Hom}(B, A)$. The rest of proof is a routine verification of the commutativity of various diagrams required by the definitions. □

The first of several important consequences of this lemma is the following embedding theorem. This theorem is obtained by taking F in the Lemma to be $\mathrm{Hom}(A, -)$, where A is an object of $\mathscr{C}$; this results in the statement that there is a natural bijection between arrows $g: A \longrightarrow B$ and natural transformations from $\mathrm{Hom}(B, -)$ to $\mathrm{Hom}(A, -)$.

6.3. Theorem (Yoneda Embeddings).

(1) The map which takes $f: A \longrightarrow B$ *to the induced natural transformation*

$$\mathrm{Hom}(B, -) \longrightarrow \mathrm{Hom}(A, -)$$

is a full and faithful contravariant functor from $\mathscr{C}$ *to* $\mathrm{Func}(\mathscr{C}, \mathsf{Set})$.

(2) The map taking f *to the natural transformation*

$$\mathrm{Hom}(-, A) \longrightarrow \mathrm{Hom}(-, B)$$

is a full and faithful functor from $\mathscr{C}$ *to* $\mathrm{Func}(\mathscr{C}^{\mathrm{op}}, \mathsf{Set})$.

Proof. It is easy to verify that the maps defined in the Theorem are functors. The fact that the first one is full and faithful follows from the Yoneda Lemma with $\mathrm{Hom}(A, -)$ in place of F. The other proof is dual. □

The induced maps in the Theorem deserve to be spelled out. If $f: S \longrightarrow T$, the natural transformation corresponding to f given by (i) has component $\mathrm{Hom}(f, A): \mathrm{Hom}(T, A) \longrightarrow \mathrm{Hom}(S, A)$ at an object A of $\mathscr{C}$—this is composing by f on the right. If $x \in^T A$, the action of $\mathrm{Hom}(f, A)$ "changes the parameter" in A along f. The other natural transformation corresponding to f is $\mathrm{Hom}(T, f): \mathrm{Hom}(T, A) \longrightarrow \mathrm{Hom}(T, B)$; since the Yoneda embedding is faithful, we can say that f is essentially the same as $\mathrm{Hom}(-, f)$. If x is an element of A based on T, then $\mathrm{Hom}(T, f)(x) = f \circ x$. Since "$f$ is essentially the same as $\mathrm{Hom}(-, f)$", this justifies the notation $f(x)$ for $f \circ x$ introduced in Section 5. The fact that the Yoneda embedding is full means that *any natural transformation* $\mathrm{Hom}(-, A) \longrightarrow \mathrm{Hom}(-, B)$ determines a morphism $f: A \longrightarrow B$, namely the image of id_A under the component of the transformation at A. Spelled out, this says that if f is any function which assigns to every element $x: T \longrightarrow A$ an element $f(x): T \longrightarrow B$ with the property that for all $T: S \longrightarrow T$, $f(x \circ t) = f(x) \circ t$ (this is the "Section 5) then f "morphism, also called f to conform to our conventions, from A to B. One says such an arrow exists "by Yoneda".

In the same vein, if $g: 1 \longrightarrow A$ is a morphism of $\mathscr{C}$, then for any object T, g determines an element $g(\ )$ of A defined on T by composition with the unique element from T to 1, which we denote $(\,)$. This notation captures the perception that a global element depends on no arguments. We will extend the functional notation to more than one variable in Section 8.

6.4. Universal elements. Another special case of the Yoneda Lemma occurs when one of the elements of F defined on $\mathrm{Hom}(A, -)$ is a natural isomorphism. If $\beta\colon \mathrm{Hom}(A, -) \longrightarrow F$ is such a natural isomorphism, the (ordinary) element $u \in FA$ corresponding to it is called a **universal element** for F, and F is called a **representable functor**, represented by A. It is not hard to see that if F is also represented by A', then A and A' are isomorphic objects of $\mathscr{C}$. (See Exercise 3, which actually says more than that.)

The following lemma gives a characterization of universal elements which in many books is given as the definition.

6.5. LEMMA. *Let $F\colon \mathscr{C} \longrightarrow \mathsf{Set}$ be a functor. Then $u \in FA$ is a universal element for F if and only if for every object B of $\mathscr{C}$ and every element $t \in FB$ there is exactly one arrow $g\colon A \longrightarrow B$ such that $Fg(u) = t$.*

PROOF. If u is such a universal element corresponding to a natural isomorphism $\beta\colon \mathrm{Hom}(A, -) \longrightarrow F$, and $t \in FB$, then the required arrow g is the element $(\beta^{-1}B)(t)$ in $\mathrm{Hom}(A, B)$. Conversely, if $u \in FA$ satisfies the conclusion of the Lemma, then it corresponds to some natural transformation $\beta\colon \mathrm{Hom}(A, -) \longrightarrow F$ by the Yoneda Lemma. It is routine to verify that the map which takes $t \in FB$ to the arrow $g \in \mathrm{Hom}(A, B)$ given by the assumption constitutes an inverse in $\mathrm{Func}(\mathscr{C}, \mathsf{Set})$ to βB. □

In this book, the phrase "$u \in FA$ is a universal element for F" carries with it the implication that u and A have the property of the lemma. (It is possible that u is also an element of FB for some object B but not a universal element in FB.)

As an example, let G be a free group on one generator g. Then g is the "universal group element" in the sense that it is a universal element for the underlying set functor $U\colon \mathsf{Grp} \longrightarrow \mathsf{Set}$ (more precisely, it is a universal element in UG). This translates into the statement that for any element x in any group H there is a unique group homomorphism $F\colon G \longrightarrow H$ taking g to x, which is exactly the definition of "free group on one generator g".

Another example which will play an important role in this book concerns the contravariant powerset functor $\mathbf{P}\colon \mathsf{Set} \longrightarrow \mathsf{Set}$ defined in Section 3. It is straightforward to verify that a universal element for $\mathbf{P}$ is the subset $\{1\}$ of the set $\{0, 1\}$; the function required by the Lemma for a subset B_0 of a set B is the characteristic function of B_0. (A universal element for a contravariant functor, as here—meaning a universal element for $\mathbf{P}\colon \mathsf{Set}^{\mathrm{op}} \longrightarrow \mathsf{Set}$—is often called a "couniversal element".)

6.6. Exercises.

1. Find a universal element for the functor

$$\mathrm{Hom}(-, A) \times \mathrm{Hom}(-, B)\colon \mathsf{Set}^{\mathrm{op}} \longrightarrow \mathsf{Set}$$

for any two sets A and B. (If $h\colon U \longrightarrow V$, this functor takes a pair (f, g) to $(h \circ f, h \circ g)$.)

(a) Show that an action of a group G on a set A is essentially the same thing as a functor from G regarded as a category to Set.

(b) Show that such an action has a universal element if and only if for any pair x and y of elements of A there is exactly one element g of G for which $gx = y$.

2. Are either of the covariant powerset functors defined in Exercise 6 of Section 3 representable?

3. Let $F: \mathscr{C} \longrightarrow \mathsf{Set}$ be a functor and $u \in FA$, $u' \in FA'$ be universal elements for F. Show that there is a unique isomorphism $\phi: A \longrightarrow A'$ such that $F\phi(u) = u'$.

4. Let $U: \mathsf{Grp} \longrightarrow \mathsf{Set}$ be the underlying set functor, and $F: \mathsf{Set} \longrightarrow \mathsf{Grp}$ the functor which takes a set A to the free group on A. Show that for any set A, the covariant functor $\mathrm{Hom}_{\mathsf{Set}}(A, U(-))$ is represented by FA, and for any group G, the contravariant functor $\mathrm{Hom}_{\mathsf{Grp}}(F(-), G)$ is represented by UG.

7. Pullbacks

7.1. The set P of composable pairs of arrows used in Section 1.1 in the alternate definition of category is an example of a "fibered product" or "pullback". A pullback is a special case of "limit", which we treat in Section 8. In this section, we discuss pullbacks in detail.

Let us consider the following diagram D in a category $\mathscr{C}$.

(7)

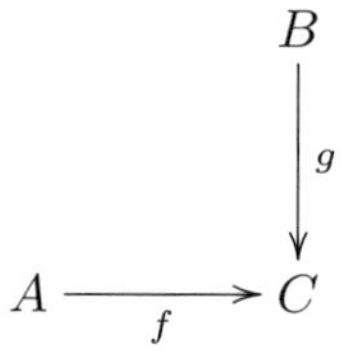

We would like to *objectify* the set $\{(x, y) \mid f(x) = g(y)\}$ in $\mathscr{C}$; that is, find an object of $\mathscr{C}$ whose elements are those pairs (x, y) with $f(x) = g(y)$. Observe that for a pair (x, y) to be in this set, x and y must be elements of A and B respectively defined over the same object T.

The set of composable pairs of arrows in a category (see Section 2) are a special case in Set of this, with $A = B$ being the set of arrows and $f = d^0$, $g = d^1$.

Thus we must consider commutative diagrams like

(8)

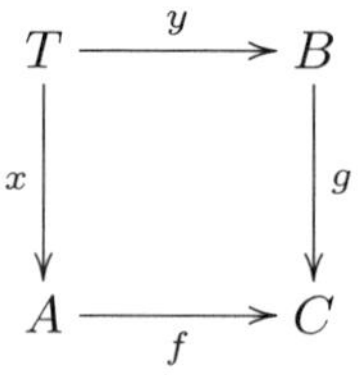

In this situation, (T, x, y) is called a **commutative cone over D based on T**. We denote by $\mathrm{Cone}(T, D)$ the set of commutative cones over D based on T. A commutative cone based on T over D may usefully be regarded as an element of D defined on T. In Section , we will see that a commutative cone is actually an arrow in a certain category, so that this idea fits with our usage of the word "element".

Our strategy will be to turn $\mathrm{Cone}(-, D)$ into a functor; then we will say that an object represents (in an informal sense) elements of D, in other words pairs (x, y) for which $f(x) = g(y)$, if that object represents (in the precise technical sense) the functor $\mathrm{Cone}(-, D)$.

We make will make $\mathrm{Cone}(-,D)$ into a contravariant functor to Set: If $h\colon W \longrightarrow T$ is an arrow of $\mathscr{C}$ and (T,x,y) is a commutative cone over (1), then

$$\mathrm{Cone}(h,D)(T,x,y) = (W, x\circ h, y\circ h)$$

which it is easy to see is a commutative cone over D based on W.

An element (P,p_1,p_2) of D which is a universal element for $\mathrm{Cone}(-,D)$ (so that $\mathrm{Cone}(-,D)$ is representable) is called the **pullback** or the **fiber product** of the diagram D. The object P is often called the pullback, with p_1 and p_2 understood. As the reader can verify, this says that (P,p_1,p_2) is a pullback if

(9)

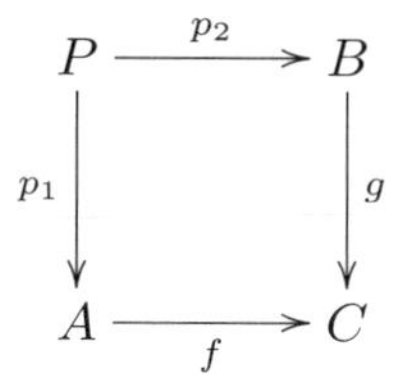

commutes and for any element of D based on T, there is a unique element of $z \in P$ based on T which makes

(10)

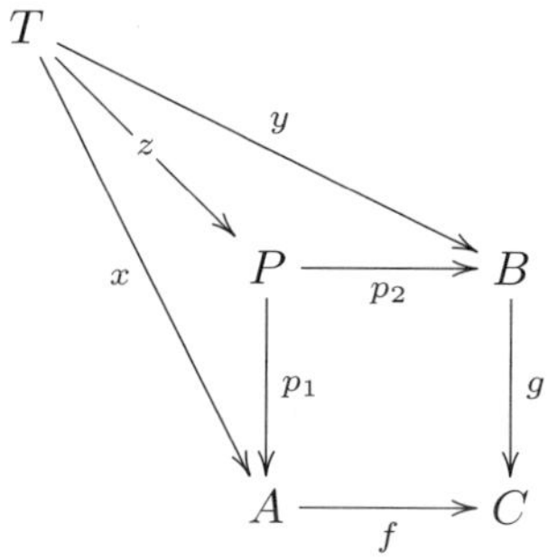

commute. Thus there is a bijection between the elements of the diagram D defined on T and the elements of the fiber product P defined on T. When a diagram like 10 has this property it is called a **pullback diagram**.

The Cone functor exists for any category, but a particular diagram of the form 7 need not have a pullback.

7.2. Proposition. *If Diagram* 9 *is a pullback diagram, then the cone in Diagram* 8 *is also a pullback of Diagram* 7 *if and only if the unique arrow from T to P making everything in Diagram* 10 *commute is an isomorphism.*

Proof. (This theorem actually follows from Exercise 3 of Section 6, but I believe a direct proof is instructive.) Assume that (2) and (3) are both pullback diagrams. Let $u\colon T \longrightarrow P$ be the unique arrow given because 9 is a pullback diagram, and let $v\colon P \longrightarrow T$ be the unique arrow given because 8 is a pullback diagram. Then both for $g = u\circ v\colon P \longrightarrow P$ and $g = \mathrm{id}_P$ it is true that $p_1 \circ g = p_1$ and $p_2 \circ g = p_2$. Therefore by the uniqueness part of the definition of universal element, $u \circ v = \mathrm{id}_P$. Similarly, $v\circ u = \mathrm{id}_T$, so that u is an isomorphism between T and P making everything commute. The converse is easy. □

The preceding argument is typical of many arguments making use of the uniqueness part of the definition of universal element. We will usually leave arguments like this to the reader.

A consequence of Proposition 1 is that a pullback of a diagram in a category is not determined uniquely but only up to a "unique isomorphism which makes everything commute". This is an instance of a general fact about constructions defined as universal elements which is made precise in Proposition 1 of Section 8.

7.3. Notation for pullbacks. We have defined the pullback P of Diagram 7 so that it objectifies the set $\{(x,y) \mid f(x) = g(y)\}$. This fits nicely with the situation in Set, where one pullback of (1) *is* the set $\{(x,y) \mid f(x) = g(y)\}$ together with the projection maps to A and B, and any other pullback is in one to one correspondence with this one by a bijection which commutes with the projections. This suggest the introduction of a setlike notation for pullbacks: We let $[(x,y) \mid f(x) = g(y)]$ denote a pullback of (1). In this notation, $f(x)$ denotes $f \circ x$ and $g(y)$ denotes $g \circ y$ as in Section 5, and (x,y) denotes the unique element of P defined on T which exists by definition of pullback. It follows that $p_1(x,y) = x$ and $p_2(x,y) = y$, where we write $p_1(x,y)$ (not $p_1((x,y))$) for $p_1 \circ (x,y)$.

The idea is that square brackets around a set definition denotes an object of the category which represents the set of arrows listed in curly brackets—"represents" in the technical sense, so that the set in curly brackets has to be turned into the object map of a set-valued functor. The square bracket notation is ambiguous. Proposition 1 spells out the ambiguity precisely.

We could have defined a commutative cone over (1) in terms of *three* arrows, namely a cone (T,x,y,z) based on T would have $x{:}\,T \longrightarrow A$, $y{:}\,T \longrightarrow B$ and $z{:}\,T \longrightarrow C$ such that $f \circ x = g \circ y = z$. Of course, z is redundant and in consequence the Cone functor defined this way would be naturally isomorphic to the Cone functor defined above, and so would have the same universal elements. (The component of the natural isomorphism at T takes (T,x,y) to $(T,x,y,f \circ x)$). Thus the pullback of (1) also represents the set $\{(x,y,z) \mid f(x) = g(y) = z\}$, and so could be denoted $[(x,y,z) \mid f(x) = g(y) = z]$. Although this observation is inconsequential here, it will become more significant when we discuss more general constructions (limits) defined by cones.

There is another way to construct a pullback in Set when the map g is monic. In general, when g is monic, $\{(x,y) \mid f(x) = g(y)\} \cong \{x \mid f(x) \in g(B)\}$, which in Set is often denoted $f^{-1}(B)$. In general, a pullback along a subobject can be interpreted as an inverse image which as we will see is again a subobject.

The pullback Diagram 9 is often regarded as a sort of generalized inverse image construction even when g is not monic. In this case, it is called the "pullback of g along f". Thus when P is regarded as the fiber product, the notion of pullback is symmetrical in A and B, but when it is regarded as the generalized inverse image of B then the diagram is thought of as asymmetrical.

A common notation for the pullback of (1) reflecting the perception of a pullback as fiber product is "$A \times_C B$".

7.4. The subobject functor. In this section, we will turn the subobject construction into a contravariant functor, by using the inverse image construction described above. To do this, we need to know first that the inverse image of a monomorphism is a monomorphism:

7.5. LEMMA. *In any category $\mathscr{C}$, in a pullback diagram (3), if f is monic then so is p_2.*

PROOF. Consider the diagram below, in which the square is a pullback.

(11)

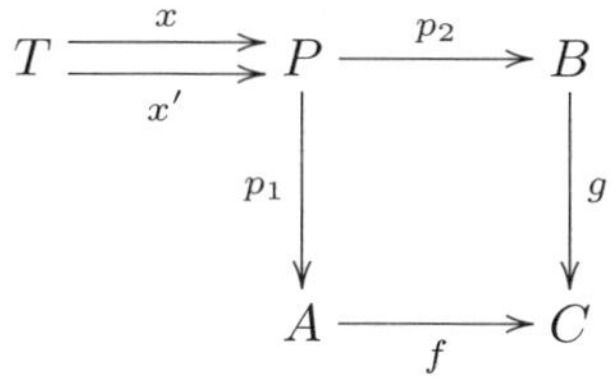

Then $P = [(a,b) \mid fa = gb]$. Showing that p_2 is monic is the same as showing that if $x = (a,b) \in^T P$ and $x' = (a',b') \in^T P$ for which $p_2(x) = p_2(x')$, then $x = x'$. But then $b = b'$ and it follows that $f(a) = g(b) = g(b') = f(a')$. Since f is monic, we see that $a = a'$ and hence that $x = x'$. □

To turn the subobject construction into a functor, we need more than that the pullback of monics is monic. We must know that the pullback of a subobject is a well-defined subobject. In more detail, for A in $\mathscr{C}$, Sub$A$ will be the set of subobjects of A. If $f: B \longrightarrow A$, then for a subobject represented by a monic $g: U \longrightarrow A$, $\mathrm{Sub}(f)(g)$ will be the pullback of g along f. To check that $\mathrm{Sub}(f)$ is well-defined, we need:

7.6. THEOREM. *If $g: U \rightarrowtail A$ and $h: V \rightarrowtail A$ determine the same subobject, then the pullbacks of g and h along $f: B \longrightarrow A$ represent the same subobjects of B.*

PROOF. This follows because the pullback of g is $[y \mid f(y) \in_A^P U]$ and the pullback of h is $[y \mid f(y) \in_A^P V]$, which has to be the same since by definition a subobject is entirely determined by its elements. □

The verification that $\mathrm{Sub}(f)$ is a functor is straightforward and is omitted.

7.7. Exercises.

1. Show how to describe the kernel of a group homomorphism $f: G \longrightarrow H$ as the pullback of f along the map which takes the trivial group to the identity of H.

2. Give an example of a pullback of an epimorphism which is not an epimorphism.

3. Prove that an arrow $f: A \longrightarrow B$ is monic if and only if the diagram

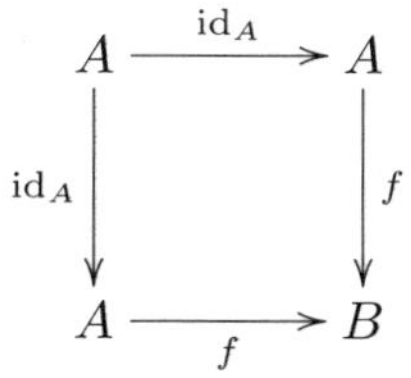

is a pullback.

4. **(a)** Suppose that

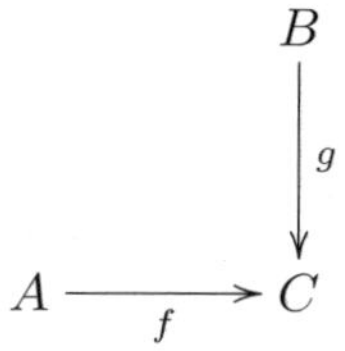

is a diagram in Set with g an inclusion. Construct a pullback of the diagram as a fiber product and as an inverse image of A along f, and describe the canonical isomorphism between them.

(b) Suppose that g is injective, but not necessarily an inclusion. Find two ways of constructing the pullback in this case, and find the isomorphism between them.

(c) Suppose f and g are both injective. Construct the pullback of Diagram 8 in four different ways: (i) fiber product, (ii) inverse image of the image of g along f, (iii) inverse image of the image of f along g, (iv) and the intersection of the images of f and g. Find all the canonical isomorphisms.

(d) Investigate which of the constructions in (c) coincide when one or both of f and g are inclusions.

5. When g is monic in diagram (1), redefine "Cone" so that

(a) $\mathrm{Cone}(T, D) = \{(x, z) \mid z \in B \text{ and } f(x) = z\}$, or equivalently

(b) $\mathrm{Cone}(T, D) = \{x \mid f(x) \in B\}$.

Show that each definition gives a functor naturally isomorphic to the Cone functor originally defined.

6. Identify pullbacks in a poset regarded as a category. Apply this to the powerset of a set, ordered by inclusion.

7. For two subobjects $g\colon U \longrightarrow A$ and $h\colon V \longrightarrow A$, say that $U \leq V$ (or $g \leq h$) if g factors through h. Show that this makes the set of subobjects of A a partially ordered set with a maximum element.

8. In a diagram

$$\begin{array}{ccccc} A & \longrightarrow & B & \longrightarrow & C \\ \downarrow & & \downarrow & & \downarrow \\ D & \longrightarrow & E & \longrightarrow & F \end{array}$$

(a) Show that if both small squares are pullbacks, so is the outer square.

(b) Show that if the outer square and right hand square are pullbacks, so is the left hand square.

8. Limits and colimits

8.1. Graphs. A limit is the categorical way of defining an object by means of equations between elements of given objects. The concept of pullback as described in Section 7 is a special case of limit, but sufficiently complicated to be characteristic of the general idea. To give the general definition, we need a special notion of "graph". What we call a graph here is what a graph theorist would probably call a "directed multigraph with loops".

Formally, a **graph G** consists of two sets, a set O of **objects** and a set A of **arrows**, and two functions $d^0, d^1\colon A \longrightarrow O$. Thus a graph is a "category without composition" and we will use some of the same terminology as for categories: O is the set of **objects** (or sometimes **nodes**) and A is the set of **arrows** of the graph; if f is an arrow, $d^0(f)$ is the **source** of f and $d^1(f)$ is the **target** of f.

A **homomorphism** $F\colon \mathbf{G} \longrightarrow \mathscr{H}$ from a graph $\mathbf{G}$ to a graph $\mathscr{H}$ is a function taking objects to objects and arrows to arrows and preserving source and target; in other words, if $f\colon A \longrightarrow B$ in $\mathbf{G}$, then $F(f)\colon F(A) \longrightarrow F(B)$ in $\mathscr{H}$.

It is clear that every category $\mathscr{C}$ has an **underlying graph** which we denote $|\mathscr{C}|$; the objects, arrows, source and target maps of $|\mathscr{C}|$ are just those of $\mathscr{C}$. Moreover, any functor $F\colon \mathscr{C} \longrightarrow \mathscr{D}$ induces a graph homomorphism $|F|\colon |\mathscr{C}| \longrightarrow |\mathscr{D}|$. It is easy to see that this gives an **underlying graph functor** from the category of categories and functors to the category of graphs and homomorphisms. A **diagram** in a category $\mathscr{C}$ (or in a graph $\mathbf{G}$—the definition is the same) is a graph homomorphism $D\colon \mathscr{I} \longrightarrow |\mathscr{C}|$ for some graph $\mathscr{I}$. $\mathscr{I}$ is the **index graph** of the diagram. Such a diagram is called a **diagram of type** $\mathscr{I}$. For example, a diagram of the form of 7 of Section 7 (which we used to define pullbacks) is a diagram of type $\mathscr{I}$ where $\mathscr{I}$ is the graph

$$1 \longrightarrow 2 \longleftarrow 3$$

D is called a **finite diagram** if the index category has only a finite number of nodes and arrows.

We will write $D\colon \mathscr{I} \longrightarrow \mathscr{C}$ instead of $D\colon \mathscr{I} \longrightarrow |\mathscr{C}|$; this conforms to standard notation.

Observe that any object A of $\mathscr{C}$ is the image of a constant graph homomorphism $K\colon \mathscr{I} \longrightarrow \mathscr{C}$ and so can be regarded as a degenerate diagram of type $\mathscr{I}$.

If D and E are two diagrams of type $\mathscr{I}$ in a category $\mathscr{C}$, a **natural transformation** $\lambda\colon D \longrightarrow E$ is defined in exactly the same way as a natural transformation of functors (which does not involve the composition of arrows in the domain category anyway); namely, λ is a family of arrows

$$\lambda i\colon D(i) \longrightarrow E(i)$$

of $\mathscr{C}$, one for each object i of $\mathscr{I}$, for which

(12)
$$\begin{array}{ccc} D(i) & \xrightarrow{\lambda i} & E(i) \\ {\scriptstyle D(e)}\downarrow & & \downarrow{\scriptstyle E(e)} \\ D(j) & \xrightarrow[\lambda j]{} & E(j) \end{array}$$

commutes for each arrow $e\colon i \longrightarrow j$ of $\mathscr{I}$.

A **commutative cone** with vertex W over a diagram $D\colon \mathscr{I} \longrightarrow \mathscr{C}$ is a natural transformation α from the constant functor with value W on $\mathscr{I}$ to D. We will refer to it as the "cone $\alpha\colon W \longrightarrow D$". This amounts to giving a compatible family $\{\alpha i\}$ of elements of the vertices $D(i)$ based on W. This commutative cone α is an element (in the category of diagrams of type $\mathscr{I}$) of the diagram D based on the constant diagram W. The individual elements αi (elements in $\mathscr{C}$) are called the **components** of the element α.

Thus to specify a commutative cone with vertex W, one must give for each object i of $\mathscr{I}$ an element αi of $D(i)$ based on W (that is what makes it a cone) in such a way that if $e\colon i \longrightarrow j$ is an arrow of $\mathscr{I}$, then $D(e)(\alpha i) = \alpha j$ (that makes it commutative). This says that the following diagram must commute for all $e: i \longrightarrow j$.

(13)

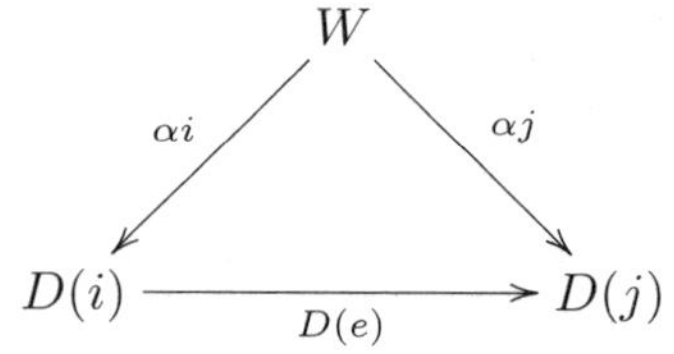

The definition of commutative cone for pullbacks in Section 7 does not fit our present definition, since we give no arrow to C in Diagram 13. Of course, this is only a technicality, since there is an *implied* arrow to C which makes it a commutative cone. This is why we gave an alternative, but equivalent construction in terms of three arrows in Section 7.

Just as in the case of pullbacks, an arrow $W' \longrightarrow W$ defines a commutative cone over D with vertex W' by composition, thus making $\mathrm{Cone}(-, D): \mathscr{C} \longrightarrow \mathsf{Set}$ a contravariant functor. ($\mathrm{Cone}(W, D)$ is the set of commutative cones with vertex W.) Then a **limit** of D, denoted $\lim D$, is a universal element for $\mathrm{Cone}(-, D)$.

Any two limits for D are isomorphic via a unique isomorphism which makes everything commute. This is stated precisely by the following proposition, whose proof is left as an exercise.

8.2. PROPOSITION. *Suppose $D: \mathscr{I} \longrightarrow \mathscr{C}$ is a diagram in a category $\mathscr{C}$ and $\alpha: W \longrightarrow D$ and $\beta: V \longrightarrow D$ are both limits of D. Then there is a unique isomorphism $u: V \longrightarrow W$ such that for every object i of $\mathscr{I}$, $\alpha i \circ u = \beta i$.*

The limit of a diagram D objectifies the set

$$\{x \mid x(i) \in D(i) \text{ and for all } e: i \longrightarrow j, D(e)(x(i)) = x(j)\}$$

and so will be denoted

$$[x \mid x(i) \in D(i) \text{ and for all } e: i \longrightarrow j, D(e)(x(i)) = x(j)]$$

As in the case of pullbacks, implied arrows will often be omitted from the description. In particular, when $y \in^T B$ and $g: A \longrightarrow B$ is a monomorphism we will often write "$y \in A$" or if necessary $\exists x(g(x) = y)$ when it is necessary to specify g.

By taking limits of different types of diagrams one obtains many well known constructions in various categories. We can recover subobjects, for example, by noting that the limit of the diagram $g: A \longrightarrow B$ is the commutative cone with vertex A and edges id_A and g. Thus the description of this limit when g is monic is $[(x, y) \mid gx = y] = [y \mid y \in A]$, which is essentially the same as the subobject determined by g since a subobject is determined entirely by its elements. In other words, the monomorphisms which could be this limit are precisely those equivalent to (in the same subobject as) g in the sense of Section 7.

A category $\mathscr{C}$ is **complete** if every diagram in the category has a limit. It is **finitely complete** if every finite diagram has a limit. Set, Grp and Top are all complete.

8.3. Products. A **discrete** graph is a graph with no arrows. If the set $\{1, 2\}$ is regarded as a discrete graph $\mathscr{I}$, then a diagram of type $\mathscr{I}$ in a category $\mathscr{C}$ is simply an ordered pair of objects of $\mathscr{C}$. A commutative cone over the diagram

(A, B) based on T is simply a pair (x, y) of elements of A and B. Commutativity in this case is a vacuous condition.

Thus a limit of this diagram represents the set $\{(x, y) \mid x \in A, y \in B\}$ and is called the **product** of A and B. It is denoted $A \times B = [(x, y) \mid x \in A, y \in B]$. The object $B \times A = [(y, x) \mid y \in B, x \in A]$ is differently defined, but it is straightforward to prove that it must be isomorphic to $A \times B$.

It follows from the definition that $A \times B$ is an object P together with two arrows $p_1 : P \longrightarrow A$ and $p_2 : P \longrightarrow B$ with the property that for any elements x of A and y of B based on T there is a unique element (x, y) of $A \times B$ based on T such that $p_1(x, y) = x$ and $p_2(x, y) = y$. These arrows are conventionally called the **projections**, even though they need not be epimorphisms. Conversely, any element h of $A \times B$ based on T must be of the form (x, y) for some elements of A and B respectively based on T: namely, $x = p_1(h)$ and $y = p_2(h)$. In other words, there is a canonical bijection between $\mathrm{Hom}(T, A \times B)$ and $\mathrm{Hom}(T, A) \times \mathrm{Hom}(T, B)$ (this is merely a rewording of the statement that $A \times B$ represents $\{(x, y) : x \in A, y \in B\}$).

Note that (x, x') and (x', x) are distinct elements of $A \times A$ if x and x' are distinct, because $p_1(x, x') = x$, whereas $p_1(x', x) = x'$. In fact, $(x, x') = (p_2, p_1) \circ (x, x')$.

If $f : A \longrightarrow C$ and $g : B \longrightarrow D$, then we define

$$f \times g = (f \circ p_1, g \circ p_2) : A \times B \longrightarrow C \times D$$

Thus for elements x of A and y of B defined on the same object, $(f \times g)(x, y) = (f(x), g(y))$.

It should be noted that the notation $A \times B$ carries with it the information about the arrows p_1 and p_2. Nevertheless, one often uses the notation $A \times B$ to denote the object P; the assumption then is that there is a well-understood pair of arrows which make it the genuine product. We point out that in general there may be no canonical choice of which object to take be $X \times Y$, or which arrows as projections. There is apparently such a canonical choice in Set but that requires one to choose a canonical way of defining ordered pairs.

In a poset regarded as a category, the product of two elements is their infimum, if it exists. In a group regarded as a category, products do not exist unless the group has only one element. The direct product of two groups is the product in Grp and the product of two topological spaces with the product topology is the product in Top. There are similar constructions in a great many categories of sets with structure.

The product of any indexed collection of objects in a category is defined analogously as the limit of the diagram $D : \mathscr{I} \longrightarrow \mathscr{C}$ where $\mathscr{I}$ is the index set considered as the objects of a graph with no arrows and D is the indexing function. This product is denoted $\prod_{i \in \mathscr{I}} Di$, although explicit mention of the index set is often omitted. Also, the index is often subscripted as D_i if that is more convenient. There is a general associative law for products which holds up to isomorphism.

There is certainly no reason to expect two objects in an arbitrary category to have a product. A category **has products** if any indexed set of objects in the category has a product. It has **finite products** if any finite indexed set of objects has a product. By an obvious induction, it is sufficient for finite products to assume an empty product and that any pair of objects has a product. Similar terminology is used for other types of limits; in particular, a category $\mathscr{C}$ **has finite limits** if every diagram $D : \mathscr{I} \longrightarrow \mathscr{C}$ in which $\mathscr{I}$ is a finite graph, has a limit.

8.4. Equalizers. The **equalizer** of two arrows $f, g\colon A \longrightarrow B$ (such arrows are said to be **parallel**) is the object $[x \in A \mid f(x) = g(x)]$. As such this does not describe a commutative cone, but the equivalent expression $[(x, y) \mid x \in A, y \in B, f(x) = g(x) = y]$ does describe a commutative cone, so the equalizer of f and g is the limit of the diagram

$$A \underset{g}{\overset{f}{\rightrightarrows}} B$$

We will also call it $\mathrm{Eq}(f, g)$. In Set, the equalizer of f and g is of course the set $\{x \in A \mid f(x) = g(x)\}$. In Grp, the kernel of a homomorphism $f\colon G \longrightarrow H$ is the equalizer of f and the constant map at the group identity.

8.5. THEOREM. *A category has finite limits if and only if it has equalizers, binary products and a terminal object.*

PROOF (SKETCH). With a terminal object and binary products, we get, by induction, all finite products. Given a diagram $D\colon \mathscr{I} \longrightarrow \mathscr{C}$, with $\mathscr{I}$ a non-empty finite graph, we let $A = \prod_{i \in \mathrm{Ob}\,\mathscr{I}} Di$ and $B = \prod_{\alpha \in \mathrm{Ar}\,\mathscr{I}} \operatorname{cod} \alpha$. We define two arrows $f, g\colon A \longrightarrow B$ by $p_\alpha \circ f = \alpha \circ p_{\operatorname{dom} \alpha}$ and $p_\alpha \circ g = p_{\operatorname{cod} \alpha}$. This means that the following diagrams commute.

$$\begin{array}{ccc} \prod_{i \in \mathrm{Ob}\,\mathscr{I}} Di & \xrightarrow{f} & \prod_{\alpha \in \mathrm{Ar}\,\mathscr{I}} \operatorname{cod} \alpha \\ {\scriptstyle p_{\operatorname{dom}\alpha}}\downarrow & & \downarrow{\scriptstyle p_\alpha} \\ D \operatorname{dom} \alpha & \xrightarrow[\alpha]{} & D \operatorname{cod} \alpha \end{array} \qquad \begin{array}{ccc} \prod_{i \in \mathrm{Ob}\,\mathscr{I}} Di & \xrightarrow{g} & \prod_{\alpha \in \mathrm{Ar}\,\mathscr{I}} \operatorname{cod} \alpha \\ & {\scriptstyle p_{\operatorname{cod}\alpha}}\searrow \quad \swarrow{\scriptstyle p_\alpha} & \\ & D \operatorname{cod} \alpha & \end{array}$$

If $E \xrightarrow{h} \prod Di$ is an equalizer of f and g, then $f \circ h = g \circ h$ expresses the fact that $h\colon E \longrightarrow D$ is a cone, while the universal mapping property into the equalizer expresses the universality of that cone. As for the empty cone, its limit is the terminal object. $\square$

With exactly the same argument one shows that the existence of arbitrary limits is equivalent to the existence of equalizers and arbitrary products. The theorems of this book (as opposed to some of the constructions used) depend only on finite limits and finite colimits (see below for the latter).

By suitable modifications of this argument, we can show that a functor preserves finite limits if and only if it preserves binary products, the terminal object and equalizers.

Another version of this theorem asserts that a category has finite limits if and only if it has a terminal object and pullbacks.

8.6. Preservation of limits. Let $D\colon \mathscr{I} \longrightarrow \mathscr{C}$ be a diagram and $F\colon \mathscr{C} \longrightarrow \mathscr{B}$ be a functor. Let $d\colon \lim D \longrightarrow D$ be a universal element of D. We say that F preserves $\lim D$ if $Fd\colon F(\lim D) \longrightarrow FD$ is a universal element of FD.

8.7. Colimits. A **colimit** of a diagram is a limit of the diagram in the opposite category. Spelled out, a commutative cocone from a diagram $D\colon \mathscr{I} \longrightarrow \mathscr{C}$ with vertex W is a natural transformation from D to the constant diagram with value W. The set of commutative cocones from D to an object A is $\mathrm{Hom}(D, A)$ and becomes a covariant functor by composition. A colimit of D is a universal element for $\mathrm{Hom}(D, -)$.

For example, let us consider the dual notion to "product". If A and B are objects in a category, their **sum** (also called **coproduct**) is an object Q together with two arrows $i_1: A \longrightarrow Q$ and $i_2: B \longrightarrow Q$ for which if $f: A \longrightarrow C$ and $g: B \longrightarrow C$ are any arrows of the category, there is a unique arrow $\langle f, g\rangle: Q \longrightarrow C$ for which $\langle f, g\rangle \circ i_1 = f$ and $\langle f, g\rangle \circ i = 2 = g$. The arrows i_1 and i_2 are called the coproduct injections although they need not be monic. Since $\mathrm{Hom}(A+B, C) \cong \mathrm{Hom}(A, C) \times \mathrm{Hom}(B, C)$, $\langle f, g\rangle$ represents an ordered pair of maps, just as the symbol (f, g) we defined when we treated products in Section 8.

The sum of two sets in Set is their disjoint union, as it is in Top. In Grp the categorical sum of two groups is their free product; on the other hand the sum of two abelian groups *in the category of abelian groups* is their direct sum with the standard inclusion maps of the two groups into the direct sum. The categorical sum in a poset regarded as a category is the supremum. The categorical sum of two posets in the category of posets and non-decreasing maps is their disjoint with no element of the one summand related to any element of the second.

The **coequalizer** of two arrows $f, g: A \longrightarrow B$ is an arrow $h: B \longrightarrow C$ such that

(i) $h \circ f = h \circ g$, and
(ii) if $k: B \longrightarrow W$ and $k \circ f = k \circ g$, then there is a unique arrow $u: C \longrightarrow W$ for which $u \circ h = k$.

The coequalizer of any two functions in Set exists but is rather complicated to construct. If K is a normal subgroup of a group G, then the coequalizer of the inclusion of K into G and the constant map at the identity is the canonical map $G \longrightarrow G/K$.

The dual concept to "pullback" is "pushout", which we leave to the reader to formulate.

The notion of a functor creating or preserving a colimit, or a class of colimits, is defined analogously to the corresponding notion for limits. A functor that preserves finite colimits is called **right exact**. In general, a categorical concept that is defined in terms of limits and/or colimits is said to be defined by "exactness conditions".

8.8. Regular monomorphisms and epimorphisms. A map that is the equalizer of two arrows is automatically a monomorphism and is called a **regular** monomorphism. For let $h: E \longrightarrow A$ be an equalizer of $f, g: A \longrightarrow B$ and suppose that $k, l: C \longrightarrow E$ are two arrows with $h \circ k = h \circ l$. Call this common composite m. Then $f \circ m = f \circ h \circ k = g \circ h \circ k = g \circ m$ so that, by the universal mapping property of equalizers, there is a unique map $n: C \longrightarrow E$ such that $h \circ n = m$. But k and l already have this property, so that $k = n = l$.

The dual property of being the coequalizer of two arrows is called **regular monomorphism**. In many familiar categories (monoids, groups, abelian groups, rings, ...) the regular epimorphisms are the surjectives mappings, but it is less often the case that the injective functions are regular monomorphisms. Of the four categories mentioned above, two (groups and abelian groups) have that property, but it is far from obvious for groups.

8.9. Regular categories. A category $\mathscr{A}$ will be called **regular** if every finite diagram has a limit, if every parallel pair of arrows has a coequalizer and if whenever

$$\begin{array}{ccc} A & \xrightarrow{f} & B \\ {\scriptstyle g}\downarrow & & \downarrow{\scriptstyle h} \\ C & \xrightarrow[k]{} & D \end{array}$$

is a pullback square, then h a regular epimorphism implies that g is a regular epimorphism. In **Set** and in many other familiar category groups, abelian groups, rings, categories of modules, etc., the regular epics are characterized as the surjective homomorphisms and these are closed in this way under pulling back. However, many familiar categories are not regular. For example neither the category of topological spaces and continuous maps, nor the category of posets and order preserving maps, is regular. If you know what an equational theory is, it is useful to know that the category of models of any equational theory is always regular (and exact, see below for the definition).

8.10. PROPOSITION. *In a regular category, every arrow f can be written as $f = m \circ e$ where m is a monomorphism and e is a regular epimorphism.*

PROOF. The obvious way to proceed is to begin with an arrow $f: A \longrightarrow A'$ and form the kernel pair of f, which can be described symbolically as $\{(a, b) \mid fa = fb\}$. If this kernel pair is $K(f) \overset{d^0}{\underset{d^1}{\rightrightarrows}} A$, then let $g: A \longrightarrow B$ be the coequalizer of d^0 and d^1. Since $f \circ d^0 = f \circ d^1$, the universal mapping property of coequalizers implies there is a unique $h: B \longrightarrow A'$ such that $h \circ g = f$. Now g is a regular epimorphism by definition. If you try this construction in the category of sets or groups or, ..., you will discover that h is always monic and then $f = h \circ g$ is the required factorization. There are, however, categories in which such an h is not always monic. We will now show that in a regular category, it is. Actually, a bit less than regularity suffices. It is sufficient that a pullback of a regular epimorphism be an epimorphism. Call an arrow a weakly regular epimorphism if it is gotten as a composite of arrows, each of which is gotten by pulling back a regular epimorphism. Since a pullback stacked on top of a pullback is a pullback, it follows that weakly regular epimorphisms are both closed under pullback (Exercise 8) and under composition and since a pullback of a regular epimorphism is an epimorphism, every weakly regular epimorphism is an epimorphism. Next note that since $A \twoheadrightarrow B$ is a regular epimorphism, $f \times 1: A \times A \longrightarrow B \times A$ is a weakly regular epimorphism since

$$\begin{array}{ccc} A \times A & \xrightarrow{p_1} & A \\ {\scriptstyle f\times 1}\downarrow & & \downarrow{\scriptstyle f} \\ B \times A & \xrightarrow[p_1]{} & B \end{array}$$

is a pullback. Similarly, $1 \times f: B \times A \longrightarrow B \times B$ and hence $f \times f: A \times A \longrightarrow B \times B$ is a weakly regular epimorphism. Let $K(h) \overset{e^0}{\underset{e^1}{\rightrightarrows}} A$ be the kernel pair of g. The fact that $h \circ g \circ d^0 = f \circ d^0 = f \circ d^1 = h \circ g \circ d^1$, together with the universal mapping property of $K(h)$ implies the existence of an arrow $k: K(f) \longrightarrow K(h)$ such that the left hand square in the diagram

$$\begin{array}{ccccc} K(f) & \xrightarrow{k} & K(h) & \longrightarrow & A' \\ {\scriptstyle (d^0,d^1)}\downarrow & & \downarrow{\scriptstyle (e^0,e^1)} & & \downarrow{\scriptstyle (1,1)} \\ A \times A & \xrightarrow[g\times g]{} & B \times B & \xrightarrow[h\times h]{} & A' \times A' \end{array}$$

commutes. The right hand square and the outer squares are pullbacks by definition—they have the universal mapping properties of the kernel pairs. By a standard property of pullbacks, the left hand square is also a pullback. But $g \times g$ is a weakly regular epimorphism and hence so is k. Now in the square

$$\begin{array}{ccc} K(f) & \xrightarrow{k} & K(h) \\ {\scriptstyle d^0}\downarrow\downarrow{\scriptstyle d^1} & & {\scriptstyle e^0}\downarrow\downarrow{\scriptstyle e^1} \\ A & \xrightarrow[g]{} & B \end{array}$$

we have $e^0 \circ k = g \circ d^0 = g \circ d^1 = e^1 \circ k$ and k is epic and therefore $e^0 = e^1$. But that means that h is monic, which finishes the argument. □

8.11. Equivalence relations and exact categories. Let $\mathscr{A}$ be a category with finite limits. If A is an object, a subobject $(d^0, d^1): E \longrightarrow A \times A$ is called an **equivalence relation** if it is

ER–1. reflexive: there is an arrow $r: a \longrightarrow E$ such that $d^0 \circ r = d^1 \circ r = \mathrm{id}$;

ER–2. symmetric: there is an arrow $s: E \longrightarrow E$ such that $s \circ d^0 = d^1$ and $s \circ d^1 = d^0$;

ER–3. transitive: if

$$\begin{array}{ccc} T & \xrightarrow{q_1} & E \\ {\scriptstyle q_2}\downarrow & & \downarrow{\scriptstyle p_2} \\ E & \xrightarrow[p_1]{} & A \end{array}$$

is a pullback, there is an arrow $t: T \longrightarrow E$ such that $p_1 \circ t = p_1 \circ q_1$ and $p_2 \circ t = p_2 \circ q_2$.

The interpretation of the last point is that $E \subseteq A \times A$, so is a set of ordered pairs (a_1, a_2); $T \subseteq E \times E$, so T is a set of ordered 4-tuples (a_1, a_2, a_3, a_4) such that $(a_1, a_2) \in E$ and $(a_3, a_4) \in E$ and the condition $p_1 \circ q_2 = p_2 \circ q_1$ simply expresses $a_3 = a_4$. Then the condition $p_1 \circ t = p_1 \circ q_1$ means that $t(a_1, a_2, a_3, a_4)$ has first coordinate a_1 and $p_2 \circ t = p_2 \circ q_2$ means that the second coordinate is a_4. So taken all together, this says that when $(a_1, a_2) \in E$, $(a_3, a_4) \in E$ and $a_2 = a_3$, then $(a_1, a_4) \in E$, which is just transitivity in the usual sense.

If $f: A \longrightarrow A'$ is an arrow, then the kernel pair of f is an equivalence relation. It is internally the relation $a_1 \sim a_2$ if and only if $fa_1 = fa_2$. We say that an equivalence relation is **effective** if it is the kernel pair of some arrow. Another term for effective equivalence relation is **congruence**.

A category is called **exact** if it is regular and if every equivalence relation is effective.

The following will be needed for 2.2.4

8.12. PROPOSITION. *Suppose $\mathscr{A}$ is a regular, respectively exact, category. Then for any object A the slice $\mathscr{A}/A$ is regular, respectively exact.*

PROOF. Let us write $[b\colon B \to A]$ for an object of $\mathscr{A}/A$. Suppose $f\colon [b\colon B \to A] \longrightarrow [b'\colon B' \to A]$ is an arrow such that $f\colon B \longrightarrow B'$ is a regular epimorphism in $\mathscr{A}$. Then there is a pair of arrows $B'' \underset{d^1}{\overset{d^0}{\rightrightarrows}} B$ whose coequalizer is f. Then we have the diagram

$$[b \circ d^0 = b \circ d^1\colon B'' \to A] \underset{d^1}{\overset{d^0}{\rightrightarrows}} [b\colon B \to A] \xrightarrow{f} [b'\colon B' \to A]$$

which is a coequalizer in $\mathscr{A}/A$ so that f is a regular epimorphism there. Conversely, suppose that $f\colon [b\colon B \to A] \longrightarrow [b'\colon B' \to A]$ is a regular epimorphism in $\mathscr{A}/A$. Then we have a coequalizer

$$[b''\colon B'' \to A] \underset{d^1}{\overset{d^0}{\rightrightarrows}} [b\colon B \to A] \xrightarrow{f} [b'\colon B' \to A]$$

Given $g\colon B \longrightarrow C$ such that $g \circ d^0 = g \circ d^1$, it is easy to see that we have a morphism $(g, b)\colon [b\colon B \to A] \longrightarrow [p_2, C \times A]$. Moreover,

$$(g, b) \circ d^0 = (g \circ d^0, b \circ d^0) = (g \circ d^1, b'') = (g \circ d^1, b \circ d^1) = (g, b) \circ d^1$$

so that there is a unique $(h, k)\colon [b'\colon B' \to A] \longrightarrow [p_2\colon C \times A \to A]$ with $(h, k) \circ f = (g, b)$. This implies that $h \circ f = g$ and $k \circ f = b$. Thus $h\colon B' \longrightarrow C$ satisfies $h \circ f = g$. If h' were a different map for which $h' \circ f = g$, then (h', k) would be a second map for which $(h', k) \circ f = (g, b)$, contradicting uniqueness. Thus far we have shown that f is a regular epic in $\mathscr{A}$ if and only if it is so in $\mathscr{A}/A$. If we have $[b\colon B \to A] \xrightarrow{f} [b'\colon B' \to A] \xleftarrow{g} [c', C']$ and if

$$\begin{array}{ccc} C & \xrightarrow{g'} & B \\ {\scriptstyle f'}\downarrow & & \downarrow{\scriptstyle f} \\ C' & \xrightarrow[g]{} & B' \end{array}$$

is a pullback, then it is immediate that for $c = c' \circ f' = b' \circ g \circ f' = b' \circ f \circ g' = b \circ g'$ the square

$$\begin{array}{ccc} [c\colon C \to A] & \xrightarrow{g'} & [b\colon B \to A] \\ {\scriptstyle f'}\downarrow & & \downarrow{\scriptstyle f} \\ [c'\colon C' \to A] & \xrightarrow[g]{} & [b'\colon B' \to A] \end{array}$$

is a pullback in $\mathscr{A}$. If f is regular epic in $\mathscr{A}/A$ it is so in $\mathscr{A}$; hence f' is regular epic in $\mathscr{A}$ and therefore is so in $\mathscr{A}/A$. This proves it for regular categories.

For exact categories, the argument is similar. The previous discussion amounts to showing that pullbacks and coequalizers are the same in $\mathscr{A}$ and $\mathscr{A}/A$. As a matter of fact, the full story is that all colimits are the same. Not all limits are; however all pullbacks are and that is all that is used in the definition of exact category. For example, the terminal object in $\mathscr{A}/A$ is [id: $A \to A$] and that is not the terminal object of $\mathscr{A}$ (unless $A = 1$, in which case $\mathscr{A}/A$ is equivalent to $\mathscr{A}$). See Exercise 2 below. □

8.13. Exercises.

1. Suppose that the category $\mathscr{A}$ has finite limits. Show that the kernel pair of any arrow is an equivalence relation. Hint: you will have to use the universal mapping properties of limits.

2. Call a graph **connected** if it is not the disjoint union of two non-empty subgraphs. Show that the forgetful functor $\mathscr{A}/A \longrightarrow \mathscr{A}$ preserves the limits of diagrams over connected graphs (which are called **connected diagrams**).

3. Suppose

$$\begin{array}{ccc} A & \xrightarrow{\ e\ } & B \\ {\scriptstyle f}\downarrow & & \downarrow{\scriptstyle g} \\ C & \xrightarrow[\ m\ ]{} & D \end{array}$$

is a commutative square in a regular category and that e is a regular epimorphism and m is a monomorphism. Show there is a unique $h: B \longrightarrow C$ making both (actually either) triangles commute. This is called the *diagonal fill-in*.

9. Adjoint functors

9.1. Adjunction of group underlying function. Let A be a set and G be a group. We have noted that for any function from A to G, in other words for any element of $\mathrm{Hom}_{\mathsf{Set}}(A, UG)$, there is a unique group homomorphism from the free group FA with basis A to G which extends the given function. This is thus a bijection

$$\mathrm{Hom}_{\mathsf{Grp}}(FA, G) \longrightarrow \mathrm{Hom}_{\mathsf{Set}}(A, UG)$$

The inverse simply restricts a group homomorphism from FA to G to the basis A. Essentially the same statement is true for monoids instead of groups (replace FA by the free monoid A^*) and also for the category of abelian groups, with FA the free abelian group with basis A.

The bijection just mentioned is a natural isomorphism β of functors of two variables, in other words a natural isomorphism from the functor $\mathrm{Hom}_{\mathsf{Grp}}(F(-), -)$ to $\mathrm{Hom}_{\mathsf{Set}}(-, U(-))$. This means precisely that for all functions $f: B \longrightarrow A$ and all group homomorphisms $g\colon G \longrightarrow H$,

$$
\begin{array}{ccc}
\mathrm{Hom}_{\mathsf{Grp}}(FA, G) & \xrightarrow{\beta(A,G)} & \mathrm{Hom}_{\mathsf{Set}}(A, UG) \\
\Big\downarrow{\scriptstyle \mathrm{Hom}_{\mathsf{Grp}}(Ff,g)} & & \Big\downarrow{\scriptstyle \mathrm{Hom}_{\mathsf{Set}}(f,Ug)} \\
\mathrm{Hom}_{\mathsf{Grp}}(FB, H) & \xrightarrow[\beta(B,H)]{} & \mathrm{Hom}_{\mathsf{Set}}(B, UH)
\end{array} \tag{14}
$$

commutes.

9.2. Unit and counit. The free group functor and the underlying set functor are a typical pair of "adjoint functors". Formally, if $\mathscr{A}$ and $\mathscr{D}$ are categories and $L\colon \mathscr{A} \longrightarrow \mathscr{D}$ and $R\colon \mathscr{D} \longrightarrow \mathscr{A}$ are functors, then L is **left adjoint** to R and R is right adjoint to L if for every objects A of $\mathscr{A}$ and B of $\mathscr{D}$ there is an isomorphism

$$\mathrm{Hom}_{\mathscr{A}}(A, RB) \cong \mathrm{Hom}_{\mathscr{D}}(LA, B)$$

which is natural in the sense of diagram 14. Informally, elements of RB defined on A are essentially the same as element of B defined on LA.

In particular, if L is left adjoint to R and A is an object of $\mathscr{A}$, then corresponding to id_{LA} in $\mathrm{Hom}_{\mathscr{A}}(LA, LA)$ there is an arrow $\eta A\colon A \longrightarrow RLA$; the arrows ηA form a natural transformation from the identity functor on $\mathscr{A}$ to $R \circ L$. This natural transformation η is the **unit** of the adjunction of L to R. A similar trick also produces a natural transformation $\epsilon\colon L \circ R \longrightarrow \mathrm{id}_{\mathscr{D}}$ called the **counit** of the adjunction. The unit and counit essentially determine the adjunction completely.

9.3. Examples. We give a number of examples that will be needed later in this book.

1. The underlying functor $\mathsf{Ab} \longrightarrow \mathsf{Set}$. The adjoint takes a set S to the set of all finite sums
$$\sum_{s \in S} n_s s$$
where for each $s \in S$, n_s is an integer, but in any given sum, only finitely many of them are non-zero. The abelian group structure is just term-wise addition (and subtraction).
2. The underlying functor **CommMon** $\longrightarrow \mathsf{Set}$. This takes a set S to the set of all terms
$$\prod_{s \in S} s^{n_s}$$
where for each $s \in S$, n_s is a non-negative integer, but in any given product, only finitely many of them are non-zero. Of course, this could be written additively, but for the purpose of the next example, we prefer to do it multiplicatively.
3. The underlying set functor **CommRing** $\longrightarrow \mathsf{Set}$. Here the left adjoint can be described as the composite of the two previous examples. If S is a set, then the free commutative ring, which we will call $\mathbf{Z}[S]$ since it is, in fact the ring of polynomials in S is gotten by first forming the free commutative monoid generated by S and then the free abelian group generated by that. It is still a monoid, since the distributive law of multiplication tells us how to multiply sums of monomials. The general process by which two such free

functors can be composed was first studied by Jon Beck under the name "distributive laws" [Beck, 1969].

4. The underlying set functors on the categories of monoids (not necessarily commutative), rings (ditto) and Lie algebras all have adjoints. Lest the reader get the idea that all underlying set functors have adjoints, we mention the category of fields, whose underlying set functor does not have an adjoint. An interesting case is that of torsion abelian groups. If we fix an exponent d and look at all groups satisfying $x^d = 1$, there is an adjoint that takes a set S to the direct sum of S many copies of $\mathbf{Z}/d\mathbf{Z}$, but on the full category, there is no adjoint.

9.4. Representability and adjointness. The statement that L is left adjoint to R immediately implies that for each object A of $\mathscr{A}$, the object LA of $\mathscr{B}$ represents the functor $\text{Hom}_{\mathscr{A}}(A, R(-)): \mathscr{B} \longrightarrow \mathsf{Set}$. The universal element for this representation, which must be an element of $\text{Hom}_{\mathscr{A}}(A, RLA)$, is the unit ηA. Dually, the object RB with universal element ϵA represents the contravariant functor $\text{Hom}(L(-), B)$. The following theorem is a strong converse to these facts.

9.5. THEOREM ("Pointwise construction of adjoints"). *Let $\mathscr{A}$ and $\mathscr{B}$ be categories.*

(a) *If $R: \mathscr{B} \longrightarrow \mathscr{A}$ is a functor such that the functor $\text{Hom}_{\mathscr{A}}(A, R(-))$ is representable for every object A of $\mathscr{A}$, then R has a left adjoint.*
(b) *If $L: \mathscr{A} \longrightarrow \mathscr{B}$ is a functor such that $\text{Hom}_{\mathscr{B}}(L(-), B)$ is representable for every object B of $\mathscr{B}$, then L has a left adjoint.*

With little more work, one can prove parametrized versions of these results.

9.6. THEOREM. *Let $\mathscr{A}$, $\mathscr{B}$, and $\mathscr{X}$ be categories.*

(a) *Suppose $R: \mathscr{X} \times \mathscr{B} \longrightarrow \mathscr{A}$ is a functor such that for every pair of objects $A \in \text{Ob}(\mathscr{A})$ and $X \in \text{Ob}(\mathscr{X})$ the functor $\text{Hom}_{\mathscr{B}}(A, R(X, -)): \mathscr{B} \longrightarrow \mathsf{Set}$ is representable. Then there is a unique functor $L: \mathscr{A} \times \mathscr{X}^{\text{op}} \longrightarrow \mathscr{B}$ such that*
$$\text{Hom}_{\mathscr{A}}(-, R(-,-)) \cong \text{Hom}_{\mathscr{B}}(L(-,-),-)$$
as functors $\mathscr{A}^{\text{op}} \times \mathscr{X} \times \mathscr{B} \longrightarrow \mathsf{Set}$.
(b) *Suppose $L: \mathscr{A} \times \mathscr{X}^{\text{op}} \longrightarrow \mathscr{B}$ is a functor such that for every pair of objects $B \in \text{Ob}(\mathscr{B})$ and $X \in \text{Ob}(\mathscr{X})$ the functor $\text{Hom}_{\mathscr{B}}(L(-, X), B): \mathscr{A} \longrightarrow \mathsf{Set}$ is representable. Then there is a unique functor $R: \mathscr{X} \times \mathscr{B} \longrightarrow \mathscr{A}$ such that*
$$\text{Hom}_{\mathscr{A}}(-, R(-,-)) \cong \text{Hom}_{\mathscr{B}}(L(-,-),-)$$
as functors $\mathscr{A}^{\text{op}} \times \mathscr{X} \times \mathscr{B} \longrightarrow \mathsf{Set}$.

PROOF. The two statements are dual, so we will prove the first. Begin by choosing, for each $A \in \text{Ob}(\mathscr{A})$, $X \in \text{Ob}(\mathscr{X})$, and $B \in \text{Ob}(\mathscr{B})$ an object function $L: \text{Ob}(\mathscr{A}) \times \text{Ob}(\mathscr{X}) \longrightarrow \text{Ob}(\mathscr{B})$ such that $\text{Hom}_{\mathscr{A}}(A, R(X, B)) \cong \text{Hom}_{\mathscr{B}}(L(A, X), B)$. Now we want to make L into a functor. Choose arrows

$f: A \longrightarrow A'$ and $g: X' \longrightarrow X$. Now for any $B \in \mathrm{Ob}(\mathscr{B})$ we have a diagram

$$\begin{array}{ccc} \mathrm{Hom}_{\mathscr{A}}(A', R(X', B)) & \xrightarrow{\cong} & \mathrm{Hom}_{\mathscr{B}}(L(A', X'), B) \\ {\scriptstyle \mathrm{Hom}_{\mathscr{A}}(f, R(g,B))}\Big\downarrow & & \\ \mathrm{Hom}_{\mathscr{A}}(A, R(X, B)) & \xrightarrow[\cong]{} & \mathrm{Hom}_{\mathscr{B}}(L(A, X), B) \end{array} \tag{15}$$

There is thus a unique arrow

$$\phi(f, g, B): \mathrm{Hom}_{\mathscr{B}}(L(A', X'), B) \longrightarrow \mathrm{Hom}_{\mathscr{B}}(L(A, X), B)$$

that makes the square commute. Moreover, since both the isomorphisms and $\mathrm{Hom}_{\mathscr{A}}(f, R(g, B))$ are natural with respect to B, we conclude that $\phi(f, g, B)$ is as well. By the Yoneda lemma, there is a unique arrow we call $L(f, g): L(A, X) \longrightarrow L(A', X')$ such that $\phi(f, g, B) = \mathrm{Hom}_{\mathscr{B}}(L(f, g), B)$. If now we have $f': A' \longrightarrow A''$ and $g': X'' \longrightarrow X'$ we can stack another diagram of shape 15 on top of that one to show that $L(f, g) \circ L(f', g') = L(f \circ f', g' \circ g)$. The fact that L preserves identities is even easier. □

One of the most important properties of adjoints is their limit preservation properties.

9.7. Proposition. *Let $L: \mathscr{A} \longrightarrow \mathscr{B}$ be left adjoint to $R: \mathscr{B} \longrightarrow \mathscr{A}$. Then R preserves the limit of an any diagram in $\mathscr{B}$ that has a limit and L preserves the colimit of any diagram in $\mathscr{A}$ that has a colimit.*

Proof. Suppose that $D: \mathscr{I} \longrightarrow \mathscr{B}$ is a diagram and that $B \longrightarrow D$ is a limit cone. Given a cone $A \longrightarrow RD$, the adjunction gives a cone $LA \longrightarrow D$ by applying the adjunction to each element of the cone. The universality gives an arrow $LA \longrightarrow C$ and then the adjunction gives $A \longrightarrow UC$. We can summarize this argument as follows:

$$\mathrm{Cone}(A, RD) \cong \mathrm{Cone}(LA, D) \cong \mathrm{Hom}(LA, B) \cong \mathrm{Hom}(A, RB)$$ □

10. Categories of fractions

The definitions and results of this section are essentially those of [Gabriel & Zisman, 1967].

The main acyclic models theorem is stated in terms of a fundamental construction in category theory, called categories of fractions. This is a relatively straightforward generalization of the construction in monoids, to which we turn by way of introduction.

10.1. Monoids. Let M be a monoid and $\Sigma \subseteq M$ be a multiplicatively closed (which is understood to include the identity element—the empty product) subset. There is a monoid denoted $\Sigma^{-1}M$ and a monoid homomorphism $\phi: M \longrightarrow \Sigma^{-1}M$ with the following two properties:

1. If $\sigma \in \Sigma$, then $\phi(\sigma)$ is invertible;
2. if $f: M \longrightarrow N$ is a monoid homomorphism such that $f(\sigma)$ is invertible for all $\sigma \in \Sigma$, then there is a unique monoid homomorphism $g: \Sigma^{-1}M \longrightarrow N$ such that $g \circ \phi = f$.

This can readily be set up as an adjoint and the general adjoint functor applied. However, it is instructive to give a direct construction. Consider the free monoid F generated by the elements of M and formal inverses of elements of Σ. Write $\langle x\rangle$ and $\langle\sigma^{-1}\rangle$ for the two kinds of generators. Factor out the congruence relation generated by all pairs of the forms $(\langle x\rangle\langle y\rangle, \langle xy\rangle)$, $(\langle\sigma\rangle\langle\sigma^{-1}\rangle, 1)$, $(\langle\sigma^{-1}\rangle\langle\sigma\rangle, 1)$ and $(\langle 1\rangle, 1)$. This means that we first form the submonoid of $M \times M$ generated by all such pairs and then the equivalence relation generated by that. The result is an equivalence relation that is also a submonoid. The set of equivalence classes has a unique monoid structure for which the class map is a homomorphism and it is clear that in the quotient monoid, the classes of the elements of Σ are invertible.

One point should be noted. This process can, depending on the nature of Σ, distort M profoundly. For example, if we carry out this procedure on the multiplicative monoid of integers and $0 \in \Sigma$, then the fact that $0n = 0m$ implies, when you invert 0, that $n = m$. Thus that procedure causes the monoid to collapse to a single element.

In general, every element of $\Sigma^{-1}M$ can be written in the form

$$x_1\sigma_1^{-1}x_2\sigma_2^{-1}\cdots x_n\sigma_n^{-1}$$

Of course, it is possible that x_1 or σ_n or both is 1, so it can start with an inverse or end with an ordinary element of M. One way of seeing this is to observe that $\Sigma^{-1}M$ must contain all the elements of M as well as inverses to all elements of Σ and hence all such products. Next observe that the set of all such products forms a submonoid that contains all the elements of M and the inverses of all the elements of Σ and this submonoid clearly satisfies the universal mapping property. Since the solution of a universal mapping problem is unique, there can be no additional elements.

10.2. Calculuses of fractions: monoids. A multiplicatively closed subset $\Sigma \subseteq M$ is said to have a **calculus of right fractions** if for any $\sigma \in \Sigma$ and $x \in M$, there are $y \in M$ and $\tau \in \Sigma$ such that $\sigma y = x\tau$ and if for any x, $y \in M$ and $\sigma \in \Sigma$, $\sigma x = \sigma y$ implies there is a $\tau \in \Sigma$ for which $x\tau = y\tau$. Dually, we say that Σ has a **calculus of left fractions** if for any $\sigma \in \Sigma$ and $x \in M$, there are $y \in M$ and $\tau \in \Sigma$ such that $y\sigma = \tau x$.

10.3. PROPOSITION. *If the multiplicatively closed subset $\Sigma \subseteq M$ has a calculus of right fractions, then every element of $\Sigma^{-1}M$ can be written as $x\sigma^{-1}$ with $\sigma \in \Sigma$. Moreover, $x\sigma^{-1} = y\tau^{-1}$ if and only if there are elements a, $b \in M$ such that $\sigma a = \tau b \in \Sigma$ and $xa = yb$. Dually, if Σ has a calculus of left fractions, then every element of $\Sigma^{-1}M$ can be written as $\sigma^{-1}x$ with $\sigma \in \Sigma$. Moreover, $\sigma^{-1}x = \tau^{-1}y$ if and only if there are elements a, $b \in M$ such that $a\sigma = b\tau \in \Sigma$ and $ax = by$.*

We will leave the proof till the corresponding theorem for categories, of which this is a special case. There is no real difference between the proofs. In a sense, the one for categories is easier because there are fewer possibilities since elements cannot be composed unless the domain of one is the codomain of the other and then only in one direction.

10.4. Categories. In dealing with categories, we have a problem of size. Usually, one assumes that in a category the class of arrows between any two objects is a set. In the case of categories of fractions, this will not necessarily be true even if it is in the original category. One way of dealing with this is to suppose the original category is itself small (that is, there in all only a set of arrows in the category), in which case any fraction category is too. Another approach is to carry out the construction in general and allow the possibility of large hom classes. It may still happen in individual cases that these classes will be small. We adopt the latter approach.

Let $\mathscr{M}$ be a category and Σ denote a class of arrows closed under composition and including all the identity arrows. The category $\Sigma^{-1}\mathscr{M}$ comes with a functor $\Phi: \mathscr{M} \longrightarrow \Sigma^{-1}\mathscr{M}$ such that:

1. if $\sigma \in \Sigma$, then $\Phi(\sigma)$ is an isomorphism;
2. if $F: \mathscr{M} \longrightarrow \mathscr{N}$ is an functor with the property that for all $\sigma \in \Sigma$, the arrow $F(\sigma)$ is an isomorphism, then there is a unique functor $G: \Sigma^{-1}\mathscr{M} \longrightarrow \mathscr{N}$ such that $G \circ \Phi = F$.

Here is how to construct $\Sigma^{-1}\mathscr{M}$. The category has the same objects as those of $\mathscr{M}$. If M and M' are objects, an arrow from M to M' is an equivalence class of formal composites

$$f_n \circ \sigma_n^{-1} \circ f_{n-1} \circ \sigma_{n-1}^{-1} \circ \cdots \circ f_1 \circ \sigma_1^{-1}$$

for which

1. $\mathrm{cod}(\sigma_1) = M$;
2. $\mathrm{cod}(f_n) = M'$;
3. $\mathrm{dom}(f_i) = \mathrm{dom}(\sigma_i)$ for $i = 1, \ldots, n$;
4. $\mathrm{cod}(f_i) = \mathrm{cod}(\sigma_{i+1})$ for $i = 1, \ldots, n-1$.

We picture an arrow as follows:

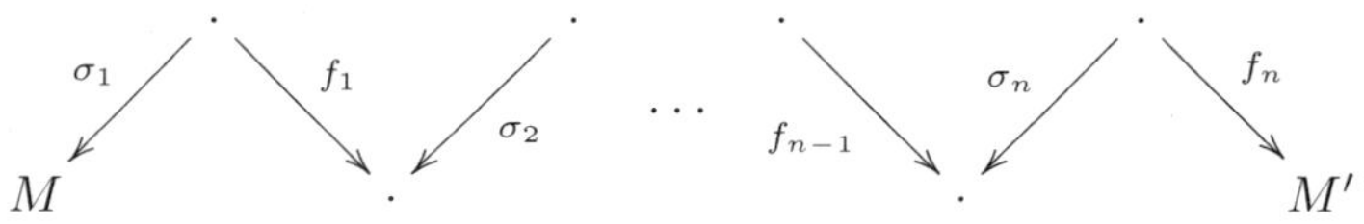

Composition is juxtaposition so that the empty string is the identity and we will unambiguously denote it by id. The equivalence relation $\sim$ is the smallest one closed under juxtaposition such that $f \circ \sigma^{-1} \sim \tau^{-1} \circ g$ whenever $\tau \circ f = g \circ \sigma$ and such that for any object C of $\mathscr{C}$, $\mathrm{id}_C \circ \mathrm{id}_C^{-1}$ is the empty string, that is, the identity of C in $\Sigma^{-1}\mathscr{C}$. Note that $\tau^{-1} \circ g$ is short for $\mathrm{id} \circ \tau^{-1} \circ g \circ \mathrm{id}^{-1}$. This equivalence relation implies, for example, that $\sigma \circ \sigma^{-1} = \mathrm{id}^{-1} \circ \mathrm{id} = \mathrm{id}_{\bullet} id^{-1} = \mathrm{id}$ and $\sigma^{-1} \circ \sigma = \mathrm{id} \circ \mathrm{id}^{-1} = \mathrm{id}$, so that σ is invertible in $\Sigma^{-1}\mathscr{C}$. Conversely, it is clear that if each of element of Σ is invertible, then $\tau \circ f = g \circ \sigma$ implies that $f \circ \sigma^{-1} = \tau^{-1} \circ g$ so that this is the least equivalence that suffices.

We will denote the equivalence classes by any element and $\sim$ by $=$ from now on.

The functor Φ is the identity on objects and $\Phi(f) = f \circ \mathrm{id}^{-1}$ which we will also denote f. It is clear that $\Phi(\sigma) = \sigma$ is invertible and if $F: \mathscr{M} \longrightarrow \mathscr{N}$ inverts every element of Σ, then we let

$$\begin{aligned} G(f_n \circ \sigma_n^{-1} \circ f_{n-1} \circ \sigma_{n-1}^{-1} \circ \cdots \circ f_1 \circ \sigma_1^{-1}) \\ = F(f_n) \circ F(\sigma_n)^{-1} \circ F(f_{n-1}) \circ F(\sigma_{n-1})^{-1} \circ \cdots \circ F(f_1) \circ F(\sigma_1)^{-1} \end{aligned}$$

This is clearly the unique functor that extends F.

10.5. Calculuses of fractions: categories. We say that Σ has a **calculus of left fractions** if, for any $\sigma \in \Sigma$ and $f \in M$ with the same domain, there is a

commutative square

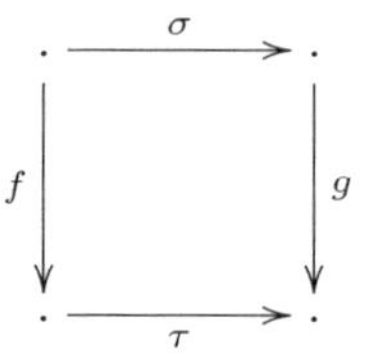

and if, for any parallel pair $f,\ g\colon M \longrightarrow N$ and $\sigma\colon N \longrightarrow N'$ in Σ such that $\sigma \circ f = \sigma \circ g$, there is a $\tau\colon M' \longrightarrow M$ belonging to Σ such that $f \circ \tau = g \circ \tau$. Dually, we say that Σ has a **calculus of right fractions** if for any $\sigma \in \Sigma$ and $f \in M$, with the same codomain, there is a commutative square

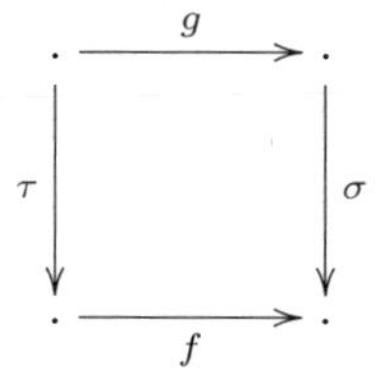

and if, for any parallel pair $f,\ g\colon M \longrightarrow N$ and $\sigma\colon M' \longrightarrow M$ in Σ such that $f \circ \sigma = g \circ \sigma$, there is a $\tau\colon N \longrightarrow N'$ belonging to Σ such that $\tau \circ f = \tau \circ g$.

10.6. PROPOSITION. *If the multiplicatively closed subset $\Sigma \subseteq \mathscr{C}$ has a calculus of right fractions, then every arrow of $\Sigma^{-1}\mathscr{C}$ can be written as $f \circ \sigma^{-1}$ with $\sigma \in \Sigma$. Moreover, if $\operatorname{dom}(f) = \operatorname{dom}(\sigma) = C'$ and $\operatorname{dom}(g) = \operatorname{dom}(\tau) = C''$, then $f \circ \sigma^{-1} = g \circ \tau^{-1}\colon C \longrightarrow D$ if and only if there is an object B and arrows $a\colon B \longrightarrow C'$ and $b\colon B \longrightarrow C''$ such that such that $\sigma \circ a = \tau \circ b \in \Sigma$ and $f \circ a = g \circ b$. Dually, if Σ has a calculus of left fractions, then every arrow of $\Sigma^{-1}\mathscr{C}$ can be written as $f \circ \sigma^{-1}$ with $\sigma \in \Sigma$. Moreover, if $\operatorname{cod}(f) = \operatorname{cod}(\sigma) = D'$ and $\operatorname{cod}(g) = \operatorname{cod}(\tau) = D''$, then $\sigma^{-1} \circ f = \tau^{-1} \circ g\colon C \longrightarrow D$ if and only if there is an object B and arrows $a\colon D' \longrightarrow E$ and $b\colon D'' \longrightarrow E$ such that such that $a \circ \sigma = b \circ \tau \in \Sigma$ and $a \circ f = b \circ g$.*

PROOF. Suppose $\mathscr{C}$ has a calculus of right fractions. Any map of the form $f \circ \sigma^{-1}$ can be written as $\tau^{-1} \circ g$ by completing the square. Composites of these maps can obviously be rewritten in this form as well. Next we consider the equivalence relation. Let R be the relation described in the theorem. The picture is

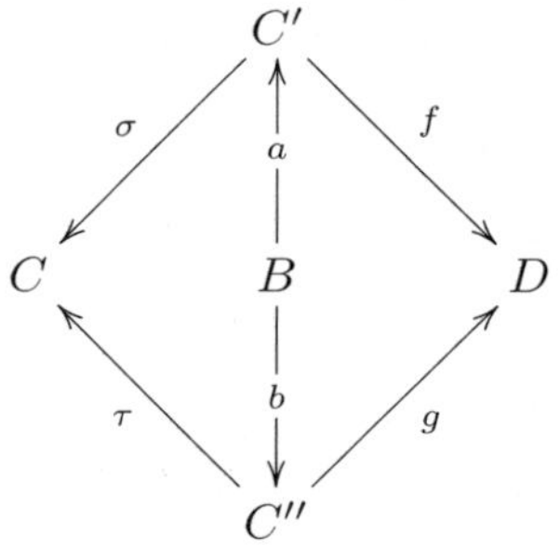

If $\sigma \circ a = \tau \circ b \in \Sigma$ and $f \circ a = g \circ b$ then

$$\begin{aligned} f \circ \sigma^{-1} &= f \circ \sigma^{-1} \circ \sigma \circ a \circ (\sigma \circ a)^{-1} = f \circ a \circ (\sigma \circ a)^{-1} \\ &= g \circ b \circ (\tau \circ b)^{-1} = g \circ \tau^{-1} \circ \tau \circ b \circ (\tau \circ b)^{-1} = g \circ \tau^{-1} \end{aligned}$$

Now suppose that $f \circ \sigma^{-1} = g \circ \tau^{-1}$ in $\Sigma^{-1}\mathscr{C}$. The equality in that category is the transitive closure of the relation S in which $(f \circ \sigma^{-1})S(g \circ \tau^{-1})$ if there is an object B and arrows $h: B \longrightarrow C$ and $\rho: B \longrightarrow D$ with $\rho \in \Sigma$ such that the diagram

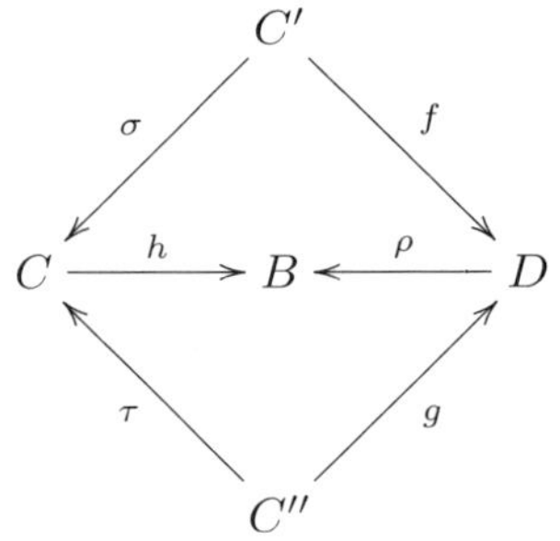

commutes. There is an object A and arrows $u: A \longrightarrow C'$ and $v: A \longrightarrow C''$ such that $\sigma \circ u = \tau \circ v$ and we may suppose that either u or v belongs to Σ, but in either case $\sigma \circ u = \tau \circ v$ does. Then

$$\rho \circ f \circ u = h \circ \sigma \circ u = h \circ \tau \circ v = \rho \circ g \circ v$$

so that there is an object A' and an arrow $\theta: A' \longrightarrow A$ in Σ such that $f \circ u \circ \theta = g \circ v \circ \theta$. Also, $\sigma \circ u \circ \theta = \tau \circ v \circ \theta \in \Sigma$ since $\sigma \circ u$ and θ do.

Next we show that R is transitive, since it is evidently reflexive and symmetric. Suppose $f \circ \sigma^{-1} = g \circ \tau^{-1}$ and $g \circ \tau^{-1} = h \circ \rho^{-1}$ with $C''' = \mathrm{dom}(h)$. Then there are objects B and B' and arrows $a: B \longrightarrow C'$, $b: B \longrightarrow C''$, $c: B' \longrightarrow C''$ and $d: B' \longrightarrow C'''$ such that $\sigma \circ a = \tau \circ b \in \Sigma$, $\tau \circ c = \rho \circ d \in \Sigma$, $f \circ a = g \circ b$ and $g \circ c = h \circ d$. Now $\tau \circ b: B \longrightarrow C$ and $\tau \circ c: B' \longrightarrow C$ belong to Σ and so there is an object A and arrows $u: A \longrightarrow B$ and $v: A \longrightarrow B'$ such that $\tau \circ b \circ u = \tau \circ c \circ v$ and we can suppose that either u or v belongs to Σ. It does not matter which one we suppose, so suppose it is u. Then since $\tau \circ b \in \Sigma$, it follows that $\tau \circ b \circ u = \tau \circ c \circ v \in \Sigma$. Since τ coequalizes $b \circ u$ and $c \circ v$, there is an object A' and $\theta \in \Sigma$ such that $b \circ u \circ \theta = c \circ v \circ \theta$. Now we have $a \circ u \circ \theta: A' \longrightarrow C'$ and $d \circ v \circ \theta: A' \longrightarrow C'''$. We see that

$$\sigma \circ a \circ u \circ \theta = \tau \circ b \circ u \circ \theta = \tau \circ c \circ v \circ \theta = \rho \circ d \circ v \circ \theta$$

Moreover, this arrow belongs to Σ because $\sigma \circ a$, u and θ do. Finally,

$$f \circ a \circ u \circ \theta = g \circ b \circ u \circ \theta = g \circ c \circ v \circ \theta = h \circ d \circ v \circ \theta$$

which shows that $f \circ \sigma^{-1}$ is related to $g \circ \rho^{-1}$ under R. □

11. The category of modules

In Chapter 7 we will need the category of all left modules. An object of the category is a pair (R, M) where R is a ring and M is a left R-module. If (R, M) and (R', M') are two such objects, a morphism $(R, M) \longrightarrow (R', M')$ is a pair (ϕ, f) where $\phi: R \longrightarrow R'$ is a ring homomorphism and $f: M \longrightarrow M'$ is an additive homomorphism such that $f(rm) = \phi(r)f(m)$ for $r \in R$ and $m \in M$. The category structure is the obvious one. We will call this category Lmod.

Here is an interesting example. It is entirely possible that $(R, M) \cong (R, M')$ without $M \cong M'$ as R-modules. Let $R = \mathbf{Z}[x, y]$ be a polynomial ring in two variables. Let $M = \mathbf{Z} \oplus \mathbf{Z} \oplus \cdots$, the direct sum of countable many copies of $\mathbf{Z}$. Both x and y act by translation of coordinates:

$$x(n_1, n_2, \ldots) = y(n_1, n_2, \ldots) = (0, n_1, n_2, \ldots)$$

We let M' be the same abelian group and the action of x is the same, while y acts as the 0 homomorphism. There can be no non-zero homomorphism between M and M' since it cannot preserve the action of y. On the other hand $R = \mathbf{Z}[x, x-y]$ as well and the action of $x-y$ on M is just like that of y on M'. Precisely, let $\phi\colon R \longrightarrow R'$ be the unique homomorphism for which $\phi(x) = x$ and $\phi(y) = x - y$. It is an isomorphism, with $\phi^{-1}(x) = x$ and $\phi^{-1}(y) = x + y$. Then (ϕ, id) is an isomorphism.

When we use this construction in Chapter 7, we will use this example, except with $2n$ variables instead of just 2.

A trivial observation is that if $\phi\colon R \longrightarrow R'$ is a ring homomorphism, then $(\phi, \phi)\colon (R, R) \longrightarrow (R, R)$ is a homomorphism in the category of all modules, since in that case the required identity is $\phi(rs) = \phi(r)\phi(s)$.

We will have need of the following proposition.

11.1. PROPOSITION. *Suppose that $f\colon M \longrightarrow M'$ is a homomorphism of R modules, $(\phi, g)\colon (R, M) \longrightarrow (S, N)$ is an isomorphism and $(\phi, h)\colon (R, M') \longrightarrow (S, N')$ is a homomorphism in* Lmod. *Then $h \circ f \circ g^{-1}\colon N \longrightarrow N'$ is a homomorphism of S-modules.*

PROOF. This can readily be done directly. Another way is to observe that the composite $(\phi, h) \circ (\mathrm{id}, f) \circ (\phi^{-1}, g^{-1})$ is $(\phi \circ \mathrm{id} \circ \phi^{-1}, h \circ f \circ g^{-1}) = (\mathrm{id}, h \circ f \circ g^{-1})$ in Lmod. □

11.2. COROLLARY. *Suppose that $(\phi, g)\colon (R, M) \longrightarrow (S, N)$ and $(\phi, h)\colon (R, M') \longrightarrow (S, N')$ are homomorphisms in the category of all modules. Then $f \mapsto h \circ f \circ g^{-1}$ defines an isomorphism $\mathrm{Hom}_R(M, M') \longrightarrow \mathrm{Hom}_S(N, N')$.* □

We note that these homsets are just abelian groups. Even when $R = S$ is commutative (which is the case we will actually be applying this) and the homsets are R-modules, respectively, the isomorphisms will not be of R-modules.

All results of this section apply, *mutatis mutandi* to the category Rmod of all right modules, whose definition is obvious.

CHAPTER 2

Abelian Categories and Homological Algebra

In order to make this book self-contained, we include a fairly brief description of additive and abelian categories and of homological algebra. There are a number of books that are devoted mostly or entirely to one or the other of these topics, so we can only hit the high spots. See, for example, Cartan & Eilenberg [1956], Freyd [1964], Mitchell [1963, 1964], and Mac Lane [1965, 1972].

1. Additive categories

1.1. Definition. A category is called **preadditive** if there is an abelian group structure on the homsets in such a way that composition on each side is a homomorphism. More precisely, all the homsets are equipped with an abelian group structure such that for $f_1, f_2: A \longrightarrow B$, $g: A' \longrightarrow A$ and $h: B \longrightarrow B'$, we have

$$h \circ (f_1 + f_2) \circ g = h \circ f_1 \circ g + h \circ f_2 \circ g$$

Of course, the addition on the left side takes place in $\mathrm{Hom}(A, B)$ while that on the right is in $\mathrm{Hom}(A', B')$.

A preadditive category is called **additive** if it has finite sums and finite products. It actually suffices that there be either finite sums or finite products. As pointed out next, the additive structure forces the finite sums and products to be isomorphic.

1.2. THEOREM. *Let $\mathscr{A}$ be a preadditive category with finite products. Then for any objects A_1 and A_2, there are arrows $u_1: A_1 \longrightarrow A_1 \times A_2$ and $u_2: A_2 \longrightarrow A_1 \times A_2$ that make the cocone*

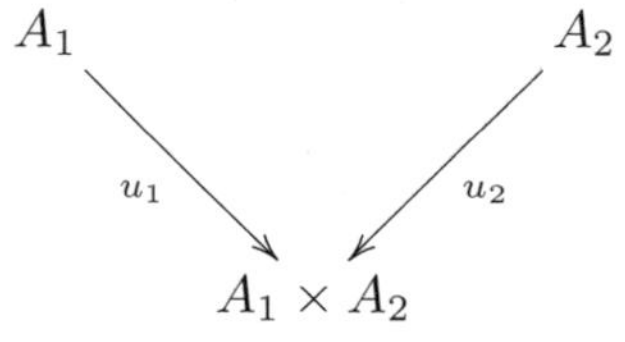

into a sum cocone.

PROOF. Since $\mathrm{Hom}(A_1, A_2)$ and $\mathrm{Hom}(A_2, A_1)$ are abelian groups, there is a 0 element for the abelian group structure. Call these elements $0^{A_1}_{A_2}: A_1 \longrightarrow A_2$ and $0^{A_2}_{A_1}: A_2 \longrightarrow A_1$. We just call them 0 since there is no ambiguity. In particular, the map $0: 1 \longrightarrow 1$ has to be the identity, since that is the only map $1 \longrightarrow 1$. There is, for each object A the map $0: 1 \longrightarrow A$ and if $f: 1 \longrightarrow A$ is any map, then $f = f \circ 0^1_1 = 0$ since composition is a group homomorphism. Thus, there is a unique map $1 \longrightarrow A$ for each object A, which means that 1 is also initial and $\mathscr{A}$ is pointed. Let $u_1 = (1, 0): A_1 \longrightarrow A_1 \times A_2$ and $u_2 = (0, 1): A_2 \longrightarrow A_1 \times A_2$. Suppose $f_1: A_1 \longrightarrow A$ and $f_2: A_2 \longrightarrow A$ are given. Let $p_1: A_1 \times A_2 \longrightarrow A_1$ and

$p_2: A_1 \times A_2 \longrightarrow A_2$ denote the product projections. Define $f = f_1 \circ p_1 + f_2 \circ p_2: A_1 \times A_2 \longrightarrow A$. Then

$$\begin{aligned} f \circ u_1 &= (f_1 \circ p_1 + f_2 \circ p_2) \circ (1, 0) = f_1 \circ p_1 \circ (1, 0) + f_2 \circ p_2 \circ (1, 0) \\ &= f_1 \circ 1 + f_2 \circ 0 = f_1 \end{aligned}$$

and similarly, $f \circ u_2 = f_2$. Suppose $g: A_1 \times A_2 \longrightarrow A$ is another arrow such that $g \circ u_1 = f_1$ and $g \circ u_2 = f_2$. Observe that the map $u_1 \circ p_1 + u_2 \circ p_2: A_1 \times A_2 \longrightarrow A_1 + A_2$ has the properties that

$$\begin{aligned} p_1 \circ (u_1 \circ p_1 + u_2 \circ p_2) &= p_1 \circ u_1 \circ p_1 + p_1 \circ u_1 \circ p_2 \\ &= p_1 \circ (1, 0) \circ p_1 + p_1 \circ (0, 1) \circ p_2 \\ &= 1 \circ p_1 + 0 \circ p_2 = p_1 \end{aligned}$$

and similarly $p_2 \circ (u_1 \circ p_1 + u_2 \circ p_2) = p_2$ and so, by the universal mapping properties of maps into a product, it follows that $u_1 \circ p_1 + u_2 \circ p_2 = \text{id}$. Then

$$f = f_1 \circ p_1 + f_2 \circ u_2 = g \circ u_1 \circ p_1 + g \circ u_2 \circ p_2 = g \circ (u_1 \circ p_1 + u_2 \circ p_2) = g$$

which shows the uniqueness of f and so that u_1, u_2 give a sum cocone. □

It is possible, given that finite sums and products exist and are isomorphic, to introduce the additive structure making the homsets into commutative monoids, but something more is needed to give group structure. Since the empty sum (the initial object) is the empty product (the terminal object), it follows that the category is pointed.

If $\mathscr{A}$ is a pointed category and also has finite products and finite sums, then for any objects A_1 and A_2 there is a canonical arrow $j: A_1 + A_2 \longrightarrow A_1 \times A_2$ as follows. Let $A_1 \xrightarrow{u} A_1 + A_2 \xleftarrow{v} A_2$ be the canonical inclusions and $A_1 \xleftarrow{p} A_1 \times A_2 \xrightarrow{q} A_2$ the canonical projections. Then using the universal mapping properties out of sums and into products, there is then a unique map $j: A_1 + A_2 \longrightarrow A_1 \times A_2$ for which

$$\begin{aligned} p_1 \circ j \circ u_1 &= \text{id} \\ p_2 \circ j \circ u_1 &= 0 \\ p_1 \circ j \circ u_2 &= 0 \\ p_2 \circ j \circ u_2 &= \text{id} \end{aligned}$$

Assuming j is the isomorphism, then the sum of two arrows $f, g: A \longrightarrow B$ is the composite

$$A \xrightarrow{\Delta} A \times A \xrightarrow{f \times g} B \times B \cong B + B \xrightarrow{\nabla} B$$

or, equivalently,

$$A \xrightarrow{\Delta} A \times A \cong A + A \xrightarrow{f + g} B + B \xrightarrow{\nabla} B$$

In both of these formulas, $\Delta: A \longrightarrow A \times A$ is the diagonal map, defined by the equations $p \circ \Delta = q \circ \Delta = \text{id}$. Dually, $\nabla: B \times B \longrightarrow B$ is determined by $\nabla \circ u = \nabla \circ v = \text{id}$.

We do not require this and do not prove it. It does imply that, under reasonable assumptions on the existence of products, the abelian group structure is unique.

It is common to introduce a new operation symbol $\oplus$, called the direct sum, to denote the simultaneous binary sum/product. Another extremely useful notation is the matrix notation for maps between sums. Since both sum and product are naturally associative, so is direct sum and thus we can write, for example, $A = A_1 \oplus A_2 \oplus \cdots \oplus A_n$, simultaneously the sum and the product. Suppose we let $u_i: A_i \longrightarrow A$ denote the injection into the sum and $p_j: B = B_1 \oplus B_2 \oplus \cdots \oplus B_n \longrightarrow B_j$ denote the projection from the product. Then a map $f: A \longrightarrow B$ is uniquely determined by the composites $f_{ji} = p_j \circ f \circ u_i$, which we will think of as a matrix and write $f = (f_{ji})$. Actually, in any category, a map from a sum to a product could be denoted by a matrix. What is different in an additive category is that we can also compose them. If we also have $g = (g_{kj}): B \longrightarrow C = C_1 \oplus C_2 \oplus \cdots \oplus C_p$, then it is easy to show (by composing with the projections and injections) that $g \circ f$ is the matrix product $(g_{kj})(f_{ji})$.

Since the addition, but not the subtraction is equivalent to the isomorphism between sums and products, it is reasonable to ask when you get an abelian group structure. A sufficient (but definitely not necessary, see Exercise 7) condition is that every arrow that is both monic and epic is an isomorphism. For one can show that the arrow $\left(\begin{smallmatrix} 1 & 1 \\ 0 & 1 \end{smallmatrix}\right): A \oplus A \longrightarrow A \oplus A$ is both monic and epic and if it is an isomorphism, its inverse is $\left(\begin{smallmatrix} 1 & -1 \\ 0 & 1 \end{smallmatrix}\right)$ and -1 is an additive inverse of the identity and, once all the identity arrows have additive inverses, all arrows do by composing (on either side!) with -1.

1.3. Additive functors. If $\mathscr{A}$ and $\mathscr{B}$ are preadditive categories, a functor $F: \mathscr{A} \longrightarrow \mathscr{B}$ is called an **additive functor** if for any $f, g: A \longrightarrow A'$ in $\mathscr{A}$, we have $F(f+g) = Ff + Fg$. If $\mathscr{A}$ has finite products, that is direct sums, then a necessary and sufficient condition that F be additive is that it preserve those products.

1.4. Abelian group objects. Let $\mathscr{A}$ be a category with finite limits. An **abelian group object** of $\mathscr{A}$ is an object A together with an arrow $m: A \times A \longrightarrow A$, an arrow $i: A \longrightarrow A$ and an arrow $z: 1 \longrightarrow A$ that satisfy the equations of abelian groups. These are that the following diagrams commute:

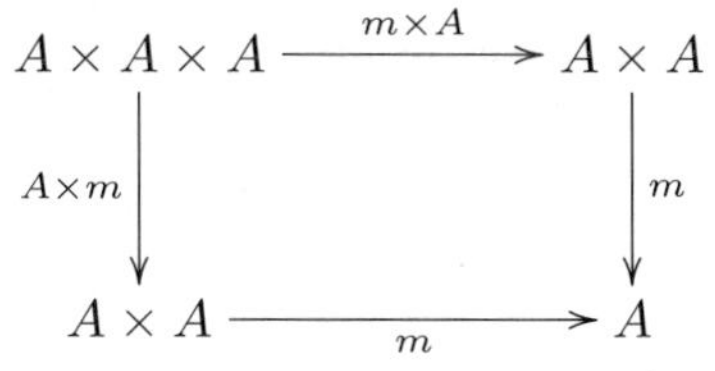

(associativity of multiplication);

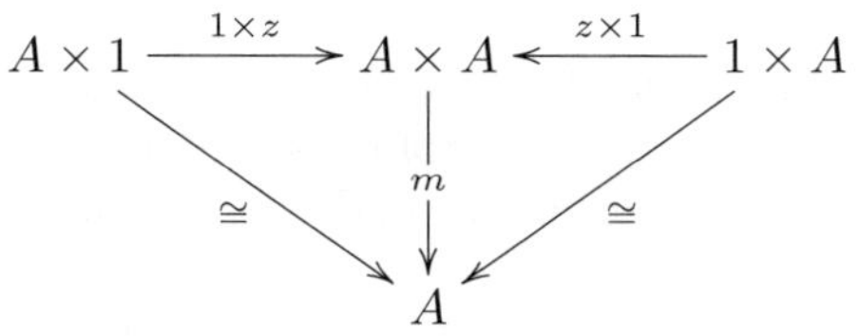

(right and left units);

$$\begin{array}{ccccc} A & \xrightarrow{(1,i)} & A\times A & \xrightarrow{(i,1)} & 1\times A \\ \big\downarrow & & \big\downarrow{\scriptstyle m} & & \big\downarrow \\ 1 & \xrightarrow[z]{} & A & \xrightarrow[z]{} & 1 \end{array}$$

(right and left inverses);

$$\begin{array}{ccc} A\times A & \xrightarrow{(p_2,p_1)} & A\times A \\ & {\scriptstyle m}\searrow \quad \swarrow{\scriptstyle m} & \\ & A & \end{array}$$

(commutativity).

If A, m, i, z and A', m', i', z' are abelian group objects of $\mathscr{A}$, then an arrow $f\colon A \longrightarrow A'$ is called a **morphism of abelian group objects** if f commutes with the structure maps in the obvious way. Just as for ordinary groups, it is sufficient that the diagram

$$\begin{array}{ccc} A\times A & \xrightarrow{m} & A \\ {\scriptstyle f\times f}\big\downarrow & & \big\downarrow{\scriptstyle f} \\ A'\times A' & \xrightarrow[m']{} & A' \end{array}$$

commute.

If A has the structure of an abelian group object, then for any object B of $\mathscr{A}$ we can give the set $\mathrm{Hom}(B, A)$ the structure of an abelian group by defining $f + g$ as the composite

$$B \xrightarrow{\nabla} B\times B \xrightarrow{f\times g} A\times A \xrightarrow{m} A$$

with zero map being $B \longrightarrow 1 \xrightarrow{z} A$ and the inverse to $f\colon B \longrightarrow A$ being $B \xrightarrow{f} A \xrightarrow{i} A$. Moreover, if $f, g\colon A' \longrightarrow A$ is a morphism of abelian group objects, so is $f + g$. One way of dealing with the last claim is to show that an abelian group morphism $A' \longrightarrow A$ can be characterized by the fact that it induces a group homomorphism $\mathrm{Hom}(B, A') \longrightarrow \mathrm{Hom}(B, A)$ for every B and then use the fact that the sum of two homomorphisms between any two abelian groups is a homomorphism, as are the negative and the 0. The product of two abelian group objects can be given the structure of an abelian group in such a way that the projections are group homomorphisms. The details can be found in virtually any book on category theory, for example Mac Lane [1971], Freyd [1964], Mitchell [1965].

The category of abelian group objects and abelian group morphisms in $\mathscr{A}$ will be denoted $\mathsf{Ab}(\mathscr{A})$.

We summarize the above statements by,

1.5. THEOREM. *For any category $\mathscr{A}$ with finite products, the category $\mathsf{Ab}(\mathscr{A})$ is an additive category.*

1.6. Exercises.

1. Assume that the category $\mathscr{A}$ has a terminal object 1 and either has equalizers or has an initial object. Show that 1 is also initial if and only if it has at least one morphism to any other object. A category with a simultaneous initial and terminal object is called pointed.

2. Show that in a pointed category, there is a unique arrow $0^A_B\colon A \longrightarrow B$ for each pair of objects A and B that factors through the initial/terminal object and that these arrows form a 2-sided ideal in the sense that whenever $f\colon A' \longrightarrow A$ and $g\colon B \longrightarrow B'$, then $g \circ 0^A_B \circ f = 0^{A'}_{B'}$. The super and subscripts are generally omitted and we write $0\colon A \longrightarrow B$.

3. Show that in a category that has a terminal object and an ideal $0^A_B\colon A \longrightarrow B$ as above, then that terminal object is also initial and the arrows are just the arrows that factor through the initial/terminal object. Of course, this shows that there is at most one such ideal if there is a terminal (or, dually, initial) object.

4. Show that in a pointed category with products and sums, given a finite number $A_1, A_2, \ldots, A_n$ of objects, there is a unique arrow $s\colon A_1 + A + 2 \cdots + A_n \longrightarrow A_1 \times A_2 \times \cdots A_n$ such that $p_i \circ s \circ u_j = \delta_{ij}$, where p_i is the projection on the ith factor, u_j is the injection of the jth summand, and δ_{ij} is the Kronecker delta, equal to the identity when $i = j$ and the 0 map otherwise.

5. Show that if s is an isomorphism, then there is a canonical structure of commutative monoids on each homset in $\mathscr{A}$ such that arrow composition distributes over the monoid structure, which is usually written as addition.

6. Conversely, show that if a category has products and a distributive commutative monoidal structure on its homsets, then those products as sums. As a consequence, there is at most one such distributive monoidal structure on the homsets.

7. Show that the category of torsion is additive, but that there is a map that is both monic and epic, but not an isomorphism.

2. Abelian categories

An additive category is said to be **abelian** if it is additive and if every arrow factors as a regular epimorphism, followed by a regular monomorphism. It is convenient to use a slightly different characterization of regular monics and epics in additive (or even pointed) categories.

In any pointed category, we can define special limits that are called kernels and, dually, special colimits that are called cokernels. The arrow $f\colon A \longrightarrow B$ is a kernel of $g\colon B \longrightarrow C$ if f is an equalizer of g and $0\colon B \longrightarrow C$. Dually, g is a cokernel of f if it is the coequalizer of f and 0.

2.1. PROPOSITION. *In any additive category, the diagram*

$$E \xrightarrow{h} A \underset{g}{\overset{f}{\rightrightarrows}} B$$

is an equalizer if and only if

$$E \xrightarrow{h} A \xrightarrow{f-g} B$$

is a kernel. Dually, $A \underset{g}{\overset{f}{\rightrightarrows}} B \xrightarrow{h} C$ *is a coequalizer if and only if*

$$A \xrightarrow{f-g} B \xrightarrow{h} C$$

is a cokernel.

PROOF. We will prove only the first. The point is that for an arrow $k: C \longrightarrow A$ the condition $f \circ k = g \circ k$ is equivalent, in an additive category, to $(f-g) \circ k = 0 \circ k$. Thus the universal mapping properties for the equalizer and the kernel are the same. □

2.2. PROPOSITION. *In any abelian category,*

1. *Every monic is regular*;
2. *every epic is regular*;
3. *every monic is a kernel of its own cokernel*;
4. *every epic is a cokernel of its own kernel*;
5. *if* $f: A \longrightarrow B$ *is arbitrary with* $k: K \longrightarrow A$ *the kernel of* f *and* $q: B \longrightarrow Q$ *the cokernel of* f*, then there is a natural arrow from the cokernel of* k *to the kernel of* q *and that is an isomorphism.*

PROOF. Let f be monic and suppose we write $f = m \circ e$ where m is regular monic and e is regular epic. Then there are two maps g, h whose codomain is the domain of e of which e is the coequalizer. Then from $e \circ g = e \circ h$ we have that $f \circ g = m \circ e \circ g = m \circ e \circ h = f \circ h$. Since f is monic, we conclude that $g = h$. But the coequalizer of a map with itself is necessarily an isomorphism and so f is an equalizer of the same pair of arrows as e. This takes care of the first assertion. The second is dual. For the third, suppose $f: A \longrightarrow B$ is the kernel of $g: B \longrightarrow C$. Let $h: B \longrightarrow D$ be a cokernel of f. From $g \circ f = 0$ and the universal mapping property of a cokernel, there is a unique arrow $k: D \longrightarrow C$ such that $k \circ h = g$. Now suppose $l: E \longrightarrow B$ is an arrow with $h \circ l = 0$. Then $g \circ l = k \circ h \circ l = k \circ 0 = 0$ and so there is a unique arrow $m: E \longrightarrow A$ such that $f \circ m = l$. Thus f has the universal mapping property of a kernel of h. The fourth assertion is dual. Finally, let $f: A \longrightarrow B$ be arbitrary. Factor it as $A \xrightarrow{e} D \xrightarrow{m} B$ with e regular epic and m regular monic. Let $k: K \longrightarrow A$ be the kernel of f and $c: B \longrightarrow C$ be the cokernel. I claim that k is also the kernel of e. In fact, with m monic, the condition $f \circ g = 0$ is equivalent to the condition $e \circ g = 0$ so that f and e must have the same kernel. But then by the previous part, $e: A \longrightarrow D$ is the cokernel of k. Dually, $m: D \longrightarrow B$ is the kernel of c. □

2.3. THEOREM. *Suppose that* $\mathscr{A}$ *is an exact category. Then* $\mathsf{Ab}(\mathscr{A})$ *is an abelian category.*

One of the consequences of this is that a category that is both additive and exact is abelian.

PROOF. This argument makes heavy use of the idea of elements discussed in 1.5. Since $\mathscr{A}$ is regular, it follows from 1.8.10 that every arrow factors as a regular epimorphism followed by a monomorphism. Therefore we must show that in an additive exact category, every regular epic is a cokernel and every monic is a kernel. Actually, we will first show that in every additive category, every regular epic is a cokernel. In fact, if $f: A' \longrightarrow A$ is the coequalizer of $g, h: A'' \longrightarrow A'$, then for an arrow $k: A' \longrightarrow B$, the condition $k \circ g = k \circ h$ is equivalent to $k \circ (g - h) = 0$. Thus f is also the cokernel of $g - h$. Now suppose $u: A' \longrightarrow A$ is monic. Here we will use elements. Map $f: A \times A' \longrightarrow A \times A$ by $f(a, a') = (a, a + ua')$. First I claim that f is injective. For if $f(a, a') = f(b, b')$, then $(a, a + ua') = (b, b + ub')$ so that $a = b$ and then $ua' = ub'$, which implies that $a' = b'$. Thus f defines a subobject of $A \times A$ and I claim it is an equivalence relation. In fact the subobject is simply $\{(a, b) \mid a - b \in \operatorname{im} u\}$. Clearly $a - a \in \operatorname{im} u$, $a - b \in \operatorname{im} u$ implies $b - a \in \operatorname{im} u$ and if $a - b$ and $b - c \in \operatorname{im} u$, then $a - c \in \operatorname{im} u$. But then there is an arrow $g: A \longrightarrow A''$ of which f is the kernel pair. This means that $ga = gb$ if and only if $a - b \in \operatorname{im} u$, which implies that $ga = 0$ if and only if $a \in \operatorname{im} u$ so that u is the kernel of g. □

Taken together with 1.8.12 this implies:

2.4. THEOREM. *Let $\mathscr{A}$ be an exact category. Then for any object A of $\mathscr{A}$, the category* $\mathsf{Ab}(\mathscr{A}/A)$ *is abelian.*

2.5. Exercise.

1. Show that any product of abelian categories is abelian.

3. Exactness

3.1. Exact sequences. If $f: A \longrightarrow B$ is an arrow of an abelian category $\mathscr{A}$, the isomorphic $\ker \operatorname{coker} f \cong \operatorname{coker} \ker f$ is called the **image** of f, denoted $\operatorname{im} f$. It is a subobject of B and a quotient object of A. Suppose

$$(*) \qquad A \xrightarrow{f} B \xrightarrow{g} C$$

is a composable pair of arrows. Then the image of f is a subobject of B as is the kernel of g. When these two subobjects are the same, the sequence $(*)$ is said to be **exact**. This can be separated into the two inclusions. The meaning of $\operatorname{im} f \subseteq \ker g$ is easy to understand: simply that $g \circ f = 0$. There is no such easy description of the opposite inclusion.

A finite or infinite sequence

$$\cdots \longrightarrow A_{n+1} \xrightarrow{d_{n+1}} A_n \xrightarrow{d_n} A_{n-1} \longrightarrow \cdots$$

is said to be **exact** if every three term subsequence $A_{n+1} \xrightarrow{d_{n+1}} A_n \xrightarrow{d_n} A_{n-1}$ is exact. An important special case is a sequence of the form

$$0 \longrightarrow A' \xrightarrow{f} A \xrightarrow{g} A'' \longrightarrow 0$$

(the two unlabeled arrows are, of course, 0), which is exact if and only if, up to isomorphism, f is the inclusion of a subobject and g is the projection on A/A'. Such a sequence is called a **short exact sequence**.

3.2. The full exact embedding theorem. Suppose that $\mathscr{A}$ and $\mathscr{B}$ are abelian categories. A functor $T: \mathscr{A} \longrightarrow \mathscr{B}$ is called **exact** if for any exact sequence $A' \longrightarrow A \longrightarrow A''$ in $\mathscr{A}$, the image $TA' \longrightarrow TA \longrightarrow TA''$ in $\mathscr{B}$ is also exact. It follows that an exact functor preserves the exactness of all exact sequences. The following theorem allows us to reduce all kinds of arguments involving exactness in abelian categories to categories of modules.

3.3. THEOREM. *Let $\mathscr{A}$ be a small abelian category. Then there is a ring R and a full exact embedding $T: \mathscr{A} \longrightarrow R$*–Mod.

This is proved in various places. See, for example, Freyd [1964], Mitchell [1964, 1965], or Popescu [1973].

The restriction to small abelian categories is not important since any small diagram we might want to know commutes or is exact or show the existence of some arrow in is in a small abelian subcategory gotten by closing the set of objects in the diagram under finite sums and then kernels and cokernels of arrows between them. If you do this countably many times you will have a small full abelian subcategory of the original category which can be embedded into a module category. Thus an exactness property of module categories is valid in any abelian category.

One of the most useful applications of this principle arises from the fact that the dual of an abelian category is an abelian category. It follows that the dual of any exactness property that is true in any abelian category is also true. This will cut in half the work needed to prove the snake lemma below.

3.4. Right and left exact functors. There are two useful variants on exact functors. A functor $F: \mathscr{A} \longrightarrow \mathscr{B}$ between abelian categories is called **left exact** if given any exact sequence $0 \longrightarrow A' \longrightarrow A \longrightarrow A'' \longrightarrow 0$ in $\mathscr{A}$, the sequence $0 \longrightarrow F(A') \longrightarrow F(A) \longrightarrow F(A'')$ is exact. Similarly, it is **right exact** if for any such short exact sequence in $\mathscr{A}$ the sequence $F(A') \longrightarrow F(A) \longrightarrow F(A'') \longrightarrow 0$ is exact.

It is evident that a left exact functor is exact if and only if it preserves epimorphisms and a right exact functor is exact if and only if it preserves monomorphisms.

3.5. PROPOSITION. *A functor between abelian categories is left exact if and only if it preserves the limits of finite diagrams; dually it is right exact if and only if it preserves the colimits of finite diagrams.*

PROOF. Suppose that F is left exact. We first show that it is additive. Given that $A = A_1 \oplus A_2$, there is an exact sequence $0 \longrightarrow A_1 \longrightarrow A \longrightarrow A_2 \longrightarrow 0$. Since F is left exact, the sequence $0 \longrightarrow TA_1 \longrightarrow TA \longrightarrow TA_2$ is exact. On the other hand, $A \longrightarrow A_2$ is a split epimorphism (that is, it has a left inverse) and hence so is $TA \longrightarrow TA_2$. Hence $0 \longrightarrow TA_2 \longrightarrow TA \longrightarrow TA_2 \longrightarrow 0$ is exact and split. But this characterizes TA as the sum of the two end terms. Hence T preserves finite products and is thus additive. Next we show it preserves equalizers. From 2.1 we know that the equalizer of two arrows $A \overset{f}{\underset{g}{\rightrightarrows}} A''$ is the kernel of $f - g$. Now factor $f - g$ as $A \xrightarrow{h} A_1 \xrightarrow{k} A''$ with h epic and k monic. Finally, let $l: A'' \longrightarrow A_1'$ be the cokernel of k. Then both sequences

$$0 \longrightarrow A' \xrightarrow{e} A \xrightarrow{h} A_1 \longrightarrow 0$$

$$0 \longrightarrow A_1 \xrightarrow{k} A'' \xrightarrow{l} A_1' \longrightarrow 0$$

are exact, whence so are

$$0 \longrightarrow FA' \xrightarrow{Fe} FA \xrightarrow{Fh} FA_1$$

$$0 \longrightarrow FA_1 \xrightarrow{Fk} FA'' \xrightarrow{Fl} FA'_1$$

Putting these together, we conclude that

$$0 \longrightarrow FA' \xrightarrow{Fe} FA \xrightarrow{F(f-g)} FA''$$

is also exact. Since F is additive, $F(f - g) = Ff - Fg$ and then it follows that Fe is the equalizer of Ff and Fg. Conversely, if F preserves finite limits, then it preserves finite products and is therefore additive. Since kernels are limits, it preserves kernels, which means that it takes any exact sequence of the form $0 \longrightarrow A' \longrightarrow A \longrightarrow A''$ to an exact sequence and is evidently left exact.

The proof for right exact functors and finite colimits is strictly dual. □

3.6. THEOREM (The snake lemma). *Suppose that in the following diagram in an abelian category the rows are exact and the squares commute*:

$$\begin{array}{ccccccc} & & A' & \xrightarrow{f} & A & \xrightarrow{f'} & A'' & \longrightarrow & 0 \\ & & \downarrow{h'} & & \downarrow{h} & & \downarrow{h''} & & \\ 0 & \longrightarrow & B' & \xrightarrow[g]{} & B & \xrightarrow[g']{} & B'' & & \end{array}$$

Then there is an arrow $\gamma\colon \ker h'' \longrightarrow \operatorname{coker} h'$ *such that the sequence*

$$\ker h' \longrightarrow \ker h \longrightarrow \ker h'' \xrightarrow{\gamma} \operatorname{coker} h' \longrightarrow \operatorname{coker} h \longrightarrow \operatorname{coker} h''$$

is exact.

PROOF. The remaining arrows in the snake are induced by f, f', g, and g'. The arrow γ can be described as being induced by the relation $g^{-1} \circ h \circ (f')^{-1}$. As mentioned, we can suppose without loss of generality that we are in a category of modules. Given an element $a'' \in \ker h''$, since f' is surjective, there is an $a \in A$ with $f'a = a''$. Since $g'ha = h''f'a = h''a'' = 0$, there is an element $b' \in B'$ with $gb' = ha$ and we define $\gamma a''$ as the class of b' modulo $\operatorname{im} h'$. We wish to show that this is independent, modulo $\operatorname{im} h'$, of the choice of a. If a_1 is another choice for a, then $a - a_1 \in \ker f' = \operatorname{im} f$ so that there is an $a' \in A'$ with $a - a_1 = fa'$. If $b'_1 \in B'$ is such that $gb'_1 = ha_1$, then $gh'a' = hfa' = h(a - a_1) = ha - ha_1 = gb' - gb'_1 = g(b' - b'_1)$. But g is injective, so we can infer that $h'a' = b' - b'_1$ which means that b' and b'_1 are in the same class modulo $\operatorname{im} h'$. Thus γ is well defined. If a''_1 and $a''_2 \in A''$ and we choose preimages a_1 and $a_2 \in A$, then we can choose $a_1 + a_2$ as a preimage of $a''_1 + a''_2$. This shows that $\gamma(a''_1 + a''_2) = \gamma a''_1 + \gamma a''_2$ and a similar argument shows that γ is a module homomorphism.

To show exactness it is sufficient to show it at the first two places; duality gives the remaining two. So suppose that $a \in \ker h$ is such that $f'a = 0$. Then exactness of the original sequence implies the existence of $a' \in A'$ such that $fa' = a$. Moreover $gh'a' = hfa' = 0$. Since g is injective, this shows that $a' \in \ker h'$ and shows exactness at the first step. If $a \in \ker h$, then we can take a as the preimage of $f'a$ and then $ha = 0$, whence $\gamma f'a = 0$ so that the image of the map induced by f' is in the kernel of γ. If $a'' \in \ker h''$ is such that $\gamma a'' = 0$, let $a \in A$ be a preimage of

a''. Then let $b' \in B'$ be such that $ha = gb'$. Since $\gamma a'' = 0$, we must have $b' \in \operatorname{im} h'$ so that there is an $a' \in A'$ with $h'a' = b'$ or $hfa' = gh'a' = gb' = ha$. But then $h(a - fa') = 0$. Since $f'(a - fa') = f'a = a''$, we see that a'' is in the image under f' of an element of $\ker h$. □

Since it is evident that if f is monic so is the induced map $\ker h' \longrightarrow \ker h$ and that if g' is epic, so is the induced map $\operatorname{coker} h \longrightarrow \operatorname{coker} h''$ we have three more forms of the snake lemma that we state for completeness

3.7. Corollary. *Suppose that in the following diagram in an abelian category the rows are exact and the squares commute:*

$$\begin{array}{ccccccccc} 0 & \longrightarrow & A' & \xrightarrow{f} & A & \xrightarrow{f'} & A'' & \longrightarrow & 0 \\ & & \downarrow{\scriptstyle h'} & & \downarrow{\scriptstyle h} & & \downarrow{\scriptstyle h''} & & \\ 0 & \longrightarrow & B' & \xrightarrow[g]{} & B & \xrightarrow[g']{} & B'' & & \end{array}$$

Then there is an arrow $\gamma\colon \ker h'' \longrightarrow \operatorname{coker} h'$ *such that the sequence*

$$0 \longrightarrow \ker h' \longrightarrow \ker h \longrightarrow \ker h'' \xrightarrow{\gamma} \operatorname{coker} h' \longrightarrow \operatorname{coker} h \longrightarrow \operatorname{coker} h''$$

is exact. □

3.8. Corollary. *Suppose that in the following diagram in an abelian category the rows are exact and the squares commute:*

$$\begin{array}{ccccccccc} & & A' & \xrightarrow{f} & A & \xrightarrow{f'} & A'' & \longrightarrow & 0 \\ & & \downarrow{\scriptstyle h'} & & \downarrow{\scriptstyle h} & & \downarrow{\scriptstyle h''} & & \\ 0 & \longrightarrow & B' & \xrightarrow[g]{} & B & \xrightarrow[g']{} & B'' & \longrightarrow & 0 \end{array}$$

Then there is an arrow $\gamma\colon \ker h'' \longrightarrow \operatorname{coker} h'$ *such that the sequence*

$$\ker h' \longrightarrow \ker h \longrightarrow \ker h'' \xrightarrow{\gamma} \operatorname{coker} h' \longrightarrow \operatorname{coker} h \longrightarrow \operatorname{coker} h'' \longrightarrow 0$$

is exact. □

3.9. Corollary. *Suppose that in the following diagram in an abelian category the rows are exact and the squares commute:*

$$\begin{array}{ccccccccc} 0 & \longrightarrow & A' & \xrightarrow{f} & A & \xrightarrow{f'} & A'' & \longrightarrow & 0 \\ & & \downarrow{\scriptstyle h'} & & \downarrow{\scriptstyle h} & & \downarrow{\scriptstyle h''} & & \\ 0 & \longrightarrow & B' & \xrightarrow[g]{} & B & \xrightarrow[g']{} & B'' & \longrightarrow & 0 \end{array}$$

Then there is an arrow $\gamma\colon \ker h'' \longrightarrow \operatorname{coker} h'$ *such that the sequence*

$$0 \longrightarrow \ker h' \longrightarrow \ker h \longrightarrow \ker h'' \xrightarrow{\gamma} \operatorname{coker} h' \longrightarrow \operatorname{coker} h \longrightarrow \operatorname{coker} h'' \longrightarrow 0$$

is exact. □

3.10. Proposition. *Suppose*

$$\begin{array}{ccccccc}
A_1 & \xrightarrow{f_1} & A_2 & \xrightarrow{f_2} & A_3 & \xrightarrow{f_3} & A_4 \\
\downarrow{\scriptstyle h_1} & & \downarrow{\scriptstyle h_2} & & \downarrow{\scriptstyle h_3} & & \downarrow{\scriptstyle h_4} \\
B_1 & \xrightarrow[g_1]{} & B_2 & \xrightarrow[g_2]{} & B_3 & \xrightarrow[g_3]{} & B_4
\end{array}$$

is a commutative diagram with exact rows in an abelian category. If h_1 is an epimorphism and h_2 and h_4 are monomorphisms, then h_3 is a monomorphism. If h_4 is a monomorphism and h_1 and h_3 are epimorphisms, then h_2 is an epimorphism.

Proof. The two statements are dual to each other, so we need prove only the first. We can replace B_1 by $B_1/\ker g_1$ and suppose that g_1 is monic. Similarly we can replace A_4 by the image of f_3 and suppose that f_3 is epic. Let $A = \operatorname{im} f_2 = \ker f_3$ and $B = \operatorname{im} g_2 = \ker g_3$. Then we have two commutative diagrams with exact rows:

$$\begin{array}{ccccccccc}
 & & A_1 & \xrightarrow{f_1} & A_2 & \longrightarrow & A & \longrightarrow & 0 \\
 & & \downarrow{\scriptstyle h_1} & & \downarrow{\scriptstyle h_2} & & \downarrow{\scriptstyle h} & & \\
0 & \longrightarrow & B_1 & \xrightarrow[g_1]{} & B_2 & \longrightarrow & B & \longrightarrow & 0
\end{array}$$

$$\begin{array}{ccccccccc}
0 & \longrightarrow & A & \longrightarrow & A_3 & \xrightarrow{f_3} & A_4 & \longrightarrow & 0 \\
 & & \downarrow{\scriptstyle h} & & \downarrow{\scriptstyle h_3} & & \downarrow{\scriptstyle h_4} & & \\
0 & \longrightarrow & B & \longrightarrow & B_3 & \xrightarrow[g_3]{} & B_4 & &
\end{array}$$

Applying the snake lemma to the first gives, in part, the exact sequence

$$\ker h_2 \longrightarrow \ker h \longrightarrow \operatorname{coker} h_1$$

which, together with h_2 monic and h_1 epic, implies that h is monic. Applying the snake lemma to the second gives, in part, the exact sequence

$$\ker h \longrightarrow \ker h_3 \longrightarrow \ker h_4$$

which, together with h and h_4 monic, implies that h_3 is monic. □

3.11. Corollary ("5 lemma"). *Suppose that*

$$\begin{array}{ccccccccc}
A_1 & \xrightarrow{f_1} & A_2 & \xrightarrow{f_2} & A_3 & \xrightarrow{f_3} & A_4 & \xrightarrow{f_4} & A_5 \\
\downarrow{\scriptstyle h_1} & & \downarrow{\scriptstyle h_2} & & \downarrow{\scriptstyle h_3} & & \downarrow{\scriptstyle h_4} & & \downarrow{\scriptstyle h_5} \\
B_1 & \xrightarrow[f_2]{} & B_2 & \xrightarrow[f_3]{} & B_3 & \xrightarrow[f_4]{} & B_4 & \xrightarrow[f_5]{} & B_5
\end{array}$$

is a commutative diagram with exact rows. If h_1 is epic, h_5 monic, and h_2 and h_4 isomorphisms, then h_3 is an isomorphism. □

Here is another immediate application of the snake lemma.

3.12. COROLLARY ("3×3 lemma"). *Suppose the columns, the middle row and either the top or bottom row of the diagram*

$$\begin{array}{ccccccccc}
 & & 0 & & 0 & & 0 & & \\
 & & \downarrow & & \downarrow & & \downarrow & & \\
0 & \longrightarrow & A' & \longrightarrow & B' & \longrightarrow & C' & \longrightarrow & 0 \\
 & & \downarrow & & \downarrow & & \downarrow & & \\
0 & \longrightarrow & A & \longrightarrow & B & \longrightarrow & C & \longrightarrow & 0 \\
 & & \downarrow & & \downarrow & & \downarrow & & \\
0 & \longrightarrow & A'' & \longrightarrow & B'' & \longrightarrow & C'' & \longrightarrow & 0 \\
 & & \downarrow & & \downarrow & & \downarrow & & \\
 & & 0 & & 0 & & 0 & &
\end{array}$$

are exact. Then the remaining row is also exact. □

3.13. Exercise.

1. Show that if $f\colon A \longrightarrow B$ and $g\colon B \longrightarrow C$ are arrows in an abelian category, then there is an exact sequence

$$0 \longrightarrow \ker f \longrightarrow \ker(g \circ f) \longrightarrow \ker g$$
$$\longrightarrow \operatorname{coker} f \longrightarrow \operatorname{coker}(g \circ f) \longrightarrow \operatorname{coker} g \longrightarrow 0$$

Hint: Show that there is a commutative square with exact rows

$$\begin{array}{ccccccccc}
0 & \longrightarrow & A & \longrightarrow & A \oplus B & \longrightarrow & B & \longrightarrow & 0 \\
 & & \downarrow{\scriptstyle f} & & \downarrow{\scriptstyle \begin{pmatrix} f & 1 \\ 0 & g \end{pmatrix}} & & \downarrow{\scriptstyle g} & & \\
0 & \longrightarrow & B & \longrightarrow & B \oplus C & \longrightarrow & C & \longrightarrow & 0
\end{array}$$

4. Homology

4.1. Differential objects. Let A be an object of the abelian category $\mathscr{A}$. A **differential** on A is an endomorphism $d\colon A \longrightarrow A$ such that $d \circ d = 0$. This is equivalent to $\operatorname{im} d \subseteq \ker d$. If, in fact, $\operatorname{im} d = \ker d$, we say that d is an **exact differential**.

A **morphism** $f\colon (A', d') \longrightarrow (A, d)$ **of differential objects** is a morphism $f\colon A' \longrightarrow A$ in $\mathscr{A}$ such that $f \circ d' = d \circ f$.

4.2. Homology. Given a differential object (A, d), we say that $B(A, d) = \operatorname{im} d$, called the object of **boundaries** and $Z(A, d) = \ker d$ is called the object of **cycles**. Since $Z(A, d) \subseteq B(A, d)$, we can form the quotient $H(A, d) = Z(A, d)/B(A, d)$, which is called the **homology object** of (A, d).

When there is no danger of confusion, we write $B(A)$, $Z(A)$, and $H(A)$ for these objects.

4.3. PROPOSITION. *Suppose* $f\colon (A',d') \longrightarrow (A,d)$ *is a morphism of differential objects over the abelian category* $\mathscr{A}$. *Then* f *induces an arrow*

$$H(f)\colon H(A',d') \longrightarrow H(A,d)\circ$$

PROOF. The equation $f \circ d' = d \circ f$ evidently implies that $f(Z(A',d')) \subseteq Z(A,d)$ and that $f(B(A',d')) \subseteq B(A,d)$, from which the conclusion readily follows. □

4.4. Another view on homology. The homology of a differential object is defined to be a quotient of a subobject. But it is always possible to take a subobject of a quotient instead. If A is a group with subgroups $A' \subseteq A''$ then A'/A'' is isomorphic to a subgroup of A/A'', namely to the subgroup consisting of those cosets $a + A''$ for which $a \in A'$. Thus if (A,d) is a differential object, the homology can be described as the subgroup of $A/(B(A,d))$ consisting of those cosets $a + B(A,d)$ for which $da = 0$.

Now write $B = B(A,d)$ and $Z = Z(A,d)$. Define $\widehat{d}\colon A/B \longrightarrow Z$ by $\widehat{d}a = da$. This makes sense since $dB = 0$. The kernel of $\widehat{d}$ is exactly those cosets modulo B that are in the kernel of d and thus $H(A,d) \cong \ker \widehat{d}$. But also the image of $\widehat{d}$ is the same as the image of d so that $H(A,d) = \operatorname{coker} \widehat{d}$. Using the exact embedding, we draw the same conclusions for any abelian category.

4.5. THEOREM. *If* (A,d) *is a differential object in an abelian category and*

$$\widehat{d}\colon A/B(A,d) \longrightarrow Z(A,d)$$

is as described above, then $H(A,d)$ *is both the kernel and the cokernel of* $\widehat{d}$. □

4.6. The homology sequence. A sequence of differential objects is called **exact** if the underlying sequence of objects is exact. Thus a short exact sequence

$$0 \longrightarrow (A',d') \longrightarrow (A,d) \longrightarrow (A'',d'') \longrightarrow 0$$

is a map of differential objects for which $0 \longrightarrow A' \longrightarrow A \longrightarrow A'' \longrightarrow 0$ is a short exact sequence.

4.7. THEOREM. *Let*

$$0 \longrightarrow (A',d') \longrightarrow (A,d) \longrightarrow (A'',d'') \longrightarrow 0$$

be exact. Then there is an arrow we denote $d\colon H(A'',d'') \longrightarrow H(A,',d')$ *such that the sequence*

$$\begin{array}{l}
\cdots \xrightarrow{d} H(A',d') \xrightarrow{H(f)} H(A,d) \xrightarrow{H(g)} H(A'',d'') \\
\qquad\qquad\qquad\qquad \swarrow_{d} \\
H(A',d') \xrightarrow[H(f)]{} H(A,d) \xrightarrow[H(g)]{} H(A'',d'') \xrightarrow[d]{} \cdots
\end{array}$$

is exact.

Notational note: The use of "d" for the so-called connecting homomorphism is hallowed by long usage. It is based on the fact that the connecting homomorphism is induced by the differential in the middle term. In the original examples, A' was simply a subgroup of A, with $dA' \subseteq A'$ and A'' the quotient. Both d' and d''

were determined uniquely by d and the inclusion and projection were ignored. The connecting homomorphism was induced from d.

PROOF. From the exactness of

$$\begin{array}{ccccccccc}
0 & \longrightarrow & A' & \xrightarrow{f} & A & \xrightarrow{f'} & A'' & \longrightarrow & 0 \\
 & & \downarrow{\scriptstyle d'} & & \downarrow{\scriptstyle d} & & \downarrow{\scriptstyle d''} & & \\
0 & \longrightarrow & A' & \xrightarrow[f]{} & A & \xrightarrow[f'']{} & A'' & \longrightarrow & 0
\end{array}$$

we conclude from 3.9 that both sequences

$$0 \longrightarrow \ker d' \longrightarrow \ker d \longrightarrow \ker d''$$

and

$$\operatorname{coker} d' \longrightarrow \operatorname{coker} d \longrightarrow \operatorname{coker} d'' \longrightarrow 0$$

are exact and we can put them together to get the commutative diagram with exact rows.

$$\begin{array}{ccccccccc}
 & & \operatorname{coker} d' & \longrightarrow & \operatorname{coker} d & \longrightarrow & \operatorname{coker} d'' & \longrightarrow & 0 \\
 & & \downarrow{\scriptstyle \widehat{d'}} & & \downarrow{\scriptstyle \widehat{d}} & & \downarrow{\scriptstyle \widehat{d''}} & & \\
0 & \longrightarrow & \ker d' & \longrightarrow & \ker d & \longrightarrow & \ker d'' & &
\end{array}$$

to conclude that the sequence

$$\ker \widehat{d'} \longrightarrow \ker \widehat{d} \longrightarrow \ker \widehat{d''} \longrightarrow \operatorname{coker} \widehat{d'} \longrightarrow \operatorname{coker} \widehat{d} \longrightarrow \operatorname{coker} \widehat{d''}$$

is exact. But in light of 4.5, this is equivalent to the exactness of

$$H(A',d') \longrightarrow H(A,d) \longrightarrow H(A'',d'') \longrightarrow H(A',d') \longrightarrow H(A,d) \longrightarrow H(A'',d'')$$

which is the homology exact sequence. □

This exact homology sequence is often called the **homology triangle** and drawn as

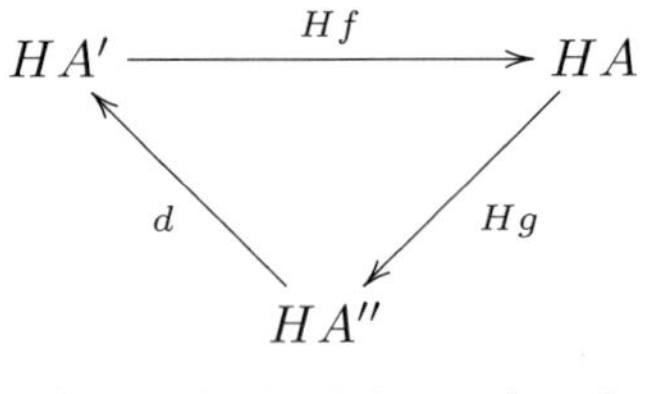

4.8. Graded objects. A **graded object** $A_\bullet$ of $\mathscr{A}$ is a **Z**-indexed sequence

$$\cdots,\ A_{n+1},\ A_n,\ A_{n-1},\ \ldots$$

of objects of $\mathscr{A}$.

There are two kinds of morphisms of graded objects. If $A'_\bullet$ and $A_\bullet$ are graded objects, a morphism $f_\bullet\colon A'_\bullet \longrightarrow A_\bullet$ is a **Z**-indexed sequence of arrows of $\mathscr{A}$, $f_n\colon A'_n \longrightarrow A_n$. This is the category we call $\mathsf{Gr}(\mathscr{A})$. For $k \in \mathbf{Z}$, a morphism of degree k is a **Z**-indexed sequence $f_n\colon A'_n \longrightarrow A_{n+k}$. We will not give this category a name and, in fact, will be interested only in the cases $k = -1, 0, 1$.

Incidentally, what we have called a graded object is sometimes called a **Z**-graded object, since other groups, monoids, or posets are also used.

4.9. Chain and cochain complexes. A **chain complex** over $\mathscr{A}$ is a differential graded object $(A_\bullet, d)$ in which the differential has degree -1 and for which there is an $m \in \mathbf{Z}$ such that $A_n = 0$ for $n < m$. We generally ignore the terms below the bottom degree and write the chain complex as

$$\cdots \xrightarrow{d_{n+1}} A_n \xrightarrow{d_n} A_{n-1} \xrightarrow{d_{n-1}} \cdots \xrightarrow{d_{m+1}} A_m$$

Most of the time, the lower bound on a chain complex will be 0 or -1, but for technical reasons (largely so that the suspension operator described below can be an isomorphism), we allow any lower bound.

A **cochain complex** over $\mathscr{A}$ is a differential graded object $(A_\bullet, d)$ in which the differential has degree -1 and for which there is an $m \in \mathbf{Z}$ such that $A_k = 0$ for $k > m$. So a cochain complex looks like

$$A_m \xrightarrow{d_m} A_{m-1} \xrightarrow{d_{m-1}} \cdots \xrightarrow{d_{n+1}} A_n \xrightarrow{d_n} \cdots$$

However, the more usual notation negates the indices and puts them as superscripts, so the cochain complex above is called $(A^\bullet, d)$ and looks like

$$A^{-m} \xrightarrow{d^{-m}} A^{-(m-1)} \xrightarrow{d^{-(m-1)}} \cdots \xrightarrow{d^{-(n+1)}} A^{-n} \xrightarrow{d^{-n}} \cdots$$

with the differential having degree $+1$. Also in this case, the bound m is likely to be 0 or -1.

The differential of a chain complex is usually called a boundary operator and is often denoted d. The differential of a cochain complex is usually called a coboundary operator and is often denoted δ.

4.10. Homology and cohomology. The homology and cohomology are as defined for differential objects. But there are some additional notational conventions that are standard. We deal first with chain complexes. If $(A_\bullet, d)$ is a chain complex, then $d: A_n \longrightarrow A_{n-1}$ for each n. Then we denote by Z_n the kernel of $d: A_n \longrightarrow A_{n-1}$ and by B_n the image of $d: A_{n+1} \longrightarrow A_n$. They are called the objects of n-cycles and n-boundaries, respectively. Then $B_n \subseteq Z_n \subseteq A_n$ and we denote the quotient Z_n/B_n by $H_n(A, d)$ or simply $H_n(A)$ if d is understood. This is called the nth homology object. The sequence H_n is simply a graded object of $\mathscr{A}$ and the sequence will usually be denoted $H_\bullet$. Analogous definitions of objects of n-cocycles, n-coboundaries and nth cohomology are made for the cohomology of a cochain complex.

An arrow $f: (A', d') \longrightarrow (A, d)$ is a map of degree 0 between the graded objects that also commutes with the differentials. We will denote the category of chain complexes over the additive category $\mathscr{A}$ by $\mathsf{CC}(\mathscr{A})$. It is immediate that a morphism of chain complexes induces a morphism of degree 0 on the graded homology objects.

4.11. PROPOSITION. *A morphism $f: A' \longrightarrow A$ of chain complexes induces a morphism $H_\bullet(f): H_\bullet(A') \longrightarrow H_\bullet(A)$.* □

A chain complex is evidently exact if and only if its homology groups are 0. In that case, we often say that the complex is **acyclic**.

4.12. Exact sequences of chain complexes. A sequence $(A', d') \xrightarrow{f} (A, d) \xrightarrow{g} (A'', d'')$ is **exact** if it is exact as a sequence of graded objects, which is equivalent to its being exact in each degree. A finite or infinite sequence

$$\cdots \longrightarrow (A_{n+1}, d_{n+1}) \longrightarrow (A_n, d_n) \longrightarrow (A_{n-1}, d_{n-1}) \longrightarrow \cdots$$

is exact if it is exact at each place. In particular, a **short exact sequence** is an exact sequence of chain complexes that looks like

$$0 \longrightarrow (A', d') \xrightarrow{f} (A, d) \xrightarrow{g} (A'', d'') \longrightarrow 0$$

where 0 here stands for the chain complex that is 0 in every degree.

4.13. THEOREM. *Suppose* $0 \longrightarrow A' \longrightarrow A \longrightarrow A'' \longrightarrow 0$ *is a short exact sequence of chain complexes. The there are arrows* $d_n : H_n(A'') \longrightarrow H_{n-1}(A')$ *such that the sequence*

$$\cdots \longrightarrow H_n(A') \xrightarrow{H_n(f)} H_n(A) \xrightarrow{H_n(g)} H_n(A'') \xrightarrow{d_n} H_{n-1}(A') \xrightarrow[H_{n-1}(f)]{} H_{n-1}(A) \xrightarrow[H_{n-1}(g)]{} H_{n-1}(A'') \longrightarrow \cdots$$

is exact.

PROOF. This is, in effect, an instance of 4.7. The only thing to note is that since the connecting morphism is induced by a composite of three relations, one of which has degree -1 and the other two have degree 0, the connecting morphism also has degree -1. Thus the homology triangle turns into the long exact sequence shown here.

4.14. Exercises.

1. Show that if $0 \longrightarrow A' \longrightarrow A \longrightarrow A'' \longrightarrow 0$ is an exact sequence of differential objects, then A' is exact if and only if the induced $H(A) \longrightarrow H(A'')$ is an isomorphism; that A is exact if and only if the connecting homomorphism $d: H(A'') \longrightarrow H(A')$ is an isomorphism; and that A'' is exact if and only the induced $H(A') \longrightarrow H(A)$ is an isomorphism.

2. Show that if

$$0 \longrightarrow A' \longrightarrow A \longrightarrow A'' \longrightarrow 0$$

is an exact sequence of differential objects and any two are acyclic, so is the third.

3. Show that if

$$\begin{array}{ccccccccc}
0 & \longrightarrow & A' & \longrightarrow & A & \longrightarrow & A'' & \longrightarrow & 0 \\
 & & \downarrow & & \downarrow & & \downarrow & & \\
0 & \longrightarrow & B' & \longrightarrow & B & \longrightarrow & B'' & \longrightarrow & 0
\end{array}$$

is a commutative diagram of differential objects with exact rows, then the diagram

$$\begin{array}{ccccccccc} HA'' & \longrightarrow & HA' & \longrightarrow & HA & \longrightarrow & HA'' & \longrightarrow & HA' \\ \downarrow & & \downarrow & & \downarrow & & \downarrow & & \downarrow \\ HB'' & \longrightarrow & HB' & \longrightarrow & HB & \longrightarrow & HB'' & \longrightarrow & HB' \end{array}$$

commutes.

5. Module categories

This section is a very short primer on Ext and Tor, two very important homology functors in module categories. There are many better sources, going back to [Cartan & Eilenberg, 1956], or [Mac Lane, 1963], any of which will give a more leisurely exposition.

5.1. Projectives. Let R be a ring. An R-module P is said to be **projective** if the homfunctor $\mathrm{Hom}_R(P,-)$ is an exact functor. This definition makes sense in any abelian category, although there is no guarantee that an arbitrary abelian category has any non-zero projective objects.

Any free module is projective, since if F is free on basis X, then for any R-module M, $\mathrm{Hom}(F,M) \cong M^X$. Thus if $0 \longrightarrow M' \longrightarrow M \longrightarrow M'' \longrightarrow 0$ is an exact sequence of R-modules, the hom sequence is $0 \longrightarrow M'^X \longrightarrow M^X \longrightarrow M''^X \longrightarrow 0$, which is readily seen to be exact.

If M is an arbitrary module, there is certainly a free module mapping surjectively on M. For example, the free module generated by M itself will do. The identify map on M extends to a unique homomorphism on the free module and that is clearly surjective, since the original arrow is. If P is projective, then for any surjection $f\colon M \twoheadrightarrow P$, the sequence $M \longrightarrow P \longrightarrow 0$ is exact, so that $\mathrm{Hom}(P,M) \longrightarrow \mathrm{Hom}(P,P) \longrightarrow 0$ is exact, meaning that $\mathrm{Hom}(P,M) \twoheadrightarrow \mathrm{Hom}(P,P)$ is surjective. In particular, this implies there is a $g \in \mathrm{Hom}(P,M)$ such that $f \circ g = \mathrm{id}$. In other words, every surjective homomorphism splits. The converse is also true. If every surjective homomorphism to P splits, then P is projective. In particular, projectives can be characterized as those modules that are retracts of free modules.

5.2. Projective resolutions. Suppose M is an R-module. By a **projective resolution** of M is meant a complex $P_\bullet = \{P_n, d \mid n \geq 0\}$ consisting of projective R-modules, together with an arrow $P_0 \longrightarrow M$ such that the augmented complex $P_\bullet \longrightarrow M \longrightarrow 0$ is exact.

5.3. PROPOSITION. *Every module has a projective resolution.*

PROOF. Let M be a module. As observed above, there is a free module, say F_0 with a surjective homomorphism $p_0\colon F_0 \twoheadrightarrow M$. Let $q_1\colon M_1 \longrightarrow F_0$ be the kernel of p_0. Suppose $p_1\colon F_1 \longrightarrow M_1$ is a surjective homomorphism with F_1 free. Continue to build the sequence $q_i\colon M_i \longrightarrow F_{i-1}$ as the kernel of p_{i-1} and $p_i\colon F_i \twoheadrightarrow M_i$ with F_i free. Then if we let $d_i = q_i \circ p_1$, the sequence $\{F_i, d_i\}$ is a projective resolution of M. □

5.4. Ext. Let M and N be left R-modules. Suppose $P_\bullet \longrightarrow M$ is a projective resolution of M. Then from the exactness of $P_1 \longrightarrow P_0 \longrightarrow M \longrightarrow 0$, we conclude that $0 \longrightarrow \mathrm{Hom}_R(M,N) \longrightarrow \mathrm{Hom}_R(P_0,N) \longrightarrow \mathrm{Hom}_R(P_1,N)$ is exact (see the proof of Proposition 3.5). Since each composite $P_{i+1} \longrightarrow P_i \longrightarrow P_{i-1}$ is 0, the same is true for each composite $\mathrm{Hom}_R(P_{i-1},N) \longrightarrow \mathrm{Hom}_R(P_i,N) \longrightarrow \mathrm{Hom}_R(P_{i+1},N)$. The result is that

$$\mathrm{Hom}_R(P_0,N) \longrightarrow \mathrm{Hom}_R(P_1,N) \longrightarrow \cdots \longrightarrow \mathrm{Hom}_R(P_n,N) \longrightarrow \cdots$$

is a cochain complex whose zeroth cohomology group is $\mathrm{Hom}_R(M,N)$. The nth cohomology group of this complex is denoted $\mathrm{Ext}^n_R(M,N)$.

5.5. THEOREM. *$\mathrm{Ext}^n_R(-,-)$ does not depend on the choice of the projective resolution. It is a contravariant functor in the first argument and a covariant functor in the second. Moreover, $\mathrm{Ext}^0_R = \mathrm{Hom}_R$.*

The last sentence has already been done and the second one is trivial. We defer the proof of the first to the next chapter, 3.6.5.

5.6. Injectives and injective resolutions. A left R-module is said to be **injective** if it is projective in the dual of the module category. Thus Q is injective if and only if for whenever $M' \rightarrowtail M$ is monomorphism of left R-modules, the induced $\mathrm{Hom}_R(M,Q) \longrightarrow \mathrm{Hom}_R(M',Q)$ is surjective. It follows that the functor $\mathrm{Hom}_R(-,Q)$ is an exact functor and thus $\mathrm{Ext}^n_R(M,Q) = 0$ for all $n > 0$.

The proof of the following theorem will be given after the discussion of tensor products, see 5.18. It can be proved without tensors, but the proofs are more complicated.

5.7. THEOREM. *Every left R-module can be embedded into an injective left R-module.*

By an **injective resolution** of the module M we mean an exact cochain complex

$$Q_0 \xrightarrow{\delta} Q_1 \xrightarrow{\delta} Q_2 \xrightarrow{\delta} \cdots$$

for which the kernel of $\delta\colon Q_0 \longrightarrow Q_1$ is M. The preceding theorem, together with the dual of Theorem 5.3 allows us to see that,

5.8. COROLLARY. *Every module has an injective resolution.* □

5.9. THEOREM. *Suppose*

$$Q_0 \xrightarrow{\delta} Q_1 \xrightarrow{\delta} Q_2 \xrightarrow{\delta} \cdots$$

is an injective resolution of the left R-module M. Then for any left R-module N, the cohomology of the cochain complex

$$0 \longrightarrow \mathrm{Hom}(N,Q_0) \longrightarrow \mathrm{Hom}(N,Q_1) \longrightarrow \mathrm{Hom}(N,Q_2) \longrightarrow \cdots$$

is $\mathrm{Ext}^\bullet_R(N,M)$.

The proof will be carried out in the next chapter, 3.6.5.

5.10. Bilinear maps and tensor products. Let R be a ring, M a left R-module and N a right R-module. For any abelian group A, an R-bilinear map $N \times M \longrightarrow A$ is a function (*not* a homomorphism) $f: N \times M \longrightarrow A$ that satisfies, for all $n, n_1, n_2 \in N$, $m, m_1, m_2 \in M$ and $r \in R$,

Bilin–1. $f(n_1 + n_2, m) = f(n_1, m) + f(n_2, m)$;
Bilin–2. $f(n, m_1 + m_2) = f(n, m_1) + f(n, m_2)$;
Bilin–3. $f(nr, m) = f(n, rm)$.

5.11. THEOREM. *Let R be a ring, M a left R-module and N a right R-module. Then there is an abelian group $N \otimes_R M$ and a bilinear map $f: N \times M \longrightarrow N \otimes_R M$ such that for any abelian group A and bilinear map $g: N \times M \longrightarrow A$ there is a unique group homomorphism $h: N \otimes_R M \longrightarrow A$ such that $h \circ f = g$.*

PROOF. Let F be the free abelian group generated by the underlying set of $N \times M$. Let E be the subgroup generated by all elements of F of the form

1. $(n_1 + n_2, m) - (n_1, m) - (n_2, m)$;
2. $(n, m_1 + m_2) - (n, m_1) - (n, m_2)$;
3. $(nr, m) - (n, rm)$.

for all $n, n_1, n_2 \in N$, $m, m_1, m_2 \in M$ and $r \in R$. Denote by $N \otimes_R M$ the quotient F/E and $n \otimes m$ the coset containing (n, m). Define $f: N \times M \longrightarrow N \otimes_R M$ by $f(m, n) = m \otimes n$. It is a triviality to see that f is a bilinear map. A bilinear map $g: N \times M \longrightarrow A$ extends, since F is freely generated by $N \times M$, to a unique homomorphism $F \longrightarrow A$. Bilinearity obviously implies that the extension vanishes on E and hence induces a unique $h: F/E \longrightarrow A$ with $h(n \otimes m) = f(n, m)$, as required. □

We call $N \otimes_R M$ the **tensor product** of N with M over R.

If N is a right R-module and A an abelian group, the group $\mathrm{Hom}(N, A)$ of additive homomorphisms $N \longrightarrow A$ has the structure of a left R-module via the formula $(rf)n = f(nr)$. Basically, it is the contravariance of the Hom functor that turns the right module structure into a left module structure. Similarly, when M is a left R-module, $\mathrm{Hom}(M, A)$ becomes a right R-module by $(fr)m = f(rm)$. Then we have the following result. Note that $\mathrm{Hom}_{R^{\mathrm{op}}}(-, -)$ denotes the *right* R homomorphisms between two right modules.

5.12. PROPOSITION. *For any right R-module N, left R-module M and abelian group A, there are natural isomorphisms*

$$\mathrm{Hom}(N \otimes_R M, A) \cong \mathrm{Hom}_R(M, \mathrm{Hom}(N, A)) \cong \mathrm{Hom}_{R^{\mathrm{op}}}(N, \mathrm{Hom}(M, A))$$

PROOF. We define $\phi: \mathrm{Hom}(N \otimes_R M, A) \longrightarrow \mathrm{Hom}_R(M, \mathrm{Hom}(N, A))$ by $(\phi f)(m)(n) = f(n \otimes m)$. It is evident how Bilin–1 and 2 make ϕf additive in N and, for each $n \in N$, make $(\phi f)n$ additive in M. The only thing missing is showing that ϕf is left R-linear. We have

$$((\phi f)(rm))n = f(n, rm) = f(nr, m) = ((\phi f)m)(nr) = (r(\phi f)m)n$$

so that $(\phi f)(rm) = r((\phi f)m)$, as required. In the other direction, we will define $\psi: \mathrm{Hom}_R(M, \mathrm{Hom}(N, A)) \longrightarrow \mathrm{Hom}(N \otimes M, A)$. Given an R-linear $g: M \longrightarrow \mathrm{Hom}(N, A)$, let $(\psi g)(n \otimes m) = (gm)n$. Bilin–1 and 2 follow from the fact that g is additive in M and, for each $m \in M$, gm is additive in N. As for Bilin–3, we have that

$$(\psi g)(n \otimes rm) = (g(rm))n = (r(gm))n = (gm)(nr) = (\psi g)(nr \otimes m)$$

It is evident that ϕ and ψ are inverse to each other, so the first isomorphism follows. The second isomorphism is similar. □

5.13. COROLLARY. *For a fixed right R-module N, the functor $N \otimes_R -$ preserves colimits. In particular, it is right exact.*

PROOF. The functor $N \otimes -: R\text{-}\mathsf{Mod} \longrightarrow \mathsf{Ab}$ is left adjoint to $\mathrm{Hom}(N, -)$. □

5.14. Existence of injectives. We begin the proof of the existence of an injective container of each module with the case of abelian groups. An abelian group A is said to be **divisible** if for all $a \in A$ and $n \in \mathbf{N}$ there is an $a' \in A$ with $na' = a$.

5.15. PROPOSITION. *An abelian group is injective as a $\mathbf{Z}$-module if and only if it is divisible.*

PROOF. The easy way is gotten by looking at the injective homomorphism of multiplication by n. Assuming that Q is an injective $\mathbf{Z}$-module, the induced $\mathrm{Hom}(\mathbf{Z}, Q) \longrightarrow \mathrm{Hom}(\mathbf{Z}, Q)$ is just multiplication by n, while $\mathrm{Hom}(\mathbf{Z}, Q) \cong Q$ and so multiplication by n is a surjective endomorphism of Q and so Q is divisible.

For the converse, suppose that Q is divisible. Suppose that A_0 is a subgroup of A. We will show that any homomorphism $f_0: A_0 \longrightarrow Q$ has an extension to a homomorphism $f: A \longrightarrow Q$. Since any monomorphism is, up to isomorphism, the inclusion of a subgroup, the conclusion will follow. Consider the poset of pairs (A_i, f_i) where A_i is a subgroup of A that contains A_0 and $f_i: A_i \longrightarrow Q$ is a homomorphism that extends f_0. This poset is closed under increasing sup and so there is a maximal element. Suppose (A_1, f_1) is a maximal element. If $A_1 \neq A$, suppose $a \in A - A_1$. If $na \notin A_1$ for all $n > 0$, we can extend f_1 to the subgroup A' generated by A_1 and a by letting $f'a = 0$. Otherwise, let n be the least positive integer for which $na \in A_1$ and choose $q \in Q$ such that $nq = f_1(na)$ and then we can extend f_1 by letting $f'a = q$. In either case, this contradicts the maximality of A_1, so that we must have $A_1 = A$ and then f_1 is the required extension. □

5.16. PROPOSITION. *Every abelian group can be embedded into a divisible abelian group.*

PROOF. Let A be an abelian group and let $A \cong F/K$ where F is free. Then $\mathcal{Q} \otimes F$ is divisible and K is isomorphic to a subgroup of it via the composite $K \longrightarrow F \longrightarrow \mathcal{Q} \otimes F$ and then A is isomorphic to a subgroup of $(\mathcal{Q} \otimes F)/K$. Since a quotient of a divisible abelian group is easily seen to be divisible, we conclude that $(\mathcal{Q} \otimes F)/K$ is divisible. □

A right R-module P is *flat* if $- \otimes_R P$ is exact, for which it is sufficient that it preserve monics. There is, of course, a similar definition for left modules.

5.17. PROPOSITION. *Suppose that P is a flat right R-module and Q is an injective abelian group. Then $\mathrm{Hom}(P, Q)$ with the induced left R-module structure is an injective left R-module.*

PROOF. Suppose $f: M' \longrightarrow M$ is a monomorphism of left R-modules. Since P is flat, the induced $f \otimes P: M' \otimes_R P \longrightarrow M' \otimes_R P$ is still monic. The diagram

$$\begin{array}{ccc} \mathrm{Hom}_R(M, \mathrm{Hom}(P,Q)) & \xrightarrow{\cong} & \mathrm{Hom}(M \otimes_R P, Q) \\ \Big\downarrow{\scriptstyle \mathrm{Hom}(f, \mathrm{Hom}(P,Q))} & & \Big\downarrow{\scriptstyle \mathrm{Hom}(f \otimes P, Q)} \\ \mathrm{Hom}_R(M', \mathrm{Hom}(P,Q)) & \xrightarrow[\cong]{} & \mathrm{Hom}(M' \otimes_R P, Q) \end{array}$$

commutes and the right hand arrow is surjective since Q is injective and hence so is the left hand arrow. □

5.18. COROLLARY. *Every left R-module can be embedded into an injective.*

PROOF. If M is an R-module, then treat is as an abelian group. Embed it into an injective abelian group Q. Then

$$R \cong \mathrm{Hom}_R(R, M) \subseteq \mathrm{Hom}(R, M) \subseteq \mathrm{Hom}(R, Q)$$

embeds M into the injective module $\mathrm{Hom}(R, Q)$. □

5.19. Tor. We begin the discussion of Tor by showing that projective modules are flat.

5.20. PROPOSITION. *Projective modules are flat.*

PROOF. We begin with the obvious fact that $\mathrm{Hom}_R(R, M) \cong M$ for any left R-module M, since a homomorphism is determined uniquely by its value at 1. For any abelian group A,

$$\mathrm{Hom}_R(R, \mathrm{Hom}(N, A)) \cong \mathrm{Hom}(N \otimes_R R, A)$$

and the Yoneda Lemma (see 1.6.1), we conclude that $- \otimes_R R$ is equivalent to the identity functor, which certainly preserves monics. The ring R is the free module on one generator. The free module F generated by the set X is the direct sum of X many copies of R. Since direct sum is a colimit and tensor commutes with colimit, it follows that $N \otimes_R F$ is the direct sum of X many copies of N. Since in module categories a direct sum of an arbitrary set of monics is monic (this is not true, in general, even in abelian categories), it follows that free modules are flat. Finally, let P be projective. Then there is a free module F and maps $P \xrightarrow{f} F \xrightarrow{g} P$ with $g \circ f = \mathrm{id}$. Then for $m: N' \rightarrowtail N$, we have

$$\begin{array}{ccccc} N' \otimes P & \overset{N' \otimes f}{\rightarrowtail} & N' \otimes F & \xrightarrow{N' \otimes g} & N' \otimes P \\ \Big\downarrow{\scriptstyle m \otimes P} & & \Big\downarrow{\scriptstyle m \otimes F} & & \Big\downarrow{\scriptstyle m \otimes P} \\ N \otimes P & \xrightarrow[N \otimes f]{} & N \otimes F & \xrightarrow[N \otimes g]{} & N \otimes P \end{array}$$

The arrow $N' \otimes f$ is monic, being split by $N' \otimes g$ and $m \otimes F$ is monic because F is flat. Hence the composite $N \otimes f \circ m \otimes P$ is monic and therefore the first factor $m \otimes P$ is monic and so P is flat. □

Suppose M is an R-module. By a **flat resolution** of M is meant a complex $P_\bullet = \{(P_n, d) \mid n \geq 0\}$ consisting of flat R-modules, together with an arrow $P_0 \longrightarrow M$ such that the augmented complex $P_\bullet \longrightarrow M \longrightarrow 0$ is exact. Since every module has a projective resolution and projectives are flat, every module has a flat resolution.

Now suppose that $F_\bullet \longrightarrow M$ is a flat resolution of M. Then for a right R-module N, we can form the chain complex $N \otimes_R F_\bullet$. Using the right exactness of tensor, it is easy to show that the zeroth homology of this complex is just $N \otimes_R M$. We denote by $\mathrm{Tor}_n^R(N, M)$ the nth homology group of this complex. It is sometimes called the nth **torsion product** of N with M over R.

5.21. THEOREM. $\mathrm{Tor}_n^R(-,-)$ *does not depend on the choice of the flat resolution. It is a covariant functor in each argument Moreover,* $\mathrm{Tor}_0^R(N, M) = N \otimes_R M$.

The proof will be given in the next chapter, 3.6.5

5.22. Exercises.

1. Show that if

$$\begin{array}{ccccccccc} 0 & \longrightarrow & A' & \longrightarrow & A & \longrightarrow & A'' & \longrightarrow & 0 \\ & & \downarrow{\scriptstyle f'} & & \downarrow{\scriptstyle f} & & \downarrow{\scriptstyle f''} & & \\ 0 & \longrightarrow & B' & \longrightarrow & B & \longrightarrow & B'' & \longrightarrow & 0 \end{array}$$

is a commutative diagram with exact rows in a module category, then f' is monic, respectively epic, if and only if the induced map $A \longrightarrow A'' \times_{B''} B$ is monic, resp. epic.

2. Suppose that $0 \longrightarrow A' \longrightarrow A \longrightarrow A'' \longrightarrow 0$ is an exact sequence of modules. Show that there is an exact sequence of projective resolutions $0 \longrightarrow P'_\bullet \longrightarrow P_\bullet \longrightarrow P''_\bullet \longrightarrow 0$ of A', A, and A'', respectively so that

$$\begin{array}{ccccccccc} 0 & \longrightarrow & P'_\bullet & \longrightarrow & P_\bullet & \longrightarrow & P''_\bullet & \longrightarrow & 0 \\ & & \downarrow & & \downarrow & & \downarrow & & \\ 0 & \longrightarrow & A' & \longrightarrow & A & \longrightarrow & A'' & \longrightarrow & 0 \end{array}$$

commutes with exact rows. (Hint: The kernel of a surjection between two projectives is projective.)

3. Suppose that

$$\begin{array}{ccccccccc} 0 & \longrightarrow & A' & \longrightarrow & A & \longrightarrow & A'' & \longrightarrow & 0 \\ & & \downarrow & & \downarrow & & \downarrow & & \\ 0 & \longrightarrow & B' & \longrightarrow & B & \longrightarrow & B'' & \longrightarrow & 0 \end{array}$$

is commutative with exact rows Show that there are exact sequences of projective resolutions

$$0 \longrightarrow P'_\bullet \longrightarrow P_\bullet \longrightarrow P''_\bullet \longrightarrow 0$$

of A', A, and A'', respectively and

$$0 \longrightarrow Q'_\bullet \longrightarrow Q_\bullet \longrightarrow Q''_\bullet \longrightarrow 0$$

of B', B, and B'', respectively, together with arrows $P'_\bullet \longrightarrow Q'_\bullet$, $P_\bullet \longrightarrow Q_\bullet$, and $P''_\bullet \longrightarrow Q''_\bullet$ such that

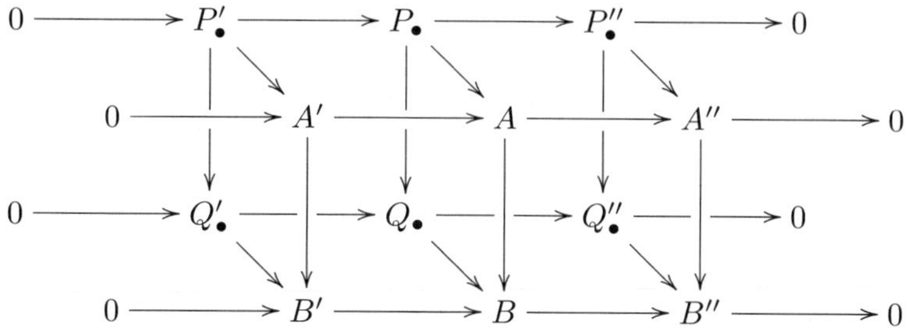

commutes with exact rows.

4. Show that any quotient of an injective abelian group is injective. Use this to show that any subgroup of a projective abelian group is projective. (Neither of these facts is true for more modules in general. Also, I should point out that every projective abelian group is actually free, again a special property of **Z**-modules.)

5. Show that if $0 \longrightarrow F' \longrightarrow F \longrightarrow F'' \longrightarrow 0$ is exact and both F and F'' are flat, then so is F'.

6. The **Z** construction

Reading of this section can be deferred till needed for Chapter 8. We put it here because it is just a construction in elementary theory of additive categories.

This construction gives the free pre-additive category associated to any category. It is not really necessary, but it really simplifies life in a few crucial places and is not at all difficult. Recall from 1.1 that a category is pre-additive if the hom sets are abelian groups in such a way that composition is distributive on the left and right.

Given a category $\mathscr{C}$, we denote by $\mathbf{Z}\mathscr{C}$ the category with the same objects as $\mathscr{C}$ and whose hom sets are the free abelian group generated by those of $\mathscr{C}$. Composition is uniquely determined by the distributive law. This means that for any finite set of arrows $\{f_i: A \longrightarrow B \mid i = 1, \ldots n\}$, any finite set of arrows $\{g_j: B \longrightarrow C \mid j = 1, \ldots, m\}$, and any finite sets of integers $\{r_i \mid i = 1, \ldots n\}$ and $\{s_j \mid j = 1, \ldots, m\}$, we have

$$\left(\sum_{j=1}^{m} s_j g_j\right) \circ \left(\sum_{i=1}^{n} r_i f_i\right) = \sum_{i,j=1,1}^{n,m} r_i s_j (g_j \circ f_i)$$

That's all there is to it. The identities are obvious and so is associativity, so that we have a category. For any (pre-)additive category $\mathscr{A}$, any functor $T: \mathscr{C} \longrightarrow \mathscr{A}$, has a unique extension to an additive functor we will still call $T: \mathbf{Z}\mathscr{C} \longrightarrow \mathscr{A}$.

6.1. Exercise.

1. Show that for any category $\mathscr{C}$, the category $\mathbf{Z}\mathscr{C}$ is preadditive.

CHAPTER 3

Chain Complexes and Simplicial Objects

In this chapter, we will develop properties of the category of differential objects and of the category of chain complexes over an abelian category $\mathscr{A}$. Since the opposite of an abelian category is also abelian, this also includes the theory of cochain complexes. For the most part, the theory is the same in the graded and ungraded case and the latter is easier to discuss. When there is a difference, we will make it clear. We let $\mathscr{C}$ denote either the category of differential objects of $\mathscr{A}$ or of chain complexes.

1. Mapping cones

1.1. Suspension. The **suspension** is one construction in which the grading matters. We take the ungraded case first. For a differential object (A, d), the suspension is simply $S(A, d) = (A, -d)$. For a chain complex $(A_\bullet, d)$ the suspension is $S(A_\bullet, d) = (A_{\bullet-1}, -d)$, meaning that an element that has degree $n-1$ in A has degree n in SA.

It is clear that in the ungraded case, $H(A) = H(SA)$. This is also true in the graded case, with a shift in dimension, so that $H_{n-1}(A) = H_n(SA)$.

Another construction that is fundamental to the theory is that of the **mapping cone** of a morphism. Suppose that $f\colon K \longrightarrow L$ is a map in $\mathscr{C}$. We define a chain complex $C = C_f$ by letting $C = L \oplus SK$ with boundary operator given by the matrix $\begin{pmatrix} d & f \\ 0 & -d \end{pmatrix}$.

1.2. PROPOSITION. *For any $f\colon K \longrightarrow L$ of $\mathscr{C}$, the mapping cone C_f is an object of $\mathscr{C}$; moreover there is an exact sequence*

$$0 \longrightarrow L \longrightarrow C_f \longrightarrow SK \longrightarrow 0$$

PROOF. The requisite commutation of the boundary operators is easy. Matrix multiplication shows that

$$\begin{pmatrix} d & f \\ 0 & -d \end{pmatrix}\begin{pmatrix} d & f \\ 0 & -d \end{pmatrix} = \begin{pmatrix} 0 & 0 \\ 0 & 0 \end{pmatrix}.$$ □

The homology exact triangle of this sequence is

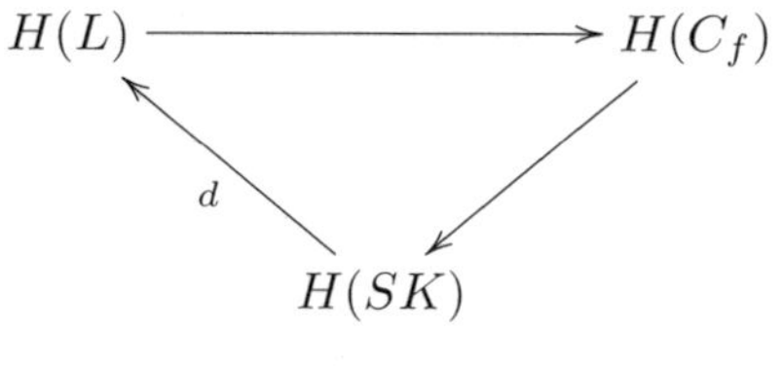

Taking account that $H(SK) = H(K)$, this triangle is

$$\begin{array}{ccc} H(L) & \longrightarrow & H(C_f) \\ & {}_{d}\nwarrow \quad \swarrow & \\ & H(K) & \end{array}$$

1.3. PROPOSITION. *The connecting homomorphism* $d: H(K) \longrightarrow H(L)$ *is just* $H(f)$.

PROOF. We do this in the case that $\mathscr{A}$ is a category of modules over some ring, using 2.3.3 to infer it for an arbitrary abelian category. The recipe for d in the proof of 2.4.7 is to represent an element of $H(K)$ by a cycle $k \in K$. Choose a preimage in C_f, which we can clearly choose as $\binom{k}{0}$. Apply the boundary to give $\binom{0}{fk}$ and choose a cycle in L mapping to that, for which choice fk clearly suffices. $\square$

Of course, in the case of a chain complex, the homology triangle unwinds to a long exact sequence

$$\cdots \longrightarrow H_n(L) \longrightarrow H_n(C_f) \longrightarrow H_{n-1}(K) \xrightarrow{H_{n-1}(f)} H_{n-1}(L) \longrightarrow H_{n-1}(C_f) \longrightarrow H_{n-2}(K) \longrightarrow \cdots$$

We will show later that when f is monic (in each degree), then C_f has homology isomorphic to that of L/K. Similarly, when f is epic, then the kernel of f has homology isomorphic to that of SC_f.

Let $U: \mathscr{C} \longrightarrow \mathsf{Gr}(\mathscr{A})$ denote the functor that forgets the boundary operator. The sequence $0 \longrightarrow L \longrightarrow C_f \longrightarrow SK \longrightarrow 0$ is said to be U-split which means that $0 \longrightarrow UL \longrightarrow UC_f \longrightarrow USK \longrightarrow 0$ is split exact. This property turns out to characterize mapping cone sequences.

1.4. PROPOSITION. *A U-split exact sequence*

$$0 \longrightarrow L \longrightarrow C \longrightarrow K \longrightarrow 0$$

is isomorphic to the mapping cone sequence of a unique map $S^{-1}K \longrightarrow L$.

PROOF. Since the sequence is split, we can suppose that as graded objects, $C = L \oplus K$ and that in degree n, the sequence is

$$0 \longrightarrow L_n \xrightarrow{\binom{1}{0}} L_n \oplus K_n \xrightarrow{(0\ 1)} K_n \longrightarrow 0$$

Now let d^K and d^L denote the differentials in K and L, respectively and suppose, using the decomposition of $C = L \oplus K$, that the differential on C has matrix

$\begin{pmatrix} d^{11} & d^{12} \\ d^{21} & d^{22} \end{pmatrix}$. The commutativity of the diagram

$$\begin{array}{ccccccccc}
0 & \longrightarrow & L & \xrightarrow{\binom{1}{0}} & L \oplus K & \xrightarrow{(0\ 1)} & K & \longrightarrow & 0 \\
 & & \downarrow{\scriptstyle d^L} & & \downarrow{\scriptstyle \begin{pmatrix} d^{11} & d^{12} \\ d^{21} & d^{22} \end{pmatrix}} & & \downarrow{\scriptstyle d^K} & & \\
0 & \longrightarrow & L & \xrightarrow[\binom{1}{0}]{} & L \oplus K & \xrightarrow[(0\ 1)]{} & K & \longrightarrow & 0
\end{array}$$

gives the equations $d^{11} = d^L$, $d^{22} = d^K$ and $d^{21} = 0$. If we write $f = d^{12}$, then the fact that

$$\begin{pmatrix} d^L & f \\ 0 & d^K \end{pmatrix} \begin{pmatrix} d^L & f \\ 0 & d^K \end{pmatrix} = 0$$

gives the equation $f \circ d^K + d^L \circ f = 0$ or equivalently, $f \circ (-d^K) = d^L \circ f$, which implies that $f\colon (S^{-1}K) = K \longrightarrow L$ is an arrow of chain complexes $f\colon \mathscr{S}^{-1}K \longrightarrow L$. □

1.5. Corollary. *Suppose $f\colon K \longrightarrow L$ is a morphism of differential objects and C_f its mapping cone. Then for any object Z of $\mathscr{A}$, the differential abelian group* $\operatorname{Hom}(Z, C_f)$ *is the mapping cone of* $\operatorname{Hom}(Z, f)\colon \operatorname{Hom}(Z, K) \longrightarrow \operatorname{Hom}(Z, L)$ *and* $\operatorname{Hom}(C_f, Z)$ *is the mapping cone of* $\operatorname{Hom}(f, Z)\colon \operatorname{Hom}(L, Z) \longrightarrow \operatorname{Hom}(K, Z)$.

Proof. If $0 \longrightarrow UL \longrightarrow UC_f \longrightarrow USK \longrightarrow 0$ is split as a sequence in $\mathscr{A}$, so is

$$0 \longrightarrow \operatorname{Hom}(Z, UL) \longrightarrow \operatorname{Hom}(Z, UC_f) \longrightarrow \operatorname{Hom}(Z, USK) \longrightarrow 0$$

which is equivalent to

$$0 \longrightarrow U\operatorname{Hom}(Z, L) \longrightarrow U\operatorname{Hom}(Z, C_f) \longrightarrow U\operatorname{Hom}(Z, SK) \longrightarrow 0$$

where we use U to denote the functor that forgets the differential for chain complexes in $\mathscr{A}$ and in Ab. Also, $\operatorname{Hom}(Z, SK)$ has the negative of the boundary of $\operatorname{Hom}(Z, K)$ and in the graded case, has the shift in degrees and is therefore the suspension of $\operatorname{Hom}(Z, K)$. From the proposition, we see therefore that $\operatorname{Hom}(Z, C_f)$ is the mapping cone of $\operatorname{Hom}(Z, f)$. □

1.6. Exercise.

1. Suppose that

$$\cdots \longrightarrow A_n \longrightarrow A_{n-1} \longrightarrow \cdots \longrightarrow A_1 \longrightarrow A_0 \longrightarrow 0$$

is a chain complex and B is an object considered as a chain complex whose only nonzero term is in degree 0. Show that if $f_\bullet\colon A_\bullet \longrightarrow B$ is a map of chain complexes, then the mapping cone is the suspension of

$$\cdots \longrightarrow A_n \longrightarrow A_{n-1} \longrightarrow \cdots \longrightarrow A_1 \longrightarrow A_0 \xrightarrow{f} B \longrightarrow 0$$

2. Show that if

$$\begin{array}{ccccccccc} 0 & \longrightarrow & A' & \longrightarrow & A & \longrightarrow & A'' & \longrightarrow & 0 \\ & & \Big\downarrow{\scriptstyle f'} & & \Big\downarrow{\scriptstyle f} & & \Big\downarrow{\scriptstyle f''} & & \\ 0 & \longrightarrow & B' & \longrightarrow & B & \longrightarrow & B'' & \longrightarrow & 0 \end{array}$$

is a commutative diagram of differential objects with exact rows and if any two of f', f, f'' are homology isomorphisms, then so is the third.

2. Contractible complexes

DEFINITION. We say that a differential object C is **contractible** if there is an arrow $s: C \longrightarrow C$ with the property that $d \circ s + s \circ d = 1$. A contractible differential object is also acyclic since $dc = 0$ implies $c = d(s(c)) + s(d(c)) = d(s(c))$ so that every cycle in a contractible differential object is a boundary.

2.1. Cycle operators. Consider now a differentiable abelian group (A, d) with $Z = \ker d$. Write $|A|$ and $|Z|$ for the underlying sets of A and Z, respectively. A necessary and sufficient condition that A be acyclic is that there be a function, not necessarily a group homomorphism, $z: |Z| \longrightarrow |A|$ such that $d \circ z$ is the inclusion $|Z| \longrightarrow |A|$. That is, for each cycle, there must be an element of which it is the boundary. Not surprisingly, something special happens when z can be chosen as an additive function. More generally, if C is a differential object in any abelian category, we say that $z: Z(C) \longrightarrow C$ is a **cycle operator** if $d \circ z$ is the inclusion $Z(C) \longrightarrow C$.

One way a cycle operator might arise is if there is a contraction s. For if $s: C \longrightarrow C$ is a contraction and we let $z = s|Z(C)$, then on $Z(C)$, $d \circ z = d \circ s = d \circ s + s \circ d = \mathrm{id}$ since $s \circ d = 0$ on $Z(C)$. Thus the restriction of s to $Z(C)$ is a cycle operator. It turns out that every cycle operator arises in this way.

2.2. PROPOSITION. *Let C be a differential object in any abelian category. Then C is contractible if and only if there is a cycle operator on C.*

PROOF. One direction has just been shown. For the other, suppose $z: Z(C) \longrightarrow C$ is a cycle operator. Since the image of d is included in the domain $Z(C)$ of z, it makes sense to form the composite $z \circ d$ and we have that $d \circ z \circ d = d$. Thus $d \circ (1 - z \circ d) = 0$ so that the image of $1 - z \circ d$ is also included in the domain of z and we can form the composite $s = z \circ (1 - z \circ d)$. Note that we cannot write $s = z - z \circ z \circ d$ since the individual terms on the right are not defined. Then we have that

$$\begin{aligned} s \circ d + d \circ s &= z \circ (1 - z \circ d) \circ d + d \circ z \circ (1 - z \circ d) \\ &= z \circ (d - d \circ d \circ d) + 1 - z \circ d = z \circ d + 1 - z \circ d = 1 \end{aligned}$$ □

In a chain complex, both cycle operators and contractions have degree $+1$ and in a cochain complex they have degree -1. But the same results are true.

2.3. Remark. One important observation is that being contractible is an additively **absolute** property: if C is contractible and F is any additive functor defined on complexes, then $F(C)$ is also contractible, since the condition is defined entirely in terms of addition, composites of arrows, and identity arrows.

Using this, we can show the following.

2.4. THEOREM. *Suppose C is a differential object in the abelian category $\mathscr{A}$. Then the following are equivalent*:

1. *C is contractible*;
2. *for each object Z of $\mathscr{A}$, the differential abelian group* $\operatorname{Hom}(Z, C)$ *is contractible*;
3. *for each object Z of $\mathscr{A}$, the differential abelian group* $\operatorname{Hom}(Z, C)$ *is acyclic*;
4. *for each object Z of $\mathscr{A}$, the differential abelian group* $\operatorname{Hom}(C, Z)$ *is contractible*;
5. *for each object Z of $\mathscr{A}$, the differential abelian group* $\operatorname{Hom}(C, Z)$ *is acyclic.*

PROOF. We will begin by proving the equivalence of the first three. If C is contractible, then there is an arrow $s: C \longrightarrow C$ such that $s \circ d + d \circ s = 1$ in which case $\operatorname{Hom}(Z, s)$ is a contracting homotopy for $\operatorname{Hom}(Z, C)$. Contractible differentiable groups are acyclic so the second condition implies the third. Assuming the third, let $Z = Z(C)$, the kernel of $d: C \longrightarrow C$, the object of cycles, and let $i: Z \longrightarrow C$ be the inclusion map. Since $d \circ i = 0$, i is a cycle in the differential group $\operatorname{Hom}(Z, C)$. Since that differential group is exact, i is also a boundary, so that there is an element $z \in \operatorname{Hom}(Z, C)$ such that $d \circ z = i$. But $z: Z \longrightarrow C$ is just a cycle operator and its existence implies that there is a contracting homotopy. This proves the equivalence of the first three parts.

As for the last two, the same argument in the dual category shows that 1, 4, and 5 are equivalent. □

2.5. Homotopies and homotopy equivalences. Suppose $f, g: C' \longrightarrow C$ are morphisms of differential objects. A **homotopy** $h: f \longrightarrow g$ is a map $h: C' \longrightarrow C$ such that $f - g = d \circ h + h \circ d$. We will say that f is homotopic to g and write $f \sim g$ if there is a homotopy $h: f \longrightarrow g$. It is easily shown that $\sim$ is an equivalence relation on morphisms. A morphism $f: C' \longrightarrow C$ is called a **homotopy equivalence** if there is a morphism $g: C \longrightarrow C'$ such that both $g \circ f \sim \mathrm{id}_{C'}$ and $f \circ g \sim \mathrm{id}_C$. It is an easy exercise to show that C is contractible if and only if either of the arrows $0 \longrightarrow C$ or $C \longrightarrow 0$ is a homotopy equivalence.

2.6. PROPOSITION. *Let $f, g: C' \longrightarrow C$ be homotopic morphisms. Then $H(f) = H(g)$.*

PROOF. Assume that $h: f \sim g$. Then restricted to $Z(C')$, $Z(f) - Z(g) = d \circ h$, which means that modulo $\operatorname{im} d$, $H(f) = H(g)$. □

A morphism of differential objects that induces an isomorphism is called a **homology equivalence**. As an immediate corollary of the preceding, we have:

2.7. COROLLARY. *A homotopy equivalence is a homology equivalence.* □

2.8. PROPOSITION. *Suppose that $f: K \longrightarrow L$ is a morphism of differential objects in the abelian category $\mathscr{A}$. Then the following are equivalent*:

1. *f is a homotopy equivalence*;
2. *for any object Z of $\mathscr{A}$, the induced* $\operatorname{Hom}(Z, f)$ *is a homotopy equivalence in the category of abelian groups*;
3. *for any object Z of $\mathscr{A}$, the induced* $\operatorname{Hom}(Z, f)$ *is a homology equivalence in the category of abelian groups*;
4. *the mapping cone of f is contractible*;

5. *for any object* Z *of* $\mathscr{A}$, *the induced* $\operatorname{Hom}(f, Z)$ *is a homotopy equivalence in the category of abelian groups*;
6. *for any object* Z *of* $\mathscr{A}$, *the induced* $\operatorname{Hom}(f, Z)$ *is a homology equivalence in the category of abelian groups.*

PROOF. If $g: L \longrightarrow K$ is a morphism of differential objects, $s: K \longrightarrow K$ is a homotopy $g \circ f \longrightarrow \mathrm{id}_K$, and $t: L \longrightarrow L$ is a homotopy $t: f \circ g \longrightarrow \mathrm{id}_L$, then for any object Z, $\operatorname{Hom}(Z, g): \operatorname{Hom}(Z, L) \longrightarrow \operatorname{Hom}(Z, K)$ is a morphism of differential abelian groups, $\operatorname{Hom}(Z, s): \operatorname{Hom}(Z, K) \longrightarrow \operatorname{Hom}(Z, K)$ is a homotopy $\operatorname{Hom}(Z, g) \circ \operatorname{Hom}(Z, f) \longrightarrow \mathrm{id}$, and $\operatorname{Hom}(Z, t)$ is a homotopy $\operatorname{Hom}(Z, f) \circ \operatorname{Hom}(Z, g) \longrightarrow \mathrm{id}$. Thus 1 implies 2. That 2 implies 3 is the previous corollary. To see that 3 implies 4, suppose that for any object Z of $\mathscr{A}$, the induced $\operatorname{Hom}(Z, f)$ is a homology equivalence. It follows from the exactness of the homology triangle and the fact that $\operatorname{Hom}(Z, f)$ is an isomorphism that $\operatorname{Hom}(Z, C_f)$ is acyclic. Then from Corollary 2.4, we see that C_f is contractible. Next we show that 4 implies 1. Let the contracting homotopy u have matrix $\left(\begin{smallmatrix} t & r \\ g & -s \end{smallmatrix}\right)$. Then the matrix of $du + ud$ is calculated to be

$$\begin{pmatrix} dt + fg + td & dr - fs + tf - rd \\ -dg + gd & ds + gf + sd \end{pmatrix}.$$

If we set this equal to the identity, we conclude that $dt + fg + td = 1$, $-dg + gd = 0$ and $ds + gf + sd = 1$ from which we see that g is a chain map and homotopy inverse to f.

A dual argument shows the equivalence of 1, 4, 5, and 6. □

This proof actually gives a method for constructing a contraction in the mapping cone out of a homotopy inverse, but the formula looks complicated and verifying directly that it is a contraction would be rather unpleasant.

2.9. COROLLARY. *Suppose* $f: K \longrightarrow L$ *is a mapping of acyclic (resp. contractible) differential objects. Then the mapping cone of* f *is acyclic (resp. contractible) and* f *is a homology (resp. homotopy) equivalence.*

PROOF. If K and L are both acyclic then the acyclicity of the mapping cone follows from the exactness of the homology triangle. Evidently, the only map between the null homology groups is an isomorphism. If K and L are both contractible, then for any object Z, $\operatorname{Hom}(Z, f): \operatorname{Hom}(Z, K) \longrightarrow \operatorname{Hom}(Z, L)$ is a map between objects with null homology and hence is a homology isomorphism. It follows that the mapping cone of $\operatorname{Hom}(Z, f)$ is contractible and that f is a homotopy equivalence. □

Probably the single most important property of the mapping cone is expressed in the following theorem.

2.10. THEOREM. *A map of differential objects is a homology equivalence if and only if its mapping cone is acyclic and it is a homotopy equivalence if and only if its mapping cone is contractible.*

PROOF. For homology, both directions are immediate consequences of the exactness of the homology triangle and the fact that an object in a homology sequence is 0 if and only if the preceding arrow is epic and the succeeding one is monic. The homotopy was dealt with in the previous theorem. □

2.11. PROPOSITION. *If* $0 \longrightarrow L \xrightarrow{f} C \xrightarrow{g} K \longrightarrow 0$ *is a* U*-split exact sequence of differential objects, then* K *is homotopic to the mapping cone* C_f *and* L *is homotopic to* SC_g.

PROOF. Except for an unavoidable arbitrariness whether to suspend one or desuspend the other term in a mapping cone, the two parts are dual; we need prove only one. Let $u: UC \longrightarrow UL$ and $v: UK \longrightarrow UC$ be such that $uf = 1$, $gv = 1$, $fu + vg = 1$ and $uv = 0$. The last equation actually follows from the first three. I claim that $\left(\begin{smallmatrix} v \\ -udv \end{smallmatrix}\right): K \longrightarrow C_f$ is a chain map. In fact,

$$\begin{aligned}
\begin{pmatrix} d & f \\ 0 & -d \end{pmatrix}\begin{pmatrix} v \\ -udv \end{pmatrix} &= \begin{pmatrix} dv - fudv \\ dudv \end{pmatrix} = \begin{pmatrix} dv - (1-vg)dv \\ ufdudv \end{pmatrix} = \begin{pmatrix} vgdv \\ udfudv \end{pmatrix} \\
&= \begin{pmatrix} vdgv \\ ud(1-vg)dv \end{pmatrix} = \begin{pmatrix} vd \\ -udvgdv \end{pmatrix} = \begin{pmatrix} vd \\ -udvdgv \end{pmatrix} \\
&= \begin{pmatrix} vd \\ -udvd \end{pmatrix} = \begin{pmatrix} v \\ -udv \end{pmatrix} d
\end{aligned}$$

It is clear that $\left(\begin{smallmatrix} g & 0 \end{smallmatrix}\right)\left(\begin{smallmatrix} v \\ -udv \end{smallmatrix}\right) = 1$. The other composite is

$$\begin{pmatrix} v \\ -udv \end{pmatrix}\begin{pmatrix} g & 0 \end{pmatrix} = \begin{pmatrix} vg & 0 \\ -udvg & 0 \end{pmatrix}$$

Thus if we let $G = \left(\begin{smallmatrix} g & 0 \end{smallmatrix}\right)$, $F = \left(\begin{smallmatrix} v \\ -udv \end{smallmatrix}\right)$, $U = \left(\begin{smallmatrix} 0 & 0 \\ u & 0 \end{smallmatrix}\right)$, and $D = \left(\begin{smallmatrix} d & f \\ 0 & -d \end{smallmatrix}\right)$, we have $GF = 1$ and

$$\begin{aligned}
DU + UD &= \begin{pmatrix} d & f \\ 0 & -d \end{pmatrix}\begin{pmatrix} 0 & 0 \\ u & 0 \end{pmatrix} + \begin{pmatrix} 0 & 0 \\ u & 0 \end{pmatrix}\begin{pmatrix} d & f \\ 0 & -d \end{pmatrix} \\
&= \begin{pmatrix} fu & 0 \\ -du & 0 \end{pmatrix} + \begin{pmatrix} 0 & 0 \\ ud & uf \end{pmatrix} = \begin{pmatrix} fu & 0 \\ -du + ud & uf \end{pmatrix} \\
&= \begin{pmatrix} 1 - vg & 0 \\ ud - ufdu & 1 \end{pmatrix} = \begin{pmatrix} 1 - vg & 0 \\ ud - udfu & 1 \end{pmatrix} \\
&= \begin{pmatrix} 1 - vg & 0 \\ udvg & 1 \end{pmatrix} = \begin{pmatrix} 1 & 0 \\ 0 & 1 \end{pmatrix} - \begin{pmatrix} vg & 0 \\ -udvg & 0 \end{pmatrix} \\
&= 1 - FG
\end{aligned}$$

□

2.12. THEOREM. *Suppose* $f: K \longrightarrow L$ *is a morphism of differential objects with mapping cone* $C = C_f$. *If* f *is monic, there is a map* $C \longrightarrow L/K$ *that induces an isomorphism on homology. Dually, if* f *is epic, there is a map* $\ker(f) \longrightarrow C$ *that induces an isomorphism on homology.*

PROOF. By duality, we need prove only one of these, say the first. Let $p: L \longrightarrow L/K$ be the projection. Then $(p \quad 0): L \oplus K \longrightarrow L/K$ is a chain map by a simple computation and thus induces a map $H(C) \longrightarrow H(K/L)$. It would be sufficient to show that we have commutative squares in the homology sequence

$$\begin{array}{ccccccccccc}
\cdots \longrightarrow & H(K) & \longrightarrow & H(L) & \longrightarrow & H(C) & \longrightarrow & H(K) & \longrightarrow & H(L) & \longrightarrow \cdots \\
& \downarrow & & \downarrow & & \downarrow & & \downarrow & & \downarrow & \\
\cdots \longrightarrow & H(K) & \longrightarrow & H(L) & \longrightarrow & H(L/K) & \longrightarrow & H(K) & \longrightarrow & H(L) & \longrightarrow \cdots
\end{array}$$

The first and last squares obviously commute and one easily checks that the second one does since in each direction the homology class of an element of K goes to the class of fk. The third square does not commute, however; it anticommutes instead. This does not really matter since it will commute if we negate the arrow $H(C) \longrightarrow H(K)$ in the upper sequence, which does not affect exactness. There is perhaps some explanation as to why this step is necessary in terms of suspension, but it is not evident. To show it anticommutes, take a cycle $\binom{l}{k} \in C$. To be a cycle means that $dl + fk = 0$ and $dk = 0$. Going clockwise around the square gives us the homology class of k. In going the other way, we first take the class of l mod K and then apply the connecting homomorphism. The recipe for doing this is to apply the boundary to get dl and then choose an element of k mapping to it. We can choose $-k$ since $dl + fk = 0$. Thus we get the homology class of $-k$, which shows that the square anticommutes. The upshot is that the arrows $H(C) \longrightarrow H(L/K)$ are trapped between isomorphisms and must be isomorphisms by the five lemma, see 2.3.11. □

3. Simplicial objects

There are two equivalent definitions of a simplicial object in a category. One, as a functor category, is useful for deriving certain formal properties. The other definition, the one we will use, is much easier for seeing what a simplicial object in a category actually is.

DEFINITION. A **simplicial object** in a category $\mathscr{X}$ is given by a sequence of objects X_0, X_1, ..., X_n, ... together with two doubly indexed family of arrows of $\mathscr{X}$. The first, called the **face operators**, are arrows $d^i_n: X_n \longrightarrow X_{n-1}$, $0 \le i \le n$, $1 \le n < \infty$; the second kind, called **degeneracy operators**, are arrows $s^i_n: X_n \longrightarrow X_{n+1}$, $0 \le i \le n$, $0 \le n < \infty$. These are subject to the following rules.

$$
\begin{aligned}
d^i_n \circ d^j_{n+1} &= d^{j-1}_n \circ d^i_{n+1} \quad \text{if } 0 \le i < j \le n+1\\
s^j_n \circ s^i_{n-1} &= s^i_n \circ s^{j-1}_{n-1} \quad \text{if } 0 \le i < j \le n\\
d^i_{n+1} \circ s^j_n &= \begin{cases} s^{j-1}_{n-1} \circ d^i_n & \text{if } 0 \le i < j \le n\\ 1 & \text{if } 0 \le i = j \le n \text{ or } 0 \le i-1 = j < n\\ s^j_{n-1} \circ d^{i-1}_n & \text{if } 0 < j < i-1 \le n \end{cases}
\end{aligned}
$$

From now on, we will usually omit the lower indices. Thus the first rule above is written $d^i \circ d^j = d^{j-1} \circ d^i$ for all values of the indices $i < j$ that make sense. Incidentally, this rule implies that $d^i \circ d^j = d^j \circ d^{i+1}$ when $i \ge j$.

We often denote by $X_\bullet$ the simplicial object consisting of objects X_n, $n \ge 0$ and the attendant faces and degeneracies. If $X_\bullet$ and $Y_\bullet$ are simplicial objects, a morphism $f_\bullet: X_\bullet \longrightarrow Y_\bullet$ consists of a family $f_n: X_n \longrightarrow Y_n$ for all $n \ge 0$ such that $d^i \circ f_n = f_{n-1} \circ d^i$ and $f_n \circ s^i = s^i \circ f_{n-1}$ whenever the indices make sense.

We will sometimes use the following symbolic notation for a simplicial set, suppressing the degeneracies:

$$
\cdots \; \vdots \; X_n \; \vdots \; X_{n-1} \; \vdots \; \cdots \; \overset{\longrightarrow}{\underset{\longrightarrow}{\longrightarrow}} \; X_1 \; \rightrightarrows \; X_0
$$

DEFINITION. The alternate definition is as follows. Let Δ denote the category whose objects are non-zero ordinals and arrows are the order preserving functions. A simplicial object in $\mathscr{X}$ can be described as a contravariant functor $\Delta^{\text{op}} \longrightarrow \mathscr{X}$.

Here is how to connect the two definitions. We will use the usual implementation of the ordinals as $n = \{0, 1, \ldots, n-1\}$. Let $\partial_i^n : n-1 \longrightarrow n$ be defined for $i \le n$ by

$$\partial_i^n(j) = \begin{cases} j & \text{for } j < i \\ j+1 & \text{for } j \ge i \end{cases}$$

Thus ∂_n^n is the inclusion of $n-1$ into n while ∂_0^n adds 1 to each ordinal. Similarly, for $0 \le i \le n$, we define $\sigma_i^n : n \longrightarrow n-1$ by

$$\sigma_i^n(j) = \begin{cases} j & \text{for } j < i \\ j-1 & \text{for } j \ge i \end{cases}$$

Then the arrows ∂_i^n and σ_i^n generate the category Δ. In fact, arrows in Δ factor as surjections followed by injections and every injection is a composite of ∂'s and every surjection is a composite of σ's. If $X : \Delta^{\mathrm{op}} \longrightarrow \mathscr{X}$ is a functor, the simplicial set that corresponds has $X_n = X(n+1)$, $d_n^i = X(\partial_i^{n+1})$ and $s_n^i = X(\sigma_i^{n+1})$. It is left as an exercise to show that this correspondence determines an equivalence between the category of simplicial objects and that of contravariant functors on Δ^{op}. The latter is often taken as the definition of simplicial object, but it is the former definition that is used in practice.

We denote the category of simplicial objects over $\mathscr{X}$ by $\mathrm{Simp}(\mathscr{X})$.

3.1. Littler fleas. Not only do we have simplicial sets as functors, with natural transformations as morphisms, but we also have arrows, called homotopies, between morphisms. These homotopies are rather complicated and used mostly in just one special case. If $X_\bullet$ and $Y_\bullet$ are simplicial objects and $f_\bullet, g_\bullet : X_\bullet \longrightarrow Y_\bullet$ are simplicial arrows, a homotopy $h_\bullet : f_\bullet \longrightarrow g_\bullet$ is given by families of arrows $h^i = h_n^i : X_n \longrightarrow Y_{n+1}$ for all $n \ge 0$ and all $0 \le i \le n$ that satisfy

$$d^i \circ h^j = \begin{cases} f_n & \text{if } i = j = 0 \\ h^{j-1} \circ d^i & \text{if } i < j \\ d^i \circ h^{j+1} & \text{if } 0 \le i-1 = j < n \\ h^j \circ d^{i-1} & \text{if } o \le j < i-1 \le n \\ g_n & \text{if } i-1 = j = n \end{cases}$$

$$s^j \circ h^i = h^i \circ s^{j-1} \qquad if 0 \le i < j \le n+1$$

We will write $h_\bullet : f_\bullet \xrightarrow{\sim} g_\bullet$ to show that $h_\bullet$ is a homotopy from $f_\bullet$ to $g_\bullet$ and $f_\bullet \xrightarrow{\sim} g_\bullet$ to indicate there is a homotopy. We choose this notation because homotopy is not symmetric. Even this notation is misleading for $\xrightarrow{\sim}$ is not transitive either. It is, however, always reflexive. In fact, if $f_\bullet : X_\bullet \longrightarrow Y_\bullet$ is a simplicial map, it is not hard to show that h given by $h^i = s^i \circ f_n = f_{n+1} \circ s^i : X_n \longrightarrow Y_{n+1}$ defines a homotopy $f_\bullet \xrightarrow{\sim} f_\bullet$.

3.2. Augmented simplicial objects. An **augmented simplicial object** in the category $\mathscr{X}$ is a simplicial object $X_\bullet$ together with an object X_{-1} together with an arrow $d : X_0 \longrightarrow X_{-1}$ such that $d \circ d^0 = d \circ d^1$.

An augmented simplicial object $X_\bullet \longrightarrow X_{-1}$ is said to be **contractible** if for each $n \ge -1$ there is a map $s_n : X_n \longrightarrow X_{n+1}$ such that $d^0 \circ s = 1$ and $d^i \circ s = s \circ d^{i-1}$, $0 < i \le n$ and $s^0 \circ s = s \circ s$ and $s^i \circ s = s \circ s^{i-1}$, $0 < i \le n+1$.

A simplicial object is called **constant** if X_n is the same object for each n and every face and degeneracy map is the identity of that object. There is a constant simplicial object for each object of the category. For an object X of $\mathscr{X}$ we will denote the corresponding simplicial object also by X.

3.3. PROPOSITION. *Suppose $X_\bullet \longrightarrow X_{-1}$ is a contractible augmented simplicial object in $\mathscr{X}$. Then there are simplicial maps $f_\bullet: X_\bullet \longrightarrow X_{-1}$ and $g_\bullet: X_{-1} \longrightarrow X_\bullet$, treating X_{-1} as a constant, such that $f \circ g = 1$ and $1 \xrightarrow{\sim} g \circ f$. Conversely, a pair of arrows involving a constant simplicial object satisfying such a condition corresponds to a contractible augmented simplicial object.*

PROOF. Let us begin with a contractible augmented simplicial object $X_\bullet \longrightarrow X_{-1}$ with contracting homotopy $s_\bullet$. At this point, we require some notational conventions. Upper indices are used on face and degeneracies and this is usually satisfactory because we cannot usually form powers. Here we will be using symbolic powers. They are not really powers, because , for example, $d^0 \circ d^0$ is really $d^0_{n+1} \circ d^0_n$ for some n, but it will be convenient to write it as a power. In order to avoid confusion, we will write it as $(d^0)^2$. In other words a true exponent will always be marked with parentheses. It will be convenient to write $d^0 = d: X_0 \longrightarrow X_{-1}$ and $s^{-1} = s: X_{n-1} \longrightarrow X_n$ for $n \geq 0$. In each case it because the equations satisfied are those appropriate to d^0, respectively s^{-1}. Now let $f_n = (d^0)^{n+1}: X_n \longrightarrow X_{-1}$ and $g_n = (s^{-1})^{n+1}: X_{-1} \longrightarrow X_n$. Before continuing, we need a lemma.

3.4. LEMMA.

$$(d^0)^j \circ d^i = \begin{cases} (d^0)^{j+1} & \text{if } i \leq j \\ d^{i-j} \circ (d^0)^j & \text{if } i < j \end{cases}$$

$$(d^0)^j \circ s^i = \begin{cases} (d^0)^{j-1} & \text{if } i < j \\ s^{i-j} \circ (d^0)^j & \text{if } i \geq j \end{cases}$$

$$d^i \circ (s^{-1})^j = \begin{cases} (s^{-1})^{j-1} & \text{if } i < j \\ (s^{-1})^j \circ d^{i-j} & \text{if } i \geq j \end{cases}$$

$$s^i \circ (s^{-1})^j = \begin{cases} (s^{-1})^{j+1} & \text{if } i < j \\ (s^{-1})^j \circ s^{i-j} & \text{if } i \geq j \end{cases}$$

PROOF. We will prove the first of these. The remaining ones are similar. When $j = 0$, there is nothing to prove. Suppose then it is true for $j - 1$. Now suppose $i \leq j - 1$. Then

$$(d^0)^j \circ d^i = d^0 \circ (d^0)^{j-1} \circ d^i = d^0 \circ (d^0)^j = (d^0)^{j+1}$$

If $i = j$, then

$$(d^0)^j \circ d^j = d^0 \circ (d^0)^{j-1} \circ d^j = d^0 \circ d^1 \circ (d^0)^{j-1} = d^0 \circ d^0 \circ (d^0)^{j-1} = (d^0)^{j+1}$$

Finally, if $i > j$, then

$$\begin{aligned}(d^0)^j \circ d^i &= d^0 \circ (d^0)^{j-1} \circ d^i = d^0 \circ d^{i-j+1} \circ (d^0)^{j-1} \\ &= d^{i-j} \circ d^0 \circ (d^0)^{j-1} = d^{i-j} \circ (d^0)^j\end{aligned}$$

□

In particular, taking $j = n$ in the first one, we see that $(d^0)^n \circ d^i = (d^0)^{n+1}$ for $0 < i < n$ so that f commutes with the face operators. Taking $j = n + 2$ in the second, we have that $(d^0)^{n+2} \circ s^i = (d^0)^{n+1}$ so that f commutes with the degeneracies. Similarly, the third and fourth equations of the lemma have as special case formulas that say that g is a simplicial map.

No we return to the proof of the proposition. We have that

$$f_n \circ g_n = (d^0)^{n+1} \circ (s^{-1})^{n+1} = (d^0)^n \circ (s^{-1})^n = \cdots = 1$$

To see the homotopy in the other direction, we let $h^i: X_n \longrightarrow X_{n+1}$ by the formula $h^i = (s^{-1})^{i+1} \circ (d^0)^i$. We see that $d^0 \circ h^0 = d^0 \circ s^{-1} = 1$ by assumption, while

$$d^{n+1} \circ h^n = d^{n+1} \circ (s^{-1})^{n+1} \circ (d^0)^n = (s^{-1})^{n+1} \circ d^0 \circ (d^0)^n = g_n \circ f_n$$

Thus, if the h^i constitute a homotopy, it is $1 \xrightarrow{\sim} g \circ f$. For $i < j$, we have

$$d^i \circ h^j = d^i \circ (s^{-1})^{j+1} \circ (d^0)^j = (s^{-1})^j \circ (d^0)^j$$

while

$$h^{j-1} \circ d^i = (s^{-1})^j \circ (d^0)^{j-1} \circ d^i = (s^{-1})^j \circ (d^0)^j$$

If $i > j+1$, we have

$$d^i \circ h^j = d^i \circ (s^{-1})^{j+1} \circ (d^0)^j = (s^{-1})^{j+1} \circ d^{i-j-1} \circ (d^0)^j$$

while

$$h^j \circ d^{i-1} = (s^{-1})^{j+1} \circ (d^0)^j \circ d^{i-1} = (s^{-1})^{j+1} \circ d^{i-1-j} \circ (d^0)^j$$

Finally,

$$d^{i+1} \circ h^i = d^{i+1} \circ (s^{-1})^{i+1} \circ (d^0)^i = (s^{-1})^{i+1} \circ d^0 \circ (d^0)^i = (s^{-1})^{i+1} \circ (d^0)^{i+1}$$

while

$$d^{i+1} \circ h^{i+1} = d^{i+1} \circ (s^{-1})^{i+2} \circ (d^0)^{i+1} = (s^{-1})^{i+1} \circ (d^0)^{i+1}$$

which establishes the homotopy. □

There is another possible definition of contractible augmented simplicial object. One can instead suppose the existence of $s_n: X_{n-1} \longrightarrow X_n$ for all $n \geq 0$ that satisfy $d^{n+1} \circ s_n = 1$, $d^i \circ s_n = s_n \circ d^i$ for $i < n$ and $s_n \circ s^i = s^i \circ s_{n-1}$. In fact, instead of looking like degeneracies labeled -1, these look like ones labeled s^{n+1} in degree n. Not surprisingly, this kind of homotopy gives rise to a homotopy $g \circ f \xrightarrow{\sim} 1$. The two are not in general equivalent, an example of the fact that the homotopy relation is not symmetric.

The following theorem gives a useful sufficient condition for a special kind of augmented simplicial set to be contractible. First we need a definition. If $f: X \longrightarrow Y$ is a function, the nth fiber power X_Y^n of f consists of all n-tuples $\langle x_0, \ldots, x_n \rangle$ of elements of X on which f is constant, that is $fx_0 = \cdots = fx_n$. This becomes a simplicial set with X_Y^{n+1} in degree n and the ith face operator $d^i\langle x_0, \ldots, x_n \rangle = \langle x_0, \ldots, x_{i-1}, x_{i+1}, \ldots, x_n \rangle$ for $i = 0, \ldots, n$. The degeneracy operators are given by $s^i\langle x_0, \ldots, x_n \rangle = \langle x_0, \ldots, x_i, x_i, \ldots, x_n \rangle$.

3.5. PROPOSITION. *The above simplicial set, augmented by $f: X \longrightarrow Y$, is contractible if and only if f is surjective.*

PROOF The augmentation term of a contractible simplicial set is always a split surjection, so that is a necessary condition. If f is surjective, then there is a section $s: Y \longrightarrow X$ such that $f \circ s = \text{id}$. Define $s: X_Y^{n+1} \longrightarrow X_Y^{n+2}$ by $s\langle x_0, \ldots, x_n \rangle = \langle x_0, \ldots, x_n, s \circ fx_n \rangle$. Then

$$d^{n+1} \circ s\langle x_0, \ldots, x_n \rangle = d^{n+1}\langle x_0, \ldots, x_n, s \circ fx_n \rangle = \langle x_0, \ldots, x_n \rangle$$

while for $0 \le i \le n$,

$$\begin{aligned} d^i \circ s\langle x_0, \dots, x_n\rangle &= d^i\langle x_0, \dots, x_n, s \circ f x_n\rangle \\ &= \langle x_0, \dots, x_{i-1}, x_{i+1} \dots, x_n, s \circ f x_n\rangle \\ &= s\langle x_0, \dots, x_{i-1}, x_{i+1} \dots, x_n\rangle = s \circ d^i\langle x_0, \dots, x_n\rangle \end{aligned}$$

which shows that this is contraction. □

4. Associated chain complex

Suppose $A_\bullet$ is a simplicial object in an additive category $\mathscr{A}$. Then the **associated chain complex** is the complex

$$\cdots \xrightarrow{d} A_{n+1} \xrightarrow{d} A_n \xrightarrow{d} A_{n-1} \xrightarrow{d} \cdots \xrightarrow{d} A_0 \longrightarrow 0$$

where $d = \sum_{i=0}^{n}(-1)^i d^i \colon A_n \longrightarrow A_{n-1}$. The first thing that has to be verified is that it is a chain complex.

4.1. Proposition. $d \circ d = 0$.

Proof. Starting at A_n, we have

$$\begin{aligned} d \circ d &= \sum_{i=0}^{n-1}(-1)^i d^i \circ \sum_{j=0}^{n}(-1)^j d^j = \sum_{i=0}^{n-1}\sum_{j=0}^{n}(-1)^{i+j} d^i \circ d^j \\ &= \sum_{i=0}^{n-1}\sum_{j=0}^{i}(-1)^{i+j} d^i \circ d^j + \sum_{i=0}^{n-1}\sum_{j=i+1}^{n}(-1)^{i+j} d^i \circ d^j \\ &= \sum_{i=0}^{n-1}\sum_{j=0}^{i}(-1)^{i+j} d^i \circ d^j + \sum_{i=0}^{n-1}\sum_{j=i+1}^{n}(-1)^{i+j} d^{j-1} \circ d^i \\ &= \sum_{i=0}^{n-1}\sum_{j=0}^{i}(-1)^{i+j} d^i \circ d^j + \sum_{j=1}^{n}\sum_{i=0}^{j-1}(-1)^{i+j} d^{j-1} \circ d^i \\ &= \sum_{i=0}^{n-1}\sum_{j=0}^{i}(-1)^{i+j} d^i \circ d^j - \sum_{j=0}^{n-1}\sum_{i=0}^{j}(-1)^{i+j} d^j \circ d^i \\ &= 0 \end{aligned}$$

□

It is evident that a simplicial map induces a chain map on the associated chain complexes. Let us write $C(A_\bullet)$ and $C(f_\bullet)$ for the associated chain complex and chain map.

4.2. Proposition. *Let $\mathscr{A}$ be an additive category and $f_\bullet, g_\bullet \colon A_\bullet \longrightarrow B_\bullet$ be simplicial maps between simplicial objects in $\mathscr{A}$. If $h_\bullet \colon f_\bullet \xrightarrow{\sim} g_\bullet$ is a simplicial homotopy, then $C(h_\bullet)$ defined in degree n as $\sum_{i=1}^{n} h^i$ is a chain homotopy from $C(f_\bullet) \longrightarrow C(g_\bullet)$.*

PROOF. The proof is the computation:

$$\begin{aligned}
d \circ h &= \sum_{i=0}^{n+1}(-1)^i d^i \circ \sum_{j=0}^{n}(-1)^j h^j = \sum_{i=0}^{n+1}\sum_{j=0}^{n}(-1)^{i+j} d^i \circ h^j \\
&= \sum_{i=2}^{n+1}\sum_{j=0}^{i-2}(-1)^{i+j} d^i \circ h^j + \sum_{i=1}^{n+1}(-1)^{2i-1} d^i \circ h^{i-1} + \sum_{i=0}^{n}(-1)^{2i} d^i \circ h^i \\
&\quad + \sum_{i=0}^{n+1}\sum_{j=i+1}^{n}(-1)^{i+j} d^i \circ h^j \\
&= \sum_{i=2}^{n+1}\sum_{j=0}^{i-2}(-1)^{i+j} h^j \circ d^{i-1} + \sum_{i=0}^{n-1}\sum_{j=i+1}^{n}(-1)^{i+j} h^{j-1} \circ d^i \\
&\quad - \sum_{i=1}^{n} d^i \circ h^{i-1} - d^{n+1} \circ h^n + d^0 \circ h^o + \sum_{i=1}^{n}(-1)^{2i} d^i \circ h^i \\
&= f_n - g_n + \sum_{i=1}^{n}\sum_{j=0}^{i-1}(-1)^{i+j+1} h^j \circ d^i + \sum_{i=0}^{n-1}\sum_{j=i}^{n-1}(-1)^{i+j+1} h^j \circ d^i \\
&= f_n - g_n - \sum_{i=0}^{n}\sum_{j=0}^{n-1}(-1)^{i+j} h^j \circ d^i \\
&= f_n - g_n - h \circ d
\end{aligned}$$

□

4.3. COROLLARY. *If $\mathscr{A} = A_\bullet$ is a contractible augmented simplicial object in an abelian category, then $C\mathscr{A}$ is a contractible chain complex.*

PROOF. For it is then homotopic to a constant simplicial set and it is obvious that the chain complex associated to a constant simplicial set is contractible. □

4.4. Exercise.

1. Calculate the associated chain complex of a constant simplicial object in an abelian category.

5. The Dold-Puppe theorem

In [1961], A. Dold and D. Puppe published a theorem that states that if $\mathscr{A}$ is an abelian category, then the category of chain complexes in $\mathscr{A}$ is equivalent to that of simplicial objects of $\mathscr{A}$. In fact, their hypotheses are even too strong. All that is needed is an additive category with split idempotents. Although we will not prove their theorem (it is really not relevant to the subject at hand), the basic construction is interesting and the reader who wants the full proof can refer to the original paper. The functor used is not C, but either of two isomorphic, but quite distinct functors that we will call $\overline{C}$ and $\underline{C}$, either of which we will call the normalized chain complex associated with a simplicial object. If $A_\bullet$ is a simplicial object, define $\underline{C}_n(A_\bullet) = \bigcap_{i=1}^{n} \ker(d_n^i) \subseteq A_n$. This is made into a chain complex using the restriction of d^0. It turns out to be a subcomplex of $C(A_\bullet)$ since the remaining terms in the sum defining the boundary vanish on this subobject. We let $\overline{C}_n(A)$ be the quotient of $A_n / \bigvee_{i=0}^{n-1} \operatorname{im}(s_{n-1}^i)$ (the sum—non-direct—of the images of the degeneracies). The boundary is induced by that of $C(A_\bullet)$. Then the Dold-Puppe theorem asserts that the composite $\underline{C}(A_\bullet) \longrightarrow C(A_\bullet) \longrightarrow \overline{C}(A_\bullet)$ is an isomorphism and the isomorphic functors $\underline{C}$ and $\overline{C}$ induce an equivalence of categories.

The way to see you need only split idempotents is as follows. The kernel of d^i is easily seen to be the kernel of $s^{i-1} \circ d^i$ and that is idempotent. (It is also the kernel of the idempotent $s^i \circ d^i$, but we cannot use that one because a choice was made to use $\bigcap_{i=1}^n \ker(d^i)$ in the definition of $\underline{C}$. We could equally well have used $\bigcap_{i=0}^{n-1} \ker(d^i)$ with boundary $(-1)^n d^n$.) But the kernel of an idempotent e in an additive category is the image of the idempotent $1 - e$ so that in an additive category with split idempotents, idempotents have kernels.

We have to work a bit harder to get the intersection of the kernels, since these idempotents do not commute. The relevant facts are these. Let $e^i = s^{i-1} \circ d^i$. Then e^i commutes with e^j provided $|i-j| \geq 2$. While e^i does not commute with e^{i+1}, they satisfy the identities $e^i \circ e^{i+1} \circ e^i = e^{i+1} \circ e^i \circ e^{i+1} = e^{i+1} \circ e^i$. Let $c^i = 1 - e^i$. Again c^i commutes with c^j when $|i-j| \geq 2$, while $c^i \circ c^{i+1} \circ c^i = c^{i+1} \circ c^i \circ c^{i+1} = c^i \circ c^{i+1}$. Note the left/right reversal here. It then follows that $c^1 \circ c^2 \circ \cdots \circ c^n$ is idempotent and the image of that idempotent is exactly $\underline{C}_n$.

The story of $\overline{C}$ is similar; the image is s^i is the same as the image of the idempotent e^i and the braided commutation identities allow the sum of the images to be realized as the image of a single idempotent.

6. Double complexes

By a **bigraded object** of $\mathscr{A}$, we mean a $\mathbf{Z} \times \mathbf{Z}$ indexed family A_{nm} of objects. This will often be denoted $A_{\bullet\bullet}$. A family of morphisms $f_{nm}: A_{nm} \longrightarrow A_{n+k\,m+l}$ is said to have **bidegree** (k, l). A **differential bigraded object** we mean a bigraded object with two differentials d^{I} and d^{II} of bidegrees $(k, 0)$ and $(0, l)$, respectively, that satisfy, in addition to $d^{\mathrm{I}} \circ d^{\mathrm{I}} = 0$ and $d^{\mathrm{II}} \circ d^{\mathrm{II}} = 0$ the equation $d^{\mathrm{I}} \circ d^{\mathrm{II}} = -d^{\mathrm{II}} \circ d^{\mathrm{I}}$. Although $d = d^{\mathrm{I}} + d^{\mathrm{II}}$ does not preserve the grading, it is trivial to see that $d \circ d = 0$, a fact we will need later.

A **morphism of differential bigraded objects**

$$f: (A'_{\bullet\bullet}, d'^{\mathrm{I}}, d'^{\mathrm{II}}) \longrightarrow (A_{\bullet\bullet}, d^{\mathrm{I}}, d'^{\mathrm{II}})$$

is a morphism of bidegree $(0, 0)$ of the bigraded objects that commutes with both boundary operators.

A sequence of morphisms of (differential) bigraded objects

$$(A'_{\bullet\bullet}, d'^{\mathrm{I}}, d'^{\mathrm{II}}) \xrightarrow{f} (A_{\bullet\bullet}, d^{\mathrm{I}}, d^{\mathrm{II}}) \xrightarrow{g} (A''_{\bullet\bullet}, d''^{\mathrm{I}}, d''^{\mathrm{II}})$$

is exact if for each n, m the sequence $A'_{nm} \xrightarrow{f_{nm}} A_{nm} \xrightarrow{g_{nm}} A''_{nm}$ is exact. We similarly define a short exact sequence of (differential) graded objects.

Now suppose that $\mathscr{A} = (A_{\bullet\bullet}, d^{\mathrm{I}}, d^{\mathrm{II}})$ is a differential bigraded object in which $k = l$. In that case there is associated a differential graded object called the total differential graded object. This object $T(\mathscr{A})$ has in degree n the direct sum $T_n(\mathscr{A}) = \sum_{i+j=n} A_{ij}$. For the differential, let $a = (\ldots, a_{i-1\,j+1}, a_{ij}, a_{i+1\,j-1}, \ldots)$ be an element of $T_n(\mathscr{A})$. Then $da = a^{\mathrm{I}} + a^{\mathrm{II}}$ where

$$a^{\mathrm{I}} = (\ldots, d^{\mathrm{I}} a_{i-1\,j+1}, d^{\mathrm{I}} a_{ij}, d^{\mathrm{I}} a_{i+1\,j-1}, \ldots)$$

and

$$a^{\mathrm{II}} = (\ldots, d^{\mathrm{II}} a_{i-1\,j+1}, d^{\mathrm{II}} a_{ij}, d^{\mathrm{II}} a_{i+1\,j-1}, \ldots)$$

Of course, in the element $d^{\mathrm{I}} a_{ij}$ lives in bidegree $(i + k, j)$, while $d^{\mathrm{II}} a_{ij}$ is in bidegree $(i, j + k)$, but they are both in $T_{n+k}(\mathscr{A})$.

The cases we are interested in will be those for which $k = l = \pm 1$. A **double chain complex** is a differential bigraded object whose differentials have bidegrees

$(-1,0)$ and $(0,-1)$ and, moreover, there are integers n_0 and m_0 for which $A_{nm} = 0$ unless $n \geq n_0$ and $m \geq m_0$. In this case, the total complex will have $T_n(\mathscr{A}) = 0$ for $n < n_0 + m_0$, so it will be a chain complex. Similarly, a **double cochain complex** is a differential bigraded object whose differentials have bidegrees $(+1,0)$ and $(0,+1)$ and, moreover, there are integers n_0 and m_0 for which $A_{nm} = 0$ unless $n < n_0$ and $m \leq m_0$. In this case, the total complex will have $T_n(\mathscr{A}) = 0$ for $n > n_0 + m_0$, so it will be a cochain complex.

In the case that $k = l = -1$, we will usually denote the two boundary operators by d^{I} and d^{II} and if $k = l = 1$, we will usually denote them by δ^{I} and δ^{II}.

The way to picture a double chain complex is to imagine $m_0 = n_0 = 0$ (which is often the case) and the non-zero objects are situated at the lattice points in the first quadrant. The first boundary operator maps A_{nm} to $A_{n-1\,m}$ and so can be thought of as arrows going to the left in each row and the second boundary operator can similarly be thought of as arrows going down. When n or m is non-zero, then the picture is similar, it just does not exactly fit into or fill out the quadrant.

If $\mathscr{A}$ is a double chain complex, we write $H_\bullet(\mathscr{A})$ for the homology of the total complex $T_\bullet(\mathscr{A})$.

6.1. Homology of double complexes. Let us say that a double differential object is an object A with two anticommuting differentials d^{I} and d^{II} so that $D = d^{\mathrm{I}} + d^{\mathrm{II}}$ is a differential. It would not be utterly astonishing (although it would be false, see the examples below) if it turned out that a double differential object was exact if both differentials were. It is rather a surprise, however, that the total complex of a double complex is already exact if just d^{I} (or d^{II}) is.

We begin with,

6.2. PROPOSITION. *Suppose*

$$0 \longrightarrow \mathscr{A}' \xrightarrow{f} \mathscr{A} \xrightarrow{g} \mathscr{A}'' \longrightarrow 0$$

is a short exact sequence of double chain complexes. Then there is an exact homology sequence

$$\cdots \longrightarrow H_n(\mathscr{A}') \xrightarrow{H_n(f)} H_n(\mathscr{A}) \xrightarrow{H_n(g)} H_n(\mathscr{A}'') \xrightarrow{D_n} H_{n-1}(\mathscr{A}') \xrightarrow{H_{n-1}(f)} H_{n-1}(\mathscr{A}) \xrightarrow{H_{n-1}(g)} H_{n-1}(\mathscr{A}'') \longrightarrow \cdots$$

is exact.

PROOF. There is just one subtlety here. The object $T_n(\mathscr{A})$ is the sum of all the A_{ij} for which $i + j = n$. There are only finitely many such because there are integers m_0 and n_0 with $A_{ij} = 0$ unless $i \geq m_0$ and $j \geq n_0$. The same is true of $\mathscr{A}'$ and $\mathscr{A}''$. Thus the sequence

$$0 \longrightarrow T_n(\mathscr{A}') \longrightarrow T_n(\mathscr{A}) \longrightarrow T_n(\mathscr{A}'') \longrightarrow 0$$

is the sum of only finitely many exact sequences and it is a standard property of abelian categories that a finite sum of exact sequences is exact. □

6.3. THEOREM. *Suppose that* $\mathscr{A} = (A_{\bullet\bullet}, d^I, d^{II})$ *is a double complex and that* d^I *is exact. Then the total complex is also exact.*

PROOF. We will suppose for simplicity that the lower bounds $m_0 = n_0 = 0$. Thus we will think of the double complex as living in the first quadrant. Let $F_n(\mathscr{A})$ denote the double complex truncated above the nth row. That is, $F_n(\mathscr{A})$ is the double complex whose (i,j)th term is A_{ij} for $j \le n$ and is 0 when $j > n$. Since both d^{I} and d^{II} go down to lower indices, $F_n(\mathscr{A})$ is a subcomplex of $\mathscr{A}$. Also for $m < n$, $F_m(\mathscr{A})$ is a subcomplex of $F_n(\mathscr{A})$. Let $R_n(\mathscr{A})$ be the nth row of $\mathscr{A}$. That is, $R_n(\mathscr{A})$ is the double complex that has A_{in} in bidegree (i,n) and all other terms are 0. The boundary operator is the restriction of d^{I}. Except for $n = 0$, it is not a subcomplex of $\mathscr{A}$. However, there is an exact sequence

$$0 \longrightarrow F_{n-1}(\mathscr{A}) \longrightarrow F_n(\mathscr{A}) \longrightarrow R_n(\mathscr{A}) \longrightarrow 0$$

Our hypothesis that d^{I} is exact implies that $R_n(\mathscr{A})$ is exact for all n. The exact homology triangle then implies that the induced

$$H_\bullet(F_{n-1}(\mathscr{A})) \longrightarrow H_\bullet(F_n(\mathscr{A}))$$

is an isomorphism. Since $F_0(\mathscr{A}) = R_0(\mathscr{A})$, this implies that $H_\bullet(F_n(\mathscr{A}))$ is identically 0 for all n. But the computation of $H_n(\mathscr{A})$ uses only terms of total degree $n-1$, n, and $n+1$, which are all in $F_{n+1}(\mathscr{A})$ so that the inclusion $F_{n+1}(\mathscr{A}) \longrightarrow \mathscr{A}$ induces an isomorphism $H_n(F_{n+1}(\mathscr{A})) \longrightarrow H_n(\mathscr{A})$. Since the left hand side is 0, so is the right hand side. □

6.4. COROLLARY. *Suppose $\mathscr{A} = (A_{\bullet\bullet}, d^I, d^{II})$ is a double chain complex with $A_{mn} = 0$ for $n < -1$ or $m < -1$. Suppose for each $n \ge 0$ the single complex $(A_{\bullet n}, d^I)$ is acyclic and for each $m \ge 0$, the single complex $(A_{m\bullet}, d^{II})$ is acyclic. Then the chain complexes $(A_{\bullet\,-1}, d^I)$ and $(A_{-1\,\bullet}, d^{II})$ are homology equivalent.*

PROOF. Let $\mathscr{B}$ denote the double complex in which all the terms A_{mn} with $m = -1$ or $n = -1$ are replaced by 0 and otherwise nothing is changed. Although $\mathscr{B}$ is not a subcomplex of $\mathscr{A}$ it is a quotient complex. Let $\mathscr{B}_1$ denote the complex in which all the terms A_{mn} with $n = -1$ have been replaced by 0 and B_2 denote the complex in which all the A_{mn} with $m = -1$ have been replaced by 0. Let $\mathscr{C}_1 = (A_{\bullet\,-1}, d^{\mathrm{I}})$ and $\mathscr{C}_2 = (A_{-1\,\bullet}, d^{\mathrm{II}})$. These are single complexes, but we treat them as double complexes with all other terms 0 for the purposes of this proof. There are exact sequences

$$0 \longrightarrow \mathscr{C}_1 \longrightarrow \mathscr{A} \longrightarrow \mathscr{B}_1 \longrightarrow 0$$
$$0 \longrightarrow \mathscr{C}_2 \longrightarrow \mathscr{A} \longrightarrow \mathscr{B}_2 \longrightarrow 0$$

Moreover, the hypotheses state that $\mathscr{B}_1$ is acyclic with d^{I} as boundary and that $\mathscr{B}_2$ is acyclic with d^{II} as boundary and therefore $\mathscr{B}_1$ and $\mathscr{B}_2$ are acyclic and therefore that both $\mathscr{C}_1 \longrightarrow \mathscr{A}$ and $\mathscr{C}_2 \longrightarrow \mathscr{A}$ are homology equivalences and therefore $\mathscr{C}_1$ is homology equivalent to $\mathscr{C}_2$. □

6.5. Proof of 2.5.9. We are now ready to prove that the definitions of Ext and Tor are independent of resolutions and also that Ext can be defined by an injective resolution of the second argument. We deal first with Ext. Given left modules N and M, let $P_\bullet \longrightarrow N \longrightarrow 0$ be an exact sequence such that $P_\bullet$ is a projective resolution of N and let $0 \longrightarrow M \longrightarrow Q_\bullet$ be an exact sequence such that $Q_\bullet$ is an injective resolution of M. Form the double cochain complex

$\mathscr{A} = (A_{mn}, d^{\mathrm{I}}, d^{\mathrm{II}})$ defined by

$$A_{mn} = \begin{cases} \mathrm{Hom}_R(P_m, Q_n), & \text{if } m \geq 0 \text{ and } n \geq 0 \\ \mathrm{Hom}_R(P_m, M), & \text{if } m \geq 0 \text{ and } n = -1 \\ \mathrm{Hom}_R(N, Q_n), & \text{if } m = -1 \text{ and } n \geq 0 \\ 0, & \text{otherwise} \end{cases}$$

The complex for $n \geq 0$ is

$$0 \longrightarrow \mathrm{Hom}_R(N, Q_n) \longrightarrow \mathrm{Hom}_R(P_0, Q_n) \longrightarrow \mathrm{Hom}_R(P_1, Q_n) \longrightarrow \cdots$$

and is acyclic since hom into an injective is an exact functor. Similarly, the complex for $m \geq 0$ is

$$0 \longrightarrow \mathrm{Hom}_R(P_m, M) \longrightarrow \mathrm{Hom}_R(P_m, Q_0) \longrightarrow \mathrm{Hom}_R(P_m, Q_1) \longrightarrow \cdots$$

which is acyclic since hom out of a projective is an exact functor. The complexes on the edges are

$$0 \longrightarrow \mathrm{Hom}_R(P_0, M) \longrightarrow \mathrm{Hom}_R(P_1, M) \longrightarrow \cdots$$
$$0 \longrightarrow \mathrm{Hom}_R(N, Q_0) \longrightarrow \mathrm{Hom}_R(N, Q_1) \longrightarrow \cdots$$

whose homology groups correspond to the two definitions of Ext given in the preceding chapter.

But we can also conclude that the Ext defined by using any projective resolution of M is the same as that using Q and so any projective resolution of M gives the same value of Ext. A similar argument implies that any injective resolution of N also gives the same value to Ext.

The argument for Tor is similar. Now suppose that M is a right module and N a left module. Let $P_\bullet \longrightarrow M \longrightarrow 0$ and $Q_\bullet \longrightarrow N \longrightarrow 0$ be flat resolutions of M and N, resp. Form the double complex that has $P_n \otimes Q_m$ in bidegree n, m, for $n, m \geq -1$. The flatness of the P_n and Q_m for $n, m \geq 0$ implies that all the rows except the (-1)st and all the columns except the (-1)st are exact and so the -1st row and column are homologous. As above, this shows that you can resolve either variable and the value of Tor is independent of the resolution.

6.6. Two examples. The first example shows that a double differential object in which both differentials are exact need not be exact. Let (A, d) be any exact complex with $A \neq 0$. Then $(A, -d)$ is also exact and $d(-d) = -(-d)d = 0$, while $(A, d - d)$ is not exact.

The second example is due to Rob Milson. It is a double differential graded group in which each row and column is exact and has only finitely many non-zero terms, but it does not fit in one quadrant and the total complex is not exact. Define A_{ij} for all $i, j \in \mathbf{N}$ as follows.

$$A_{ij} = \begin{cases} \mathbf{Z} & \text{if } i = -j \text{ or } i = -j + 1 \\ 0 & \text{otherwise} \end{cases}$$

In one direction, the only non-zero boundary operator is $d^{\mathrm{I}} = \mathrm{id}\colon A_{i+1,-i} \longrightarrow A_{i,-i}$ while in the other direction we use $d^{\mathrm{II}} = -\mathrm{id}\colon A_{i+1,-i} \longrightarrow A_{i+1,-i-1}$. This is acyclic in both directions, since each such complex looks like

$$\cdots \longrightarrow 0 \longrightarrow \mathbf{Z} \longrightarrow \mathbf{Z} \longrightarrow 0 \longrightarrow \cdots$$

with either id or $-$id for the arrow. The total complex is

$$\cdots \longrightarrow 0 \longrightarrow A_1 \longrightarrow A_0 \longrightarrow 0 \longrightarrow \cdots$$

where both A_0 and A_1 are the direct sum of a Z-indexed family of copies of Z. The boundary is $D = d^{\mathrm{I}} + d^{\mathrm{II}}$. Then if we let $(a)_i^1$ denote the element $a \in A_{i,-i+1}$ and $(a)_i^0$ denote the element $a \in A_{ii}$, we have $d^{\mathrm{I}}((a)_{i+1}^1) = (a)_i^0$ and $d^{\mathrm{II}}((a)_{i+1}^1) = -(a)_{i+1}^0$ so that

$$D((a)_{i+1}^1) = (a)_i^0 - (a)_{i+1}^0$$

We claim that $D: A_1 \longrightarrow A_0$ is not surjective while $D: A_0 \longrightarrow 0$ is 0, so that the total complex is not exact. To see this, let $s: A_0 \longrightarrow \mathsf{Z}$ be the sum of the coordinates function. Clearly $s \circ D = 0$ so it is sufficient to observe that $s \neq 0$, which is obvious.

6.7. Exercise.

1. A chain complex in the category of chain complexes is almost a double chain complex object except that the squares commute instead of anticommute. Show that given such a chain complex of chain complexes, there is at least one way of negating some of the boundary maps so that you get instead a double chain complex.

7. Double simplicial objects

7.1. definition. If $\mathscr{X}$ is a category, then an object of $\mathsf{Simp}(\mathsf{Simp}(\mathscr{X}))$ is called a **double simplicial object** of $\mathscr{X}$. It consists of a doubly indexed family X_{nm}, $n \geq 0$, $m \geq 0$, arrows $d^i: X_{nm} \longrightarrow X_{n-1\,m}$ and $s^i: X_{nm} \longrightarrow X_{n+1\,m}$ for $0 \leq i \leq n$ and arrows $\partial^j: X_{nm} \longrightarrow X_{n\,m-1}$ and $\sigma^j: X_{nm} \longrightarrow X_{n\,m+1}$ for $0 \leq j \leq m$. In addition both directions must satisfy the simplicial identities separately and all the horizontal arrows commute with all vertical arrows, which is to say that all such commutation identities as $d^i \circ \partial^j = \partial^j \circ d^i$, as well as three similar kinds, must hold.

7.2. The diagonal object. If $\mathscr{X} = X_{\bullet\bullet}$ together with all the requisite faces and degeneracies is a double simplicial object, the **diagonal simplicial object** is simply the simplicial object that has X_{nn} in degree n. The ith face operator is $\partial^i \circ d^i = d^i \circ \partial^i$ and the ith degeneracy is similarly $\sigma^i \circ s^i = s^i \circ \sigma$. The commutation laws imply that this gives a simplicial set, called the **diagonal simplicial set**, which we will denote by $\Delta\mathscr{X}$. There is an associated chain complex that has X_{nn} in degree n and whose boundary operator is given by $\sum(-1)^i d^i \partial^i$. Let us call the functor so defined $K_\bullet$.

7.3. The double complex. On the other hand, there is a double complex associated to each double simplicial set in an abelian category. Or rather, there are many, as we will see. If we take a double simplicial set $\mathscr{A} = A_{\bullet\bullet}$, then form the doubly graded object that has A_{nm} in bidegree nm with two boundary operators $d^{\mathrm{I}} = \sum(-1)^i d^i$ and $d^{\mathrm{II}} = \sum(-1)^i \partial^i$. There is a slight problem with this since if we take this definition, the squares will commute rather than anticommute as is necessary for a double simplicial object. There are many—uncountably many—ways of assigning minus signs to some of the arrows so that every square gets either one or three of them and winds up anti-commuting (see Exercise 1). One way is to define $d_n^{\mathrm{II}} = \sum(-1)^{n+i}\partial_n^i$ so that all boundary operators in the odd numbered columns are negated. To be perfectly definite, let us make that change and call the resultant double complex $L\mathscr{A}$, which is clearly functorial in $\mathscr{A}$. Then we can form the single complex $TL\mathscr{A}$ (previous section). Thus we have two functors, K

and TL that turn a double simplicial object into a chain complex. Remarkably, they are homotopic (see 7.4). This is even more surprising for the fact that, for example, an element in A_{nn} appears in $K\mathscr{A}$ with degree n and in $TL\mathscr{A}$ with degree $2n$. Nonetheless, simplicial sets are so tightly bound that the two constructions are homotopic.

8. Homology and cohomology of a morphism

Suppose $\mathscr{X}$ is a category with a chain complex functor $C_\bullet$. If $f: X \longrightarrow Y$ is an arrow in $\mathscr{X}$ we let $C_\bullet(f: X \longrightarrow Y)$ denote the mapping cone of $C_\bullet f: C_\bullet X \longrightarrow C_\bullet(Y)$. This means that $C_n(f: X \longrightarrow Y) = C_n(Y) \oplus C_{n-1}(X)$. The boundary operator has the matrix $\begin{pmatrix} d & f \\ 0 & -d \end{pmatrix}$. If f is understood, we will write $C_\bullet(Y, X)$, by analogy with the notation used in algebraic topology, of which this is a generalization. Now suppose that $g: Y \longrightarrow Z$ is another map in $\mathscr{X}$. Consider the sequence

$$0 \longrightarrow C_nY \oplus C_{n-1}X \xrightarrow{\begin{pmatrix} -g & 0 \\ 0 & 1 \\ 1 & 0 \\ 0 & f \end{pmatrix}} C_nZ \oplus C_{n-1}X \oplus C_nY \oplus C_{n-1}Y$$
$$\xrightarrow{\begin{pmatrix} 1 & 0 & g & 0 \\ 0 & -f & 0 & 1 \end{pmatrix}} C_nZ \oplus C_{n-1}Y \longrightarrow 0$$

in which the boundary of the middle object is

$$\begin{pmatrix} d & -gf & 0 & 0 \\ 0 & -d & 0 & 0 \\ 0 & 0 & d & 1 \\ 0 & 0 & 0 & -d \end{pmatrix}$$

Then one can check by a direct computation that the horizontal arrows are maps of chain complexes. Exactness of the sequence is readily proved using the arrows

$$\begin{pmatrix} 0 & 0 & 1 & 0 \\ 0 & 1 & 0 & 0 \end{pmatrix} : C_nZ \oplus C_{n-1}X \oplus C_nY \oplus C_{n-1}Y \longrightarrow C_nY \oplus C_{n-1}X$$

and

$$\begin{pmatrix} 1 & 0 \\ 0 & 0 \\ 0 & 0 \\ 0 & 1 \end{pmatrix} : C_nZ \oplus C_{n-1}Y \longrightarrow C_nZ \oplus C_{n-1}X \oplus C_nY \oplus C_{n-1}Y$$

The end terms are the nth terms of $C_\bullet(Y, X)$ and $C_\bullet(Z, Y)$, respectively, while the middle term is the direct sum of the mapping cone of $C_\bullet(Z, X)$ (with a change of sign, which is irrelevant) and that of $C_\bullet(Y, Y)$. Since the latter is the mapping cone of the identity functor and is therefore contractible, the result is the exact sequence of pairs

$$\cdots \to H_{n+1}(Z, Y) \to H_n(Y, X) \to H_n(Z, X) \to H_n(Z, Y) \to H_{n-1}(Y, X) \to \cdots$$

CHAPTER 4

Triples à la Mode de Kan

1. Triples and cotriples

1.1. Definition of triple. Let $\mathscr{C}$ be a category. A **triple** $\mathbf{T} = (T, \eta, \mu)$ on $\mathscr{C}$ consists of an endofunctor $T: \mathscr{C} \longrightarrow \mathscr{C}$ and natural transformations $\eta: \mathrm{Id} \longrightarrow T$ and $\mu: T^2 \longrightarrow T$ for which the following diagrams commute.

$$\begin{array}{ccc} T^3 & \xrightarrow{T\mu} & T^2 \\ {\scriptstyle \mu T}\downarrow & & \downarrow{\scriptstyle \mu} \\ T^2 & \xrightarrow[\mu]{} & T \end{array} \qquad\qquad \begin{array}{ccccc} T & \xrightarrow{T\eta} & T^2 & \xleftarrow{\eta T} & T \\ & {\scriptstyle =}\searrow & \downarrow{\scriptstyle \mu} & \swarrow{\scriptstyle =} & \\ & & T & & \end{array}$$

The two natural transformations are called the unit and the multiplication of the triple, respectively. The three diagrams that commute are called the left and right unit and associativity laws. The reason for these names comes from examples such as the following.

1.2. An example. Let R be a ring (associative with unit). There is a triple $\mathbf{T} = (T, \eta, \mu)$ on the category Ab of abelian groups for which $T(A) = R \otimes A$, $\eta A: A \longrightarrow R \otimes A$ is defined by $\eta A(a) = 1 \otimes a$ and $\mu A(r_1 \otimes r_2 \otimes a) = r_1 r_2 \otimes a$.

1.3. Another example. Here is a simple example of a rather different nature. Let T be a functor on the category of sets that adds one element to each set. We can write $T(S) = S \cup \{S\}$. If $f: S \longrightarrow S'$, then define $T(f): T(S) \longrightarrow T(S')$ by

$$Tf(s) = \begin{cases} f(s) & \text{if } s \in S \\ S' & \text{if } s = S \end{cases}$$

Then T is readily seen to be a functor. Let $\eta S: S \longrightarrow T(S)$ be the inclusion and define $\mu S: T^2(S) = S \cup \{S\} \cup \{T(S)\} \longrightarrow T(S)$ by

$$\mu S(s) = \begin{cases} s & \text{if } s \in S \\ S & \text{if } s = S \text{ or } s = T(S) \end{cases}$$

Many of the triples that arise in nature are shown to be triples by using the following result.

1.4. THEOREM. *Suppose that $F: \mathscr{C} \longrightarrow \mathscr{B}$ is left adjoint to $U: \mathscr{B} \longrightarrow \mathscr{C}$. Suppose $\eta: \mathrm{Id} \longrightarrow UF$ and $\epsilon: FU \longrightarrow \mathrm{Id}$ are the unit and counit, respectively, of the adjunction. Then $(UF, \eta, U\epsilon F)$ is a triple on $\mathscr{C}$.*

PROOF. We have

$$\mu \circ T\eta = U\epsilon F \circ UF\eta = U(\epsilon F \circ F\eta) = U(\mathrm{id}) = \mathrm{id}$$

and

$$\mu \circ \eta T = U\epsilon F \circ \eta UF = \mathrm{id}$$

Finally, we have,

$$\begin{aligned}\mu \circ T\mu &= U\epsilon F \circ UFU\epsilon F = U(\epsilon F \circ FU\epsilon F)\\ &= U(\epsilon F \circ \epsilon FUF) = U\epsilon F \circ U\epsilon FUF = \mu \circ \mu T\end{aligned}$$

The interesting step here is the fourth, which is an instance of the naturality law

$$\begin{array}{ccc} TA & \xrightarrow{Tf} & TB \\ {\scriptstyle \epsilon TA}\downarrow & & \downarrow{\scriptstyle \epsilon B} \\ A & \xrightarrow[f]{} & B \end{array}$$

with $B = F$ (really an instance of F), $A = TF$, and $f = \epsilon F$. □

1.5. Yet another example. Armed with this theorem, we can now write down as many triples as we like. For example, the free group triple on sets comes from the adjunction between the underlying set functor on groups and its left adjoint the free group functor. The endofunctor on Set assigns to each set S the underlying set of the free group generated by S, which is to say the set of all words (including the empty word) in the elements of S and their inverses, reduced by the equations $wss^{-1}w' = ws^{-1}sw' = ww'$, for arbitrary words w and w'. The unit of the triple takes an element of the set to the singleton word. The multiplication takes a word made up of words and reinterprets it as a word. For example, let us write $\langle a \rangle$ for the element $\eta S(a)$ corresponding to $a \in S$. Then a typical word in $T(S)$ might look like $\langle a\rangle\langle b\rangle^{-1}\langle c\rangle$. And μS applied to $\langle\langle a\rangle\langle b\rangle^{-1}\langle c\rangle\rangle\langle\langle c\rangle^{-1}\langle d\rangle^{-1}\langle e\rangle\rangle\langle\langle f\rangle\rangle$ produces the word $\langle a\rangle\langle b\rangle^{-1}\langle c\rangle\langle c\rangle^{-1}\langle d\rangle^{-1}\langle e\rangle\langle f\rangle = \langle a\rangle\langle b\rangle^{-1}\langle d\rangle^{-1}\langle e\rangle\langle f\rangle$. One point to note is that the associativity law does not merely correspond to the associativity law of group multiplication, but to the entire set of equations that groups satisfy. Indeed, there are similar triples involving free non-associative structures.

1.6. Cotriples. A *cotriple* in a category $\mathscr{B}$ is a triple in $\mathscr{B}^{\mathrm{op}}$. Thus $\mathbf{G} = (G, \epsilon, \delta)$ (this is standard notation) is a cotriple in $\mathscr{B}$ if G is an endofunctor of $\mathscr{B}$, and $\epsilon: G \longrightarrow \mathrm{Id}$, $\delta: G \longrightarrow G^2$ are natural transformations satisfying the duals to the diagrams of 1.1 above. (Thus a cotriple is the opposite of a triple, not the dual of a triple. The dual of a triple—in other words, a triple in $\mathsf{Cat}^{\mathrm{op}}$—is a triple.)

1.7. PROPOSITION. *Let $U: \mathscr{B} \longrightarrow \mathscr{C}$ have a left adjoint $F: \mathscr{C} \longrightarrow \mathscr{B}$ with adjunction morphisms $\eta: \mathrm{Id} \longrightarrow UF$ and $\epsilon: FU \longrightarrow \mathrm{Id}$. Then $\mathbf{G} = (FU, \epsilon, F\eta U)$ is a cotriple on $\mathscr{C}$.*

PROOF. This follows from Theorem 1 and the observation that U is left adjoint to F as functors between $\mathscr{B}^{\mathrm{op}}$ and $\mathscr{C}^{\mathrm{op}}$ with unit ϵ and counit η. □

1.8. Exercises.

1. Show that the example of 1.2 satisfies the equations to be a triple.

2. Show that the example of 1.3 satisfies the equations to be a triple.

3. Verify the associative law in the case of the example of 1.5. (Hint: Prove and make use of the fact that instances of μ are "really" group homomorphisms.)

4. Consider the category $\mathbf{N}$ whose objects are the natural numbers and there is a unique morphism $n \longrightarrow m$ if and only if $n \leq m$. Let $T: \mathbf{N} \longrightarrow \mathbf{N}$ be the functor defined by

$$T(n) = \begin{cases} n & \text{if } n \text{ is even} \\ n+1 & \text{if } n \text{ is odd} \end{cases}$$

so that $T(n)$ is least even number $\geq n$. Show that there is a unique η and μ that makes (T, η, μ) into a triple.

5. Dually, there is a cotriple (G, ϵ, δ) on $\mathbf{N}$ in which $G(n)$ is the greatest odd number $\leq n$. Show that G is left adjoint to T.

6. Let $\mathbf{P}$ denote the functor from Set to Set which takes a set to its powerset and a function to its direct image function (Section 1.2). For a set X, let ηX take an element of X to the singleton containing x, and let μX take a set of subsets of X (an element of $\mathbf{P}X$) to its union. Show that $(\mathbf{P}, \eta, \mu)$ is a triple in Set.

7. Let R be any commutative ring. For each set X, let TX be the set of polynomials in a finite number of variables with the variables in X and coefficients from R. Show that $\mathbf{T}$ is the functor part of a triple (μ is defined to "collect terms").

8. An ordered binary rooted tree (OBRT) is a binary rooted tree (assume trees are finite in this problem) which has an additional linear order structure (referred to as left/right) on each set of siblings. An X-labeled OBRT (LOBRT/X) is one together with a function from the set of terminal nodes to X. Show that the following construction produces a triple in Set: For any set X, TX is the set of all isomorphism classes of LOBRT/X. If $f: X \longrightarrow Y$, then Tf is relabelling along f (take a tree in TX and change the label of each node labeled x to $f(x)$). ηX takes $x \in X$ to the one-node tree labeled x, and μX takes a tree whose labels are trees in TX to the tree obtained by attaching to each node the tree whose name labels that node.

9. Let $\mathscr{B}$ be the category of sets with one binary operation (subject to no conditions) and functions which preserve the binary operation.

(a) Show that the triple of Exercise 8 arises from the underlying set functor $\mathscr{B} \longrightarrow$ Set and its left adjoint.

(b) Give an explicit description of the cotriple in $\mathscr{B}$ induced by the adjoint functors in (a).

10. (a) Give an explicit description of the cotriple in Grp induced by the underlying set functor and the free group functor.

(b) Give an explicit description of the model induced cotriple in Grp when $\mathscr{M}$ consists of the free group on one generator. (Recall that the sum in the category of groups is the free product.)

(c) Show that these cotriples are naturally equivalent.

11. Let M be a monoid and $G = \mathrm{Hom}(M, -): \mathsf{Set} \longrightarrow \mathsf{Set}$. If X is a set and $f: M \longrightarrow X$, let $\epsilon X(f) = f(1)$ and $[\delta X(f)](m)(n) = f(mn)$ for $m, n \in M$. Show that δ and ϵ are natural transformations making (G, ϵ, δ) a cotriple in Set.

12. Show that if $\mathbf{T}$ is any triple on $\mathscr{C}$ and A is an object of $\mathscr{C}$, and there is at least one mono $A \longrightarrow TA$, then ηA is monic. (Hint: If m is the monic, put Tm into a commutative square with η and use a unitary identity.)

2. Model induced triples

2.1. Other sources of triples. Not all triples arise from adjunctions. More precisely (since there is a theorem due to Kleisli [1965] and to Eilenberg and Moore [1965] that states that all triples do arise from adjunctions), there are other ways of getting triples besides using adjoints. Here is a very important example of a class of triples that do not naturally arise from an adjunction.

Let $\mathscr{C}$ be a category with arbitrary products and suppose that $\mathscr{M}$ is a set of objects of $\mathscr{C}$. Define $T(C) = \prod_{M\in\mathscr{M}} \prod_{C\longrightarrow M} M$. This means that $T(C)$ consists of the product of one copy of M corresponding to each $M \in \mathscr{M}$ and to each arrow $C \longrightarrow M$. Another way of writing this is $T(C) = \prod_{M\in\mathscr{M}} M^{\mathrm{Hom}(C,M)}$. For $u: C \longrightarrow M$, let $\langle u\rangle: T(C) \longrightarrow M$ denote the projection on the product corresponding to u. The universal mapping property of products implies that an arrow into $T(C)$ is determined and uniquely by specifying its composite with each $\langle u\rangle$. We use this observation first off to say how T is a functor. For $f: C' \longrightarrow C$, we define $T(f): T(C') \longrightarrow T(C)$ by $\langle u\rangle \circ T(f) = \langle u\circ f\rangle$ for $u: C \longrightarrow M$, $M \in \mathscr{M}$. If also $g: C'' \longrightarrow C'$ is an arrow, then

$$\langle u\rangle \circ T(f\circ g) = \langle u\circ f\circ g\rangle = \langle u\circ f\rangle \circ T(g) = \langle u\rangle \circ T(f)\circ T(g)$$

for any $u: C \longrightarrow M$ and $M \in \mathscr{M}$, whence by the universal mapping property of products, we conclude that $T(f\circ g) = T(f)\circ T(g)$. We can now define η by the formula $\langle u\rangle \circ \eta C = u$ and μ by $\langle u\rangle \circ \mu C = \langle\langle u\rangle\rangle$. To interpret the latter, we note that the projection $\langle u\rangle: T(C) \longrightarrow M$ is an arrow and corresponding to it is $\langle\langle u\rangle\rangle: T^2(C) \longrightarrow M$. Of course, there are usually other arrows from $T(C)$ to objects in $\mathscr{M}$ in general.

Now we prove the various laws. For $u: C \longrightarrow M$, we have that

$$\langle u\rangle \circ \mu C\circ \eta TC = \langle\langle u\rangle\rangle \circ \eta TC = \langle u\rangle$$

since the effect of η is to remove the (outermost) brackets. It follows that $\mu C\circ\eta TC =$ id. We also have

$$\langle u\rangle \circ \mu C\circ T\eta C = \langle\langle u\rangle\rangle \circ T\eta C = \langle\langle u\rangle \circ \eta C\rangle = \langle u\rangle$$

from which it follows that $\mu C\circ T\eta C =$ id. Finally, the associativity is shown by

$$\langle u\rangle \circ \mu C\circ \mu TC = \langle\langle u\rangle\rangle \circ \mu TC = \langle\langle\langle u\rangle\rangle\rangle$$

while

$$\langle u\rangle \circ \mu C\circ T\mu C = \langle\langle u\rangle\rangle \circ T\mu C = \langle\langle u\rangle \circ \mu C\rangle = \langle\langle\langle u\rangle\rangle\rangle$$

This triple is called a *model induced triple* and $\mathscr{M}$ is the set of models. This construction can be generalized to allow $\mathscr{M}$ to be an arbitrary small subcategory of $\mathscr{C}$ or indeed replaced by an arbitrary functor into $\mathscr{C}$ with small domain, see Exercise 1. In this book, we require only this version of the construction.

2.2. Model induced cotriples. There are also, of course, model induced cotriples. If $\mathscr{M}$ is a set of objects in a category with sums, then there is a cotriple $\mathbf{G} = (G, \epsilon, \delta)$ as follows:

$$GC = \sum_{M\in\mathscr{M}} \sum_{M\longrightarrow C} C$$

If $[u]: M \longrightarrow GC$ is the element of the sum corresponding to $u: M \longrightarrow C$, then $\epsilon C: GC \longrightarrow C$ is the unique arrow such that $\epsilon C\circ [u] = u$ and $\delta C: GC \longrightarrow G^2C$ is the unique arrow such that $\delta C\circ [u] = [[u]]$.

2.3. Exercises.

1. Generalize the results of 2 as follows. Let $\mathscr{M}$ be a small category and $I\colon \mathscr{M} \longrightarrow \mathscr{C}$ a functor. Define, for C an object of $\mathscr{C}$,

$$TC = \lim_{C \longrightarrow IM} IM$$

That is, TC is the limit of the diagram whose nodes are arrows $u\colon C \longrightarrow IM$ and for which an arrow from $u\colon C \longrightarrow IM$ to $u'\colon C \longrightarrow IM'$ is an arrow $\alpha\colon M \longrightarrow M'$ such that $I\alpha \circ u = u'$. Extend T to a functor and define η and μ so that (T, η, μ) is a triple that reduces to the construction of 2 when $\mathscr{M}$ is the discrete category consisting of a set of objects of $\mathscr{C}$ and I is the inclusion.

3. Triples on the simplicial category

A very simple triple can be described as follows. It is convenient to work in the category of augmented simplicial sets. So suppose $X \longrightarrow X_{-1}$ is an augmented simplicial set. Let $GX \longrightarrow (GX)_{-1}$ be the simplicial set described by $(GX)_n = X_{n+1}$, $(GX)_{-1} = X_0$, face operators $(Gd)^i_n = d^{i+1}_{n+1}$ and degeneracies $(Gs)^i_n = s^{i+1}_{n+1}$. Then $d^0_{n+1}\colon (GX)_n = X_{n+1} \longrightarrow X_n$ defines a simplicial map. In fact, the simplicial identities $d^0 \circ d^{i+1} = d^i \circ d^0$ for all $i \geq 0$ and $d^0 \circ s^{i+1} = s^i \circ d^0$ say exactly that. This is the map we call $\epsilon X\colon GX \longrightarrow X$. We also note that $s^0_{n+1}\colon (GX)_n = X_{n+1} \longrightarrow X_{n+2} = (G^2X)_n$ is the nth component of a simplicial map. This comes down to the simplicial identities $s^0 \circ d^i = d^{i+1} \circ s^0$ and $s^0 \circ s^{i+1} = s^{i+2} \circ s^0$ for $i > 0$. This is the map $\delta X\colon GX \longrightarrow G^2X$.

3.1. THEOREM. *The maps ϵX and δX are components of natural transformations and $\mathbf{G} = (G, \epsilon, \delta)$ is a cotriple on $\mathsf{Simp}(\mathscr{X})$.*

PROOF. The naturality is easy and is left to the reader. The identities $d^0 \circ s^0 = d^1 \circ s^0 = \text{id}$ imply, respectively, that $G\epsilon \circ \delta = \text{id}$ and $\epsilon G \circ \delta = \text{id}$. The reason is that applying G to ϵ changes only the lower index, not the upper, while ϵGX is the 0th face of GX, which is d^1. The identity $s^0 \circ s^0 = s^1 \circ s^0$ similarly implies $G\delta \circ \delta = \delta G \circ \delta$. □

Finally, we observe the most important property of GX.

3.2. THEOREM. *The augmented simplicial object GX is contractible.*

PROOF. Let $s = s^0\colon (GX)_n = X_{n+1} \longrightarrow X_{n+2} = (GX)_{n+1}$. The equations $d^1 \circ s^0 = \text{id}$, $d^{i+1} \circ s^0 = s^0 \circ d^i$ for $i > 0$ and $s^{i+1} \circ s^0 = s^0 \circ s^i$ for $i \geq 0$ give the result. □

We will call this the *path cotriple* since it is the simplicial version of the following cotriple on the category of locally connected locally pointed spaces. These are locally connected spaces for which a base point has been chosen in each component. This is analogous to augmented simplicial objects since the augmentation corresponds to fixing a point in each component (in the contractible case). Now consider the subset, we will call it $I \multimap X$, of the continuous maps of the unit interval I to the space X consisting of those maps f for which $f(0)$ is the base point. This is topologized by the compact/open topology. The map $H\colon I \times I \multimap X \longrightarrow I \multimap X$, defined by $H(s, f)(t) = f(st)$ is a homotopy between the identity and the discrete set of base points, thus a contraction to the set of components.

3.3. Double simplicial objects. On the category of double augmented simplicial objects over $\mathscr{X}$, we could apply the path cotriple to either the rows or columns of a double complex. Let us define cotriples $\mathbf{G}_{\mathrm{I}}$ and $\mathbf{G}_{\mathrm{II}}$ by $G_{\mathrm{I}}(X_{\bullet\bullet})_{mn} = X_{m+1\,n}$ and $G_{\mathrm{II}}(X_{\bullet\bullet})_{mn} = X_{m\,n+1}$. Of course, the appropriate definitions have to be given for the face and degeneracy operators, but these are obvious from the definition of $\mathbf{G}$. It is also obvious that G_{I} commutes with G_{II} to give a total cotriple $\mathbf{G}_T$ whose functor is $G_{\mathrm{I}} \circ G_{\mathrm{II}} = G_{\mathrm{II}} \circ G_{\mathrm{I}}$. The image of G_T includes double augmented simplicial objects that are contractible in both rows and columns.

4. Historical Notes

Adjoints were originated by Daniel Kan in [1958]. Triples were discovered by Claude Chevalley in [1959]. He called his "The standard construction" and it seems likely that this term was meant only to be descriptive. But various people used the term substantively, including, for example, Peter Huber who used "standard constructions" in his Ph. D. dissertation in [1960] and proved that every adjoint pair gives rise to a triple. He once told me that the reason he proved this theorem was that they (this presumably referred to him and his advisor, Beno Eckmann) had had a lot of trouble proving the associative rule for various triples and it occurred to them that they always did seem to come from adjoint pairs. They wondered if this was a general phenomenon and Huber proved that it was and that the associativity was then automatic. Heinrich Kleisli later showed [1965] that, conversely, every triple came from an adjoint pair. Samuel Eilenberg and John Moore also proved that converse using a construction that almost always gives a category different from that of Kleisli in [1965]. Eilenberg and Moore also invented the name "triple" by which they are known today, at least to some of us. At lunch one day in Oberwohlfach in the summer of 1966, Jean Bénabou suggested calling them "monads" and this term is in wide use today.

CHAPTER 5

The Main Acyclic Models Theorem

In this chapter, we develop the machinery necessary to state the main acyclic models theorem of which the various versions we use are special cases. The first thing we do is introduce the idea of an abstract class of acyclicity. This definition, as currently formulated, requires that we are dealing with a class of chain complexes. Cochain complexes also make sense, being chain complexes in the dual category. But ungraded complexes (or doubly infinite complexes) do not seem to work.

A word about notation should be inserted here. Till now, it has not mattered if the differential objects was graded. In this chapter, it definitely matters. We often denote by $K_\bullet$ the chain complex

$$\cdots \longrightarrow K_n \longrightarrow K_{n-1} \longrightarrow \cdots \longrightarrow K_1 \longrightarrow K_0 \longrightarrow 0$$

and by $f_\bullet\colon K_\bullet \longrightarrow L_\bullet$ a chain map between such complexes.

1. Acyclic classes

In this definition, $\mathscr{C} = \mathsf{CC}(\mathscr{A})$ is the category of chain complexes of an abelian category $\mathscr{A}$.

1.1. Acyclic classes. A class Γ of objects of $\mathscr{C}$ will be called an *acyclic class* provided:

AC–1. The 0 complex is in Γ.
AC–2. The complex $C_\bullet$ belongs to Γ if and only if $SC_\bullet$ does.
AC–3. If the complexes $K_\bullet$ and $L_\bullet$ are homotopic and $K_\bullet \in \Gamma$, then $L_\bullet \in \Gamma$.
AC–4. Every complex in Γ is acyclic.
AC–5. If $K_{\bullet\bullet}$ is a double complex, all of whose rows are in Γ, then the total complex of $C_\bullet$ belongs to Γ.

Of these conditions, the first two are routine, the third says the class is closed under homotopy, which implies, among other things that every contractible complex belongs to Γ. The fourth says that every complex in Γ is acyclic. But the real heart of the definition is the fifth condition. This is the one that does not seem to have an obvious generalization to the ungraded case. One cannot strengthen this condition to one involoving arbitrary (or even countable) filtered colimits since all the cohomology examples would fail; filtered colimits are not exact in the category $\mathsf{Ab}^{\mathrm{op}}$.

Before studying the properties of acyclic classes, we give some examples.

1.2. Acyclic complexes. Let Γ consist of the acyclic complexes. AC–1, 2, 3 and 4 are obvious, while 5 is an immediate consequence of 2.6.3.

1.3. Contractible complexes. Let Γ consist of the contractible complexes. AC–1, 2 and 4 are obvious. To see AC–3, suppose that $f: K_\bullet \longrightarrow L_\bullet$ and $g: L_\bullet \longrightarrow K_\bullet$ are chain maps and $s: K_\bullet \longrightarrow K_\bullet$ and $t: L_\bullet \longrightarrow L_\bullet$ are maps such that $1 = ds + sd$ and $1 - fg = dt + td$. The first of these equations says that $K_\bullet$ is contractible and the second that fg is homotopic to the identity. We could also suppose that gf is homotopic to the identity, but that turns out to be unnecessary. For already we have:

$$d(t + fsg) + (t + fsg)d = dt + td + f(ds + sd)g = 1 - fg + fg = 1$$

so that $t + fsg$ is a contracting homotopy for $L_\bullet$.

To prove AC–5, suppose that we are given a double complex K_{mn}, with $K_{mn} = 0$ for $m < 0$ or $n < 0$. The actual lower bounds make no real difference, but is just a convenience. In order to avoid ugly superscripts that make things harder to read, we will denote one boundary operator by $d: K_{mn} \longrightarrow K_{m\,n-1}$ and the other by $\partial: K_{mn} \longrightarrow K_{m-1\,n}$ and assume that $d\partial = -\partial d$. Suppose that for each m and n, there is a map $s: K_{mn} \longrightarrow K_{m\,n+1}$ that satisfies $ds + sd = 1$. The total complex has, in degree n, the direct sum $L_n = \sum_{i=0}^{n} K_{i\,n-i}$ and is 0 when $n < 0$. The boundary operator $D: L_n \longrightarrow L_{n-1}$ has the matrix

$$\begin{pmatrix} d & \partial & 0 & \cdots & 0 \\ 0 & d & \partial & \cdots & 0 \\ 0 & 0 & d & \cdots & 0 \\ \vdots & \vdots & \vdots & \ddots & \partial \\ 0 & 0 & 0 & \cdots & d \end{pmatrix}$$

For the rest of this proof, we will use S not for suspension, but for a contracting homotopy in the double complex, which we now define in degree n as a map $S: L_n \longrightarrow L_{n+1}$ with the matrix

$$\begin{pmatrix} s & -s\partial s & s\partial s\partial s & \cdots & (-1)^n s(\partial s)^n \\ 0 & s & -s\partial s & \cdots & (-1)^{n-1} s(\partial s)^{n-1} \\ \vdots & \vdots & \vdots & \ddots & \vdots \\ 0 & 0 & 0 & \cdots & s \\ 0 & 0 & 0 & \cdots & 0 \end{pmatrix}$$

Direct matrix multiplication shows that $SD + DS$ is upper triangular and has $sd + ds = 1$ in each diagonal entry (including the last, since in that case the $sd = 0$ so that $ds = 1$). In carrying that out, it is helpful to block D into an upper triangular matrix and a single column and S into an upper triangular matrix and a single row of zeros. In order to see that $SD + DS = 1$, we must show that the above diagonal entries vanish. First we claim that for $i > 0$, $ds(\partial s)^i = (\partial s)^i + (s\partial)^i ds$. In fact, for $i = 1$,

$$ds\partial s = (1 - sd)\partial s = \partial s - sd\partial s = \partial s + s\partial ds$$

Assuming that the conclusion is true for $i - 1$,

$$\begin{aligned} ds(\partial s)^i &= ((\partial s)^{i-1} + (s\partial)^{i-1} ds)\partial s = (\partial s)^i + (s\partial)^{i-1}(1 - sd)\partial s \\ &= (\partial s)^i + (s\partial)^{i-1}\partial s - (s\partial)^{i-1} sd\partial s = (\partial s)^i + (s\partial)^{i-1} s\partial ds \\ &= (\partial s)^i + (s\partial)^i ds \end{aligned}$$

Now suppose we choose indices $i < j$. The (i, j)th entry of SD is

$$\begin{pmatrix} 0 & \cdots & 0 & s & \cdots & (-1)^{j-i-1}s(\partial s)^{j-i-1} & (-1)^{j-i}s(\partial s)^{j-i} & \cdots \end{pmatrix} \begin{pmatrix} 0 \\ \vdots \\ \partial \\ d \\ 0 \\ \vdots \end{pmatrix}$$

$$\begin{aligned} &= (-1)^{j-i-1}s(\partial s)^{j-i-1}\partial + (-1)^{j-i}s(\partial s)^{j-i}d \\ &= (-1)^{j-i-1}((s\partial)^{j-i} - (s\partial)^{j-i}sd) \\ &= (-1)^{j-i-1}((s\partial)^{j-i} - (s\partial)^{j-i}(1 - ds)) \\ &= (-1)^{j-i-1}(s\partial)^{j-i}ds \end{aligned}$$

and the (i, j)th entry of DS is

$$\begin{pmatrix} 0 & \cdots & 0 & d & \partial & 0 & \cdots \end{pmatrix} \begin{pmatrix} (-1)^j s(\partial s)^j \\ \vdots \\ (-1)^{j-i}s(\partial s)^{j-i} \\ (-1)^{j-i-1}s(\partial s)^{j-i-1} \\ \vdots \end{pmatrix}$$

$$\begin{aligned} &= (-1)^{j-i}ds(\partial s)^{j-i} + (-1)^{j-i-1}\partial s(\partial s)^{j-i-1} \\ &= (-1)^{j-i}(ds(\partial s)^{j-i} - (\partial s)^{j-i}) \\ &= (-1)^{j-i}((\partial s)^{j-i} + (s\partial)^{j-i}ds - (\partial s)^{j-i}) \\ &= (-1)^{j-i}(s\partial)^{j-i}ds \end{aligned}$$

so that the terms cancel and $SD + DS = 1$.

1.4. Quasi-contractible complexes. For this example, we suppose that $\mathscr{A}_0$ is an abelian category and that $\mathscr{A} = \mathrm{Func}(\mathscr{X}, \mathscr{A}_0)$ is a category of functors into $\mathscr{A}_0$. Say that a chain complex functor $C_\bullet \colon \mathscr{X} \longrightarrow \mathscr{A}_0$ is *quasi-contractible* if for each object X of $\mathscr{X}$, the complex $C_\bullet X$ is contractible. Each of the previous results on contractible complexes carries over to these quasi-contractible ones, except that in each case the conclusion is object by object. Similarly we say that a map f of chain complexes is a quasi-homotopy equivalence if at each object X, fX is a is a homotopy equivalence. It is clear that f is a quasi-homotopy equivalence if and only if its mapping cone is quasi-contractible. The earlier material on contractible complexes implies that the quasi-contractible complexes constitute an acyclic class.

1.5. A general condition. Here is one way of generating acyclic classes. As we will explain, each of the three examples above is an instance. Suppose $\mathscr{A}$ is a given abelian category and Φ is a class of additive Ab-valued functors on $\mathscr{A}$. Let Γ denote the class of all acyclic chain complexes over $\mathscr{A}$ such that $\phi(C_\bullet)$ is acyclic for all $\phi \in \Phi$. Then I claim that Γ is automatically an acyclic class. Conditions AC–1 and 2 are obvious. AC–3 follows since additive functors preserve homotopies. AC–4 is clear since contractible complexes are acyclic. And AC–5 follows from the argument in 1.2 above applied to the category of abelian groups. The way this works in the three examples follows. To derive the first example, let $\Phi = \emptyset$. You get precisely the acyclic complexes. For the second, take all the covariant homfunctors

$\mathrm{Hom}(Z, -)$ for all objects Z of $\mathscr{A}$. We have seen in 3.2.4 that $C_\bullet$ is contractible if and only if $\mathrm{Hom}(Z, C_\bullet)$ is exact for all objects Z. For the third example, we are supposing that $\mathscr{A} = \mathrm{Func}(\mathscr{X}, \mathscr{A}_0)$ with $\mathscr{A}_0$ abelian. Each object X of $\mathscr{X}$ gives an evaluation functor $\mathrm{ev}_X : \mathscr{A} \longrightarrow \mathscr{A}_0$ given by $\mathrm{ev}_X(F) = F(X)$. Then $C_\bullet \in \Gamma$ in the third example if and only if $\mathrm{ev}_X(C_\bullet)$ is contractible for every object X of $\mathscr{X}$, which is true if and only if for each object Z of $\mathscr{A}_0$, the complex $\mathrm{Hom}(Z, \mathrm{ev}_X(C_\bullet))$ is acyclic in Ab.

1.6. Exercises.

1. Show that if

$$\cdots \xrightarrow{\partial} (C_n, d_n) \xrightarrow{\partial} (C_{n-1}, d_{n-1}) \xrightarrow{\partial} \cdots \xrightarrow{\partial} (C_0, d_0) \longrightarrow 0$$

is a chain complex of exact differential abelian groups with $d_{n-1} \circ \partial = -\partial \circ d_n$ and you let $C_\bullet = \sum C_n$ with boundary given by

$$d = \begin{pmatrix} d_0 & \partial & 0 & \cdots \\ 0 & d_1 & \partial & \cdots \\ 0 & 0 & d_2 & \cdots \\ \vdots & \vdots & \vdots & \ddots \end{pmatrix}$$

then $(C_\bullet, d)$ is exact. Conclude that if you have a bigraded double differential object

$$(C_{nm}, d : C_{nm} \longrightarrow C_{n-1\,m}, \partial : C_{n,m} \longrightarrow C_{n\,m-1})$$

with $\partial \circ d = -d \circ \partial$ such that $C_{nm} = 0$ for all $m < 0$ and if for each $m \geq 0$, the complex (C_{nm}, d) is exact, the double complex is exact.

2. Show that if

$$0 \longrightarrow (C_0, d_0) \xrightarrow{\delta} \cdots \xrightarrow{\delta} (C_{n-1}, d_{n-1}) \xrightarrow{\delta} (C_n, d_n) \xrightarrow{\delta} \cdots$$

is a cochain complex of exact differential abelian groups with $d_{n+1} \circ \delta = -\delta \circ d_n$ and you let $C = \prod C_n$ with boundary given by

$$d = \begin{pmatrix} d_0 & 0 & 0 & \cdots \\ \delta & d_1 & & \cdots \\ 0 & \delta & d_2 & \cdots \\ \vdots & \vdots & \vdots & \ddots \end{pmatrix}$$

then (C, d) is exact. Conclude that if you have a bigraded double differential object

$$(C_{nm}, d : C_{nm} \longrightarrow C_{n+1\,m}, \delta : C_{n,m} \longrightarrow C_{n\,m+1})$$

with $\delta \circ d = -d \circ \delta$ such that $C_{nm} = 0$ for all $m < 0$ and if for each $m \geq 0$, the complex (C_{nm}, d) is exact, the double complex is exact. Using the fact that the dual of Ab is the category of compact abelian groups (and continuous homomorphisms), show that the preceding exercise is also valid in the category of compact abelian groups.

2. Properties of acyclic classes

2.1. PROPOSITION. *If* $0 \longrightarrow L_\bullet \longrightarrow C_\bullet \longrightarrow K_\bullet \longrightarrow 0$ *is a* U*-split exact sequence of objects of* $\mathscr{C}$ *and if any two belong to* Γ*, then so does the third.*

PROOF. Suppose that $L_\bullet$ and $K_\bullet$, and hence $S^{-1}K_\bullet$ belong to Γ. We know from 1.4 that $C_\bullet$ is the mapping cone of a map $f_\bullet\colon S^{-1}K_\bullet \longrightarrow L_\bullet$. We can think of this as a double complex as in the following diagram. In this diagram, we use d for the boundary operator in $K_\bullet$ so that $-d$ is the boundary operator in $SK_\bullet$ and the squares commute as shown.

$$\begin{array}{ccccccccc}
\cdots \xrightarrow{-d} & K_{n+1} & \xrightarrow{-d} & K_n & \xrightarrow{-d} \cdots \xrightarrow{-d} & K_1 & \xrightarrow{-d} & K_0 \\
& \downarrow f_{n+1} & & \downarrow f_n & & \downarrow f_1 & & \downarrow \\
\cdots \xrightarrow[d]{} & L_n & \xrightarrow[d]{} & L_{n-1} & \xrightarrow[d]{} \cdots \xrightarrow[d]{} & L_0 & \longrightarrow & 0
\end{array}$$

If we replace the $-d$ in the upper row by d, the squares will anticommute and the resultant diagram can be considered as a double complex in which all rows belong to Γ. From AC–5 the total complex also belongs to Γ, but that is just the mapping cone of f which is isomorphic to $C_\bullet$ and hence belongs to Γ.

Now suppose that $L_\bullet$ and $C_\bullet$ belong to Γ. We have just seen that the mapping cone of $L_\bullet \longrightarrow C_\bullet$ is in Γ. It then follows from Proposition 2.11 and AC–3 that $K_\bullet \in \Gamma$. Dually, if $C_\bullet$ and $K_\bullet$ are in Γ, so is $L_\bullet$. □

2.2. Arrows determined by an acyclic class. Given an acyclic class Γ, let Σ denote the class of arrows f whose mapping cone is in Γ. It follows from AC–3 and 4 and the preceding proposition that this class lies between the class of homotopy equivalences and that of homology equivalences.

2.3. PROPOSITION. Σ *is closed under composition.*

PROOF. Suppose that $f = f_\bullet\colon K_\bullet \longrightarrow L_\bullet$ and $g = g_\bullet\colon L_\bullet \longrightarrow M_\bullet$ are each in Σ. Then $C_{f\bullet}$ and $C_{g\bullet}$ are in Γ. The nth term of $S^{-1}C_{f\bullet}$ is $M_{n+1} \oplus L_n$ and that of $C_{f\bullet}$ is $L_n \oplus K_{n-1}$. The boundary operators are

$$\begin{pmatrix} -d & -g \\ 0 & d \end{pmatrix} \text{ and } \begin{pmatrix} d & f \\ 0 & -d \end{pmatrix}$$

respectively, from which one calculates that $h = \left(\begin{smallmatrix} 0 & -1 \\ 0 & 0 \end{smallmatrix}\right)$ is a chain map from $S^{-1}C_{g\bullet} \longrightarrow C_{f\bullet}$:

$$\begin{pmatrix} 0 & -1 \\ 0 & 0 \end{pmatrix}\begin{pmatrix} -d & -g \\ 0 & d \end{pmatrix} = \begin{pmatrix} 0 & -d \\ 0 & 0 \end{pmatrix} = \begin{pmatrix} d & f \\ 0 & -d \end{pmatrix}\begin{pmatrix} 0 & -1 \\ 0 & 0 \end{pmatrix}$$

Thus there is an exact sequence $0 \longrightarrow C_{f\bullet} \longrightarrow C_{h\bullet} \longrightarrow C_{g\bullet} \longrightarrow 0$ and it follows from Proposition 2.1 that $C_{h\bullet} \in \Gamma$. The nth term of $C_{h\bullet}$ is $L_n \oplus K_{n-1} \oplus M_n \oplus L_{n-1}$ and the matrix of the boundary operator is

$$\begin{pmatrix} d & f & 0 & -1 \\ 0 & -d & 0 & 0 \\ 0 & 0 & d & g \\ 0 & 0 & 0 & -d \end{pmatrix}$$

Let $C_{-\mathrm{id}\bullet}$ be the mapping cone of the negative of the identity of L. Thus $(C_{-\mathrm{id}})_n = L_n \oplus L_{n-1}$ and the boundary operator is $\left(\begin{smallmatrix} d & -1 \\ 0 & -d \end{smallmatrix}\right)$. The mapping cone of gf has

$M_n \oplus K_{n-1}$ in degree n and boundary operator $\left(\begin{smallmatrix} d & gf \\ 0 & -d \end{smallmatrix}\right)$. I claim there is an exact sequence

$$0 \longrightarrow C_{gf\bullet} \xrightarrow{i} C_{h\bullet} \xrightarrow{j} C_{-\mathrm{id}\bullet} \longrightarrow 0$$

In fact, let i and j be the maps given by the matrixes

$$i = \begin{pmatrix} 0 & 0 \\ 0 & 1 \\ 1 & 0 \\ 0 & f \end{pmatrix}, \qquad j = \begin{pmatrix} 1 & 0 & 0 & 0 \\ 0 & -f & 0 & 1 \end{pmatrix}$$

Matrix multiplication shows that these are chain maps. The sequences are U-split exact; for example, $\left(\begin{smallmatrix} 0 & 0 & 1 & 0 \\ 0 & 1 & 0 & 0 \end{smallmatrix}\right)$ splits i and it follows from Proposition 2.1 that $C_{gf\bullet} \in \Gamma$ and hence $gf \in \Sigma$. □

The following theorem extends the main results of 3.6.1 to arbitrary acyclic classes.

2.4. THEOREM. *Suppose* $C_{\bullet\bullet} = \{C_{mn} \mid m \geq 0, n \geq 0\}$ *is a double complex that is augmented over the single complex* $C_{-1\bullet}$ *and such that for each* $n \geq 0$, *the complex*

$$\cdots \longrightarrow C_{mn} \longrightarrow C_{m-1\,n} \longrightarrow \cdots \longrightarrow C_{0n} \longrightarrow C_{-1\,n} \longrightarrow 0$$

belongs to Γ. *Then the induced map* $\mathrm{Tot}(C) \longrightarrow C_{-1\bullet}$ *is in* Σ.

PROOF. The mapping cone of the induced map is just the double complex including the augmentation term. From AC–5 it follows that that total double complex is in Γ since each row is. Thus the induced map is in Σ. □

2.5. COROLLARY. *Suppose* $C_{\bullet\bullet} = \{C_{mn} \mid m \geq 0, n \geq 0\}$ *is a double complex that is augmented in each direction over the single complexes* $C_{-1\bullet}$ *and* $C_{\bullet\,-1}$. *Suppose that for each* $n \geq 0$, *both complexes*

$$\cdots \longrightarrow C_{mn} \longrightarrow C_{m-1\,n} \longrightarrow \cdots \longrightarrow C_{0n} \longrightarrow C_{-1\,n} \longrightarrow 0$$

and

$$\cdots \longrightarrow C_{mn} \longrightarrow C_{m\,n-1} \longrightarrow \cdots \longrightarrow C_{n0} \longrightarrow C_{m\,-1} \longrightarrow 0$$

belong to Γ. *Then in* $\Sigma^{-1}\mathscr{C}$, *the two chain complexes* $C_{-1\bullet}$ *and* $C_{\bullet\,-1}$ *are isomorphic.*

2.6. COROLLARY. *Suppose*

$$\cdots \longrightarrow K_{\bullet n} \longrightarrow K_{\bullet n-1} \longrightarrow \cdots \longrightarrow K_{\bullet 0} \longrightarrow K_{\bullet -1} \longrightarrow 0$$

is an exact sequence of chain complexes such that for each $n \geq 0$, *the complex* $K_{\bullet n} \in \Gamma$. *Then* $K_{\bullet -1} \in \Gamma$.

PROOF. We may treat this as a double complex of the form treated in the theorem and the conclusion is that the arrow from the total complex made up from

$$\cdots \longrightarrow K_{\bullet n} \longrightarrow K_{\bullet n-1} \longrightarrow \cdots \longrightarrow K_{\bullet 0}$$

is in Σ. But if for each $n \geq 0$, the complex $K_{\bullet n} \in \Gamma$, then this total complex also belongs to Γ and then so does $K_{\bullet -1}$. □

2.7. COROLLARY. *Suppose we have a commutative diagram of double complexes*

$$\begin{array}{ccccccccccccc} 0 & \longrightarrow & K_{\bullet n} & \longrightarrow & K_{\bullet n-1} & \longrightarrow & \cdots & \longrightarrow & K_{\bullet 0} & \longrightarrow & K_{\bullet -1} & \longrightarrow & 0 \\ & & \downarrow f_n & & \downarrow f_{n-1} & & & & \downarrow f_0 & & \downarrow f_{-1} & & \\ 0 & \longrightarrow & L_{\bullet n} & \longrightarrow & L_{\bullet n-1} & \longrightarrow & \cdots & \longrightarrow & L_{\bullet 0} & \longrightarrow & L_{\bullet -1} & \longrightarrow & 0 \end{array}$$

such that for all $n \geq 0$, the arrow $f_n \in \Sigma$. Then $f_{-1} \in \Sigma$.

PROOF. This follows from the preceding corollary, by using mapping cones. □

Another useful property of Σ is the following.

2.8. PROPOSITION. *Suppose that*

$$\begin{array}{ccccccccc} 0 & \longrightarrow & K_\bullet & \xrightarrow{u} & L_\bullet & \xrightarrow{v} & M_\bullet & \longrightarrow & 0 \\ & & \downarrow f_\bullet & & \downarrow g_\bullet & & \downarrow h_\bullet & & \\ 0 & \longrightarrow & K'_\bullet & \xrightarrow[u']{} & L'_\bullet & \xrightarrow[v']{} & M'_\bullet & \longrightarrow & 0 \end{array}$$

is a commutative diagram with U-split exact rows. If two of the three vertical arrows belong to Σ, so does the third.

PROOF. The mapping cone sequence

$$0 \longrightarrow C_{f\bullet} \xrightarrow{\begin{pmatrix} u & 0 \\ 0 & u' \end{pmatrix}} C_{g\bullet} \xrightarrow{\begin{pmatrix} v & 0 \\ 0 & v' \end{pmatrix}} C_{h\bullet} \longrightarrow 0$$

is readily seen to be exact. The claim now follows immediately from 2.1. □

3. The main theorem

Now let us suppose we are given an acyclic class Γ on $\mathscr{C}$ and that Σ is the associated class of maps. Then $\Sigma^{-1}\mathscr{C}$ is the category of fractions gotten by inverting all the arrows in Σ. From AC–4 and Theorem 3.2.10 it follows that the homology inverts all arrows of Σ and hence that homology factors through $\Sigma^{-1}\mathscr{C}$ as described. In particular, any map in $\Sigma^{-1}\mathscr{C}$ induces a map in homology. We will see in Theorem 4.1 that although Σ does not generally have either a right or left calculus of fractions (see 1.10), it does have the weaker properties of homotopy right and left classes of fractions.

Suppose that $G\colon \mathscr{X} \longrightarrow \mathscr{X}$ is an endofunctor and that $\epsilon\colon G \longrightarrow \mathrm{Id}$ is a natural transformation. If $F\colon \mathscr{X} \longrightarrow \mathscr{A}$ is a functor, we define an augmented chain complex functor we will denote $FG^{\bullet+1} \longrightarrow F$ as the functor that has FG^{n+1} in degree n, for $n \geq -1$. Let $\partial^i = FG^i\epsilon G^{n-i}\colon FG^{n+1} \longrightarrow FG^n$. Then the boundary operator is $\partial = \sum_{i=0}^{n}(-1)^i\partial^i$. If, as usually happens in practice, G and ϵ are 2/3 of a cotriple, then this chain complex is the chain complex associated to a simplicial set built using the comultiplication δ to define the degeneracies. Next suppose that $K_\bullet \longrightarrow K_{-1}$ is an augmented chain complex functor. Then there is a double chain complex functor that has in bidegree (n, m) the term K_nG^{m+1}. This will actually commute since

$$\begin{array}{ccc} K_nG^{m+1} & \xrightarrow{dG^{m+1}} & K_{n-1}G^{m+1} \\ \big\downarrow{\scriptstyle K_nG^i\epsilon G^{m-i}} & & \big\downarrow{\scriptstyle K_{n-1}G^i\epsilon G^{m-i}} \\ K_nG^m & \xrightarrow[dG^m]{} & K_{n-1}G^m \end{array}$$

commutes by naturality for $0 \le i \le m$ and continues to commute with the sums $\sum_{i=0}^{m} K_nG^i\epsilon G^{m-i}$ on the left and $\sum_{i=0}^{m} K_{n-1}G^i\epsilon G^{m-i}$ on the right. However, the usual trick of negating every second column produces an anticommuting double complex.

This is augmented in both directions. The first is $\epsilon: K_\bullet G^{\bullet+1}: K_\bullet G \longrightarrow K_\bullet$; the second is $K_0G^{\bullet+1} \longrightarrow K_{-1}G^{\bullet+1}$. We say that $K_\bullet$ is *ϵ-presentable with respect to* Γ if for each $n \ge 0$, the augmented chain complex $K_nG^{\bullet+1} \longrightarrow K_n \longrightarrow 0$ belongs to Γ. We say that $K_\bullet$ is *G-acyclic with respect to* Γ if the augmented complex $K_\bullet G \longrightarrow K_{-1}G \longrightarrow 0$ belongs to Γ.

The main theorem of this chapter, and in fact, of this book, follows:

3.1. THEOREM. *Let Γ be an acyclic class and Σ be the associated class of arrows. Suppose $\alpha: K_\bullet \longrightarrow K_{-1}$ and $\beta: L_\bullet \longrightarrow L_{-1}$ are augmented chain complex functors. Suppose G is an endofunctor on $\mathscr{X}$ and $\epsilon: G \longrightarrow \mathrm{Id}$ a natural transformation for which $K_\bullet$ is ϵ-presentable and $L_\bullet \longrightarrow L_{-1} \longrightarrow 0$ is G-acyclic, both with respect to Γ. Then given any natural transformation $f_{-1}: K_{-1} \longrightarrow L_{-1}$ there is, in $\Sigma^{-1}\mathscr{C}$, a unique arrow $f_\bullet: K_\bullet \longrightarrow L_\bullet$ that extends f_{-1}.*

PROOF. For all $m \ge 0$, the augmented complex $K_mG^{\bullet+1} \longrightarrow K_m \longrightarrow 0$ belongs to Γ and hence, by AC–5, the total augmented complex $K_\bullet G^{\bullet+1} \longrightarrow K_\bullet \longrightarrow 0$ belongs to Γ whence, by Theorem 3.2.10 the arrow $K_\bullet\epsilon: K_\bullet G^{\bullet+1} \longrightarrow K_\bullet$ belongs to Σ. The same reasoning implies that $\beta G^{\bullet+1}: L_\bullet G^{\bullet+1} \longrightarrow L_{-1}G^{\bullet+1}$ is also in Σ. We can summarize the situation in the diagram

$$\begin{array}{ccccc} K_{-1}G^{\bullet+1} & \xleftarrow{\alpha G^{\bullet+1}} & K_\bullet G^{\bullet+1} & \xrightarrow{K_\bullet\epsilon} & K_\bullet \\ \big\downarrow{\scriptstyle f_{-1}G^{\bullet+1}} & & & & \\ L_{-1}G^{\bullet+1} & \xleftarrow[\beta G^{\bullet+1}]{} & L_\bullet G^{\bullet+1} & \xrightarrow[L_\bullet\epsilon]{} & L_\bullet \end{array}$$

with $K_\bullet\epsilon$ and $\beta G^{\bullet+1}$ in Σ. We now invert Σ to get the map

$$f_\bullet = L_\bullet\epsilon \circ (\beta G^{\bullet+1})^{-1} \circ f_{-1}G^{\bullet+1} \circ \alpha G^{\bullet+1} \circ (K_\bullet\epsilon)^{-1}: K_\bullet \longrightarrow L_\bullet$$

I claim that this map extends f_{-1} in the sense that $f_{-1} \circ \alpha = \beta \circ f_0$ and that $f_\bullet$ is unique with this property. Begin by observing that naturality of α and β imply that $\alpha \circ K_\bullet\epsilon = K_{-1}\epsilon \circ \alpha G^{\bullet+1}$ and $\beta \circ L_\bullet\epsilon = L_{-1}\epsilon \circ \beta G^{\bullet+1}$. Then the first claim follows from the diagram

$$\begin{array}{ccccccccc}
K_\bullet & \xrightarrow{(K_\bullet\epsilon)^{-1}} & K_\bullet G^{\bullet+1} & \xrightarrow{\alpha G^{\bullet+1}} & K_{-1}G^{\bullet+1} & \xrightarrow{f_{-1}G^{\bullet+1}} & L_{-1}G^{\bullet+1} & \underset{\beta G^{\bullet+1}}{\overset{(\beta G^{\bullet+1})^{-1}}{\rightleftarrows}} & L_\bullet G^{\bullet+1} \\
 & \searrow{=} & \downarrow{\scriptstyle K_\bullet\epsilon} & & \downarrow{\scriptstyle K_{-1}\epsilon} & & \downarrow{\scriptstyle L_{-1}\epsilon} & & \downarrow{\scriptstyle L_\bullet\epsilon} \\
 & & K_\bullet & \xrightarrow[\alpha]{} & K_{-1} & \xrightarrow[f_{-1}]{} & L_{-1} & \xleftarrow[\beta]{} & L_\bullet
\end{array}$$

Now suppose that $g_\bullet\colon K_\bullet \longrightarrow L_\bullet$ is another arrow in $\Sigma^{-1}\mathscr{C}$ for which $f_{-1}\circ\alpha = \beta\circ g_0$. Then

$$f_{-1}G^{\bullet+1}\circ\alpha G^{\bullet+1} = \beta G^{\bullet+1}\circ g_\bullet G^{\bullet+1}$$

which implies that

$$(\beta G^{\bullet+1})^{-1}\circ f_{-1}G^{\bullet+1}\circ\alpha G^{\bullet+1} = g_\bullet G^{\bullet+1}$$

and then

$$L_\bullet\epsilon\circ(\beta G^{\bullet+1})^{-1}\circ f_{-1}G^{\bullet+1}\circ\alpha G^{\bullet+1} = L_\bullet\epsilon\circ gG^{\bullet+1} = g_\bullet\circ K_\bullet\epsilon$$

from which we conclude that

$$g_\bullet = L_\bullet\epsilon\circ(\beta G^{\bullet+1})^{-1}\circ f_{-1}G^{\bullet+1}\circ\alpha G^{\bullet+1}\circ(K_\bullet\epsilon)^{-1} = f_\bullet \qquad \square$$

3.2. COROLLARY. *Suppose that $K_\bullet$ and $L_\bullet$ are each ϵ-presentable and G-acyclic on models with respect to Γ. Then any natural isomorphism $f_{-1}\colon K_{-1} \longrightarrow L_{-1}$ extends to a unique isomorphism $f_\bullet\colon K_\bullet \longrightarrow L_\bullet$ in $\Sigma^{-1}\mathscr{C}$. Moreover if $g_\bullet\colon K_\bullet \longrightarrow L_\bullet$ is a natural transformation for which $\beta\circ g_0 = f_{-1}\circ\alpha$, then $g_\bullet = f_\bullet$ in $\Sigma^{-1}\mathscr{C}$.*

PROOF. If f_{-1} is an isomorphism with inverse g_{-1}, then there is a map $f_\bullet\colon K_\bullet \longrightarrow L_\bullet$ that extends f_{-1} and $g_\bullet\colon L_\bullet \longrightarrow K_\bullet$ that extends g_{-1}. Then $g_\bullet\circ f_\bullet$ extends $g_{-1}\circ f_{-1} = \mathrm{id}$, as does the identity so that by the uniqueness of the preceding, we see that in $\Sigma^{-1}\mathscr{C}$, $g_\bullet\circ f_\bullet = \mathrm{id}$. Similarly, $f_\bullet\circ g_\bullet = \mathrm{id}$ in the fraction category. This shows that $K_\bullet \cong L_\bullet$. The second claim is obvious. $\square$

3.3. Other conditions. Sometimes, other conditions that are easy to verify can replace the stated ones. Here is one that is required to recover the form of the acyclic models theorem from [Barr & Beck, 1966].

3.4. THEOREM. *Suppose $G\colon \mathscr{X} \longrightarrow \mathscr{X}$ is a functor and $\epsilon\colon G \longrightarrow \mathrm{Id}$ is a natural transformation. Then for any functor $C\colon \mathscr{X} \longrightarrow \mathscr{A}$, $CG^{\bullet+1} \longrightarrow C \longrightarrow 0$ is contractible if and only if $C\epsilon$ splits.*

PROOF. The necessity of the condition is obvious. If $C\epsilon$ splits, let $\theta\colon C \longrightarrow C\epsilon$ be an arrow such that $C\epsilon\circ\theta = \mathrm{id}$. Let $s = \theta G^n\colon CG^n \longrightarrow CG^{n+1}$. Then $\partial^0\circ s = C\epsilon G^n\circ\theta G^n = \mathrm{id}$ and for $i > 0$,

$$\begin{aligned}
\partial^i\circ s &= CG^i\epsilon\circ G^{n-i}\circ\theta G^n = (CG^i\epsilon\circ\theta G^i)G^{n-i} \\
&= (\theta G^{i-1}\circ CG^{i-1}\epsilon)G^{n-i} = \theta G^{n-1}\circ CG^{i-1}\epsilon G^{n-i} = s\circ\partial^{i-1}
\end{aligned}$$

using naturality of θ. Then

$$\begin{aligned}\partial \circ s + s \circ \partial &= \sum_{i=0}^{n}(-1)^i \partial^i \circ s + \sum_{i=0}^{n-1} s \circ \partial^i \\ &= \mathrm{id} + \sum_{i=1}^{n}(-1)^i s \circ \partial^{i-1} + \sum_{i=0}^{n-1}(-1)^i s \circ \partial^i = 1\end{aligned}$$

□

3.5. COROLLARY. *Let $K_\bullet \longrightarrow K_{-1} \longrightarrow 0$ and $L_\bullet \longrightarrow L_{-1} \longrightarrow 0$ be augmented chain complex functors such that $GK_n \longrightarrow K_n$ is split epic for all $n \geq 0$ and $L_\bullet \longrightarrow L_{-1} \longrightarrow 0$ is G-contractible. Then any natural transformation $f_{-1}: K_{-1} \longrightarrow L_{-1}$ extends to a natural chain transformation $f_\bullet: K_\bullet \longrightarrow L_\bullet$ and any two extensions of f_{-1} are naturally homotopic.* □

If $\mathbf{G} = (G, \epsilon, \delta)$ is actually a cotriple, then it can be used to build a resolution that is automatically both acyclic on models and presentable with respect to homotopy.

3.6. THEOREM. *Let $E: \mathscr{X} \longrightarrow \mathscr{A}$ be any functor and $\mathbf{G} = (G, \epsilon, \delta)$ be cotriple on $\mathscr{X}$. Then the chain complex $EG^{\bullet+1}$ that has EG^{n+1} in degree n and boundary $\sum_{i=0}^{n}(-1)^i E\partial^i: EG^{n+1} \longrightarrow EG^n$ is G-acyclic on models and G-presentable with respect to homotopy.*

PROOF. To verify the presentability, it is sufficient, by Theorem 3.3 above, to give a map $\theta_n: EG^{n+1} \longrightarrow EG^{n+2}$ for each n such that $EG^{n+1}\epsilon \circ \theta_n =$ id. Evidently, $EG^n\delta$ is such a map. As for the acyclicity, again the arrows $\theta_n: EG^{n+1} \longrightarrow EG^{n+2}$ give a contracting homotopy in the complex $EG^{\bullet+1}G \longrightarrow EG \longrightarrow 0$. See the proof of 3.3 for details. □

4. Homotopy calculuses of fractions

We saw in Section 10 of Chapter 1 what a calculus of fractions is. In the cases considered here there is no calculus of fractions (left or right), but there is the next best thing, homotopy left and right calculuses of fractions. We will write $f \sim g$ when f and g are homotopic and $C \simeq D$ to mean that there are chain maps $f: C \longrightarrow D$ and $g: D \longrightarrow C$ such that $g \circ f \sim \mathrm{id}_C$ and $f \circ g \sim \mathrm{id}_D$. Of course, C is contractible if and only if $C \simeq 0$.

We will say that Σ has a **homotopy left calculus of fractions** if

1. Σ is closed under composition;
2. whenever $\sigma \in \Sigma$ and f are arrows with the same domain, there is a not necessarily commutative square

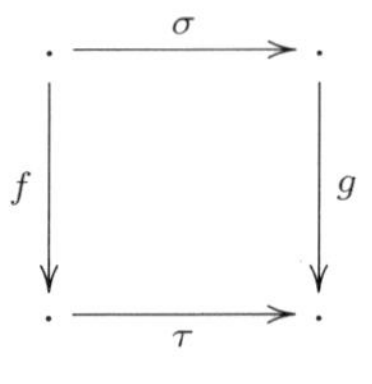

with $\tau \in \Sigma$ and for which $\tau \in \Sigma$ and $g \circ \sigma \sim \tau \circ f$;

3. for any diagram $\cdot \xrightarrow{\sigma} \cdot \underset{g}{\overset{f}{\rightrightarrows}} \cdot$ with $\sigma \in \Sigma$ such that $f \circ \sigma \sim g \circ \sigma$, there is a diagram $\cdot \underset{g}{\overset{f}{\rightrightarrows}} \cdot \xrightarrow{\tau}$ with $\tau \in \Sigma$ such that $\tau \circ f \sim \tau \circ g$.

Dually, we will say that the composition closed class Σ has a **homotopy right calculus of fractions** if, whenever $\sigma \in \Sigma$ and f are arrows with the same codomain, there is a square

$$\begin{array}{ccc} \cdot & \xrightarrow{g} & \cdot \\ {\scriptstyle\tau}\big\downarrow & & \big\downarrow{\scriptstyle\sigma} \\ \cdot & \xrightarrow[f]{} & \cdot \end{array}$$

with $\tau \in \Sigma$ for which $f \circ \tau \sim \sigma \circ g$ and if, for any $\cdot \underset{g}{\overset{f}{\rightrightarrows}} \cdot \xrightarrow{\sigma} \cdot$ with $\sigma \in \Sigma$ such that $\sigma \circ f \sim \sigma \circ g$, there is a diagram $\cdot \xrightarrow{\tau} \cdot \underset{g}{\overset{f}{\rightrightarrows}} \cdot$ with $\tau \in \Sigma$ such that $f \circ \tau \sim g \circ \tau$.

4.1. THEOREM. *Every acyclic class has homotopy left and right calculuses of fractions.*

The presence of the homotopy left calculus of fractions will be demonstrated by a series of propositions. The homotopy right calculus of fractions is dual.

4.2. PROPOSITION. *Suppose $L_\bullet \xleftarrow{\sigma} N_\bullet \xrightarrow{f} M_\bullet$ are maps of chain complexes with $\sigma \in \Sigma$. Then there is a homotopy commutative square*

$$\begin{array}{ccc} N_\bullet & \xrightarrow{f} & M_\bullet \\ {\scriptstyle\sigma}\big\downarrow & & \big\downarrow{\scriptstyle\tau} \\ L_\bullet & \xrightarrow[g]{} & K_\bullet \end{array}$$

with $\tau \in \Sigma$.

PROOF. Let $K_\bullet$ be the mapping cone of $\binom{f}{\sigma}: N_\bullet \longrightarrow M_\bullet \oplus L_\bullet$. Then $K_\bullet$ is the chain complex whose nth term is $K_n = M_n \oplus L_n \oplus N_{n-1}$, with boundary operator given by the matrix $D = \begin{pmatrix} d & 0 & f \\ 0 & d & \sigma \\ 0 & 0 & -d \end{pmatrix}$. Let $\tau = \begin{pmatrix} 1 \\ 0 \\ 0 \end{pmatrix}: M_\bullet \longrightarrow K_\bullet$, $g = \begin{pmatrix} 0 \\ -1 \\ 0 \end{pmatrix}: L_\bullet \longrightarrow K_\bullet$, and $H = \begin{pmatrix} 0 \\ 0 \\ 1 \end{pmatrix}: UN_\bullet \longrightarrow UK_\bullet$. Then it is immediate that D is a boundary operator and that g and τ are chain maps. We compute that

$$\begin{aligned} D \circ H + H \circ d &= \begin{pmatrix} d & 0 & f \\ 0 & d & \sigma \\ 0 & 0 & -d \end{pmatrix} \begin{pmatrix} 0 \\ 0 \\ 1 \end{pmatrix} + \begin{pmatrix} 0 \\ 0 \\ 1 \end{pmatrix} d \\ &= \begin{pmatrix} 0 \\ 0 \\ d \end{pmatrix} + \begin{pmatrix} f \\ \sigma \\ -d \end{pmatrix} = \begin{pmatrix} f \\ \sigma \\ 0 \end{pmatrix} = \begin{pmatrix} f \\ 0 \\ 0 \end{pmatrix} - \begin{pmatrix} 0 \\ -\sigma \\ 0 \end{pmatrix} \\ &= \tau \circ f - g \circ \sigma \end{aligned}$$

We still have to show that $\tau \in \Sigma$. But there is obviously a U-split exact sequence

$$0 \longrightarrow M_\bullet \xrightarrow{\tau} K_\bullet \xrightarrow{\left(\begin{smallmatrix} 0 & 1 & 0 \\ 0 & 0 & 1 \end{smallmatrix}\right)} C_{\sigma\bullet} \longrightarrow 0$$

which can readily be calculated to consist of chain morphisms. It follows from Proposition 3.2.11 that $C_{\tau\bullet}$ is homotopic to $C_{\sigma\bullet}$, so that AC–3 implies that $C_{\tau\bullet} \in \Gamma$. □

We note, for future reference, that since f and σ enter this proof symmetrically, if $f \in \Sigma$, then also $g \in \Sigma$.

4.3. PROPOSITION. *Given any diagram*

$$N_\bullet \xrightarrow{\sigma} L_\bullet \underset{g}{\overset{f}{\rightrightarrows}} M_\bullet$$

with $\sigma \in \Sigma$ and $f \circ \sigma$ homotopic to $g \circ \sigma$, there is a diagram

$$L_\bullet \underset{g}{\overset{f}{\rightrightarrows}} M_\bullet \xrightarrow{\tau} K_\bullet$$

with $\tau \in \Sigma$ and $\tau \circ f$ homotopic to $\tau \circ g$.

PROOF. We can use the additivity of the category to replace f by $f - g$ and reduce the assertion to the case that $g = 0$. Then our hypotheses are that $\sigma \in \Sigma$ and $f \circ \sigma$ is null homotopic. We show that there is a $\tau \colon L_\bullet \longrightarrow K_\bullet$ in Σ such that $\tau \circ f$ is null homotopic. Since $f \circ \sigma$ is null homotopic, there is an $h \colon USN_\bullet \longrightarrow UM_\bullet$ such that $f \circ \sigma = h \circ d + d \circ h$. Since $\sigma \in \Sigma$, the mapping cone $C_{\sigma\bullet} \in \Gamma$ by Theorem 2.10. One easily sees by direct computation that the square

$$\begin{array}{ccc} L_\bullet \oplus SN_\bullet & \xrightarrow{(\,f \;\; h\,)} & M_\bullet \\ {\scriptstyle \left(\begin{smallmatrix} d & \sigma \\ 0 & -d \end{smallmatrix}\right)}\Big\downarrow & & \Big\downarrow{\scriptstyle d} \\ L_\bullet \oplus SN_\bullet & \xrightarrow[(\,f \;\; h\,)]{} & M_\bullet \end{array}$$

commutes. Thus $u = (\,f \quad h\,) \colon C_{\sigma\bullet} \longrightarrow M_\bullet$ is a map of chain complexes. Let $K_\bullet$ be the mapping cone of u. This gives us a U-split exact sequence

$$0 \longrightarrow M_\bullet \xrightarrow{\tau} K_\bullet \longrightarrow SC_{\sigma\bullet} \longrightarrow 0$$

Since $C_{\sigma\bullet} \in \Gamma$ so is $SC_{\sigma\bullet}$ and it follows from the converse part of Theorem 3.2.10 that $\tau \in \Sigma$. In order to see that $\tau \circ f$ is null homotopic, we actually calculate $K_\bullet$ and τ. In fact, $UK_\bullet = UM_\bullet \oplus USC_{\sigma\bullet}$ and has boundary operator $\left(\begin{smallmatrix} d & u \\ 0 & -d \end{smallmatrix}\right)$. When we replace $C_{\sigma\bullet}$ by its components, we have $UK_\bullet = UM_\bullet \oplus USL_\bullet \oplus US^2N_\bullet$ and the boundary is

$$D = \begin{pmatrix} d & f & h \\ 0 & -d & -\sigma \\ 0 & 0 & d \end{pmatrix}$$

with $\tau = \begin{pmatrix} 1 \\ 0 \\ 0 \end{pmatrix}$. Then $\tau \circ f = \begin{pmatrix} f \\ 0 \\ 0 \end{pmatrix}$. Let $H = \begin{pmatrix} 0 \\ 1 \\ 0 \end{pmatrix}$. Then

$$H \circ D + d \circ H = \begin{pmatrix} 0 \\ 1 \\ 0 \end{pmatrix} d + \begin{pmatrix} d & f & h \\ 0 & -d & -\sigma \\ 0 & 0 & d \end{pmatrix} \begin{pmatrix} 0 \\ 1 \\ 0 \end{pmatrix} = \begin{pmatrix} f \\ -d \\ 0 \end{pmatrix} + \begin{pmatrix} 0 \\ d \\ 0 \end{pmatrix} = \begin{pmatrix} f \\ 0 \\ 0 \end{pmatrix}$$

which shows that $\tau \circ f$ is null homotopic. □

This finishes the proof of 4.1, but there is a bit more to be learned from the developments in this section.

4.4. PROPOSITION. *Homotopic maps become equal in $\Sigma^{-1}\mathscr{C}$.*

PROOF. We apply the construction used in the proof of 4.2 to $L_\bullet \xleftarrow{1} L_\bullet \xrightarrow{1} L_\bullet$ to give the homotopy commutative square

$$\begin{array}{ccc} L_\bullet & \xrightarrow{1} & L_\bullet \\ {\scriptstyle 1}\big\downarrow & & \big\downarrow{\scriptstyle \sigma} \\ L_\bullet & \xrightarrow[\tau]{} & K_\bullet \end{array}$$

with both $\sigma = \begin{pmatrix} 1 \\ 0 \\ 0 \end{pmatrix}$ and $\tau = \begin{pmatrix} 0 \\ -1 \\ 0 \end{pmatrix} \in \Sigma$. Moreover, the map $f = (\,1\ -1\ 0\,) \colon K_\bullet \longrightarrow L_\bullet$ is a chain map such that $f \circ \sigma = f \circ \tau = \mathrm{id}$. In $\Sigma^{-1}\mathscr{C}$, σ and τ are invertible, whence $f = \sigma^{-1} = \tau^{-1}$ so that $\sigma = \tau$. Now suppose that we are given two homotopic chain maps g, $h\colon L_\bullet \longrightarrow M_\bullet$. Then there is a map $s\colon UK_\bullet \longrightarrow US^{-1}L_\bullet$ such that $g - h = s \circ d + d \circ s$. Define the map $k = (\,g\ -h\ s\,) \colon K_\bullet \longrightarrow M_\bullet$. We have that

$$k \circ D = (\,g \quad -h \quad s\,) \begin{pmatrix} d & 0 & 1 \\ 0 & d & 1 \\ 0 & 0 & -d \end{pmatrix} = (\,gd \quad -hd \quad f - g - sd\,)$$
$$= (\,dg \quad -dh \quad ds\,) = d \circ k$$

so that k is a chain map. Evidently, $k \circ \sigma = g$ and $k \circ \tau = h$ so that when we invert homotopy equivalences and $\sigma = \tau$, then also $g = h$. □

4.5. PROPOSITION. *Suppose that $f \sim g$ and $g \sim h$. Then $f \sim h$.*

PROOF. Assume that $f - g = sd + ds$ and $g - h = td + dt$. Then

$$(s + t)d + d(s + t) = sd + ds + td + dt = f - g + g - h = f - h$$ □

4.6. PROPOSITION. *Suppose $K'_\bullet \xrightarrow{h} K_\bullet \overset{f}{\underset{g}{\rightrightarrows}} L_\bullet \xrightarrow{k} L'_\bullet$ are chain maps such that $f \sim g$. Then $k \circ f \circ h \sim k \circ g \circ h$.*

PROOF. Assuming that $f - g = sd + ds$, we have

$$kfh - kgh = k(f - g)h = k(sd + ds)h = ksdh + ksdh = kshd + kshd$$

since h and k commute with d. Thus ksh is the required homotopy. □

It follows that we can form the quotient category $\mathscr{C}/\sim$ with the same objects as $\mathscr{C}$ and homotopy classes of maps as arrows.

4.7. THEOREM. *Suppose $\Sigma_0 \subseteq \mathscr{C}$ is the class of homotopy equivalences. Then $\mathscr{C}/\sim$ is equivalent to $\Sigma^{-1}\mathscr{C}$. In particular, a parallel pair of maps are homotopic in $\mathscr{C}$ if and only if they become equal in $\Sigma^{-1}\mathscr{C}$.*

PROOF. Let Γ_0 denote the class of contractible complexes. We have just seen that homotopic maps become equal in $\Sigma_0^{-1}\mathscr{C}$ so that the canonical functor $\mathscr{C} \longrightarrow \Sigma_0^{-1}\mathscr{C}$ factors through $\mathscr{C}/\sim$. Conversely, if $\sigma \in \Sigma_0$, it has a homotopy inverse τ, so that $\sigma \circ \tau$ and $\tau \circ \sigma$ are homotopic to identity arrows and hence those composites are identity arrows in $\mathscr{C}/\sim$. Thus the maps in Σ become invertible in $\mathscr{C}/\sim$, which means that the canonical functor $\mathscr{C} \longrightarrow \mathscr{C}/\sim$ factors through $\Sigma_0^{-1}\mathscr{C}$. Thus the categories $\mathscr{C} \longrightarrow \mathscr{C}/\sim$ and $\Sigma_0^{-1}\mathscr{C}$ are homotopic. □

4.8. THEOREM. *For any acyclic class Σ on $\mathscr{C}$, we have that $\Sigma^{-1}\mathscr{C}$ is equivalent to $\Sigma^{-1}(\mathscr{C}/\sim)$ and, in the latter category, Σ has calculuses of right and left fractions.*
□

4.9. COROLLARY. *Every map in $\Sigma^{-1}\mathscr{C}$ has the form $f \circ \sigma^{-1}$ where $f \in \mathscr{C}$ and $\sigma \in \Sigma$. Moreover, $f \circ \sigma^{-1} = f' \circ \sigma'^{-1}$ in $\sigma^{-1}\mathscr{C}$ if and only if there is a homotopy commutative diagram in $\mathscr{C}$:*

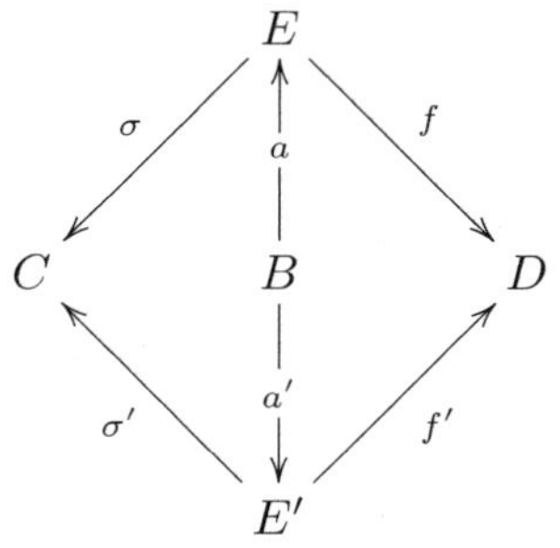

for which $a \circ \sigma$ (and therefore $a' \circ \sigma'$) belongs to Σ. □

4.10. COROLLARY. *Every map in $\Sigma^{-1}\mathscr{C}$ has the form $\tau^{-1} \circ g$ where $f \in \mathscr{C}$ and $\sigma \in \Sigma$. Moreover, $\tau^{-1} \circ g = \tau'^{-1} \circ g'$ in $\sigma^{-1}\mathscr{C}$ if and only if there is a homotopy commutative diagram*

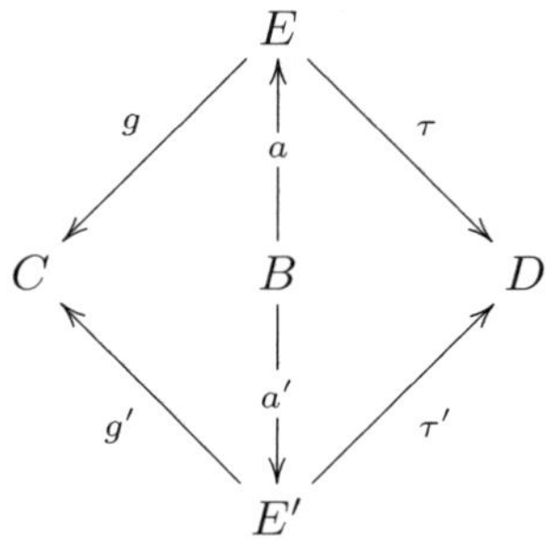

for which $a \circ \tau$ (and therefore $a' \circ \tau'$) belongs to Σ. □

These facts hold despite the fact that there is no calculus of left—or dually of right—fractions in this case. For example, in the proof of Proposition 4.4, the homotopy equivalence f coequalizes σ and τ, but only the 0 map equalizes them and that is a homotopy equivalence if and only if $K_\bullet$ is contractible.

5. Exactness conditions

In this section, we consider conditions that simplify the verification of the main hypotheses of the acyclic models theorem. If $\mathbf{G}$ is a cotriple on an additive category $\mathscr{A}$, by the standard $\mathbf{G}$-resolution of an object A of $\mathscr{A}$, we mean the chain complex

$$\cdots \longrightarrow G^{n+1}A \longrightarrow G^nA \longrightarrow \cdots \longrightarrow GA$$

with boundary operator

$$d = \sum_{i=0}^{n}(-1)^i G^i \epsilon G^{n-i}A\colon G^{n+1}A \longrightarrow G^nA$$

This is augmented over A by $\epsilon A\colon GA \longrightarrow A$ and by the augmented standard $\mathbf{G}$-resolution, we mean that augmented complex.

5.1. Homology. We do the case of homology first, since the argument for quasi-homotopy depends on it.

5.2. THEOREM. *Suppose that $\mathscr{A}$ is an abelian category and $\mathbf{G}$ is a cotriple on $\mathscr{A}$. A necessary and sufficient condition that the augmented standard complex be acyclic for each A is that ϵA be an epimorphism for each A.*

PROOF. We begin by showing that for any object A, there is an exact sequence

$$\cdots \longrightarrow GA_n \longrightarrow GA_{n-1} \longrightarrow \cdots \longrightarrow GA_0 \longrightarrow A \longrightarrow 0$$

We begin with $A_0 = A$ and $\epsilon A\colon GA \longrightarrow A$. If we have $d\colon GA_{n-1} \longrightarrow GA_{n-2}$ already constructed, let $A_n = \ker d$ and then the next term is the composite $GA_n \xrightarrow{\epsilon A_n} A_n \longrightarrow GA_{n-1}$. Since ϵA_n is an epimorphism, it follows that $GA_n \longrightarrow GA_{n-1} \longrightarrow GA_{n-2}$ is exact.

The remainder of the argument will follow from a sequence of propositions.

5.3. PROPOSITION. *For any object B, the chain complex of abelian groups*

$$\cdots \longrightarrow \mathrm{Hom}(GB, GA_n) \longrightarrow \mathrm{Hom}(GB, GA_{n-1}) \longrightarrow \cdots$$
$$\longrightarrow \mathrm{Hom}(GB, GA) \longrightarrow \mathrm{Hom}(GB, A) \longrightarrow 0$$

is contractible.

PROOF. The splitting of $\mathrm{Hom}(GB, GA) \longrightarrow \mathrm{Hom}(GB, A)$ has already been done. Suppose we have maps $s_i\colon \mathrm{Hom}(GB, GA_i) \longrightarrow \mathrm{Hom}(GB, GA_{i+1})$ for $i = 1, \ldots, n-1$ given such that $ds_i + s_{i-1}d = \mathrm{id}$ for $i = 0, \ldots, n-1$. For $f\colon GB \longrightarrow GA_n$, we define $s_nf\colon GB \longrightarrow GA_{n+1}$ as follows. Let $u_{n+1}\colon A_{n+1} \longrightarrow GA_n$ be the inclusion of the kernel of d. Since

$$d \circ (f - s_{n-1}df) = d \circ f - ds_{n-1}d \circ f = d \circ f - (d \circ f - s_{n-2}d \circ d \circ f) = 0$$

there is a unique $g\colon GB \longrightarrow A_{n+1}$ such that $u_{n+1} \circ g = f - s_{n-1}d \circ f$. Finally, let $s_nf = Gg \circ \delta B$. Note that $u_{n+1} \circ \epsilon A_{n+1} = d$. Then

$$d \circ s_nf = u_{n+1} \circ \epsilon A_{n+1} \circ Gg \circ \delta B = u_{n+1} \circ g \circ \epsilon GB \circ \delta B = f - s_{n-1}d \circ f$$

so that $d \circ s + s \circ d = \mathrm{id}$ as required. □

5.4. PROPOSITION. *For any object B, the chain complex of abelian groups*

$$\cdots \longrightarrow \operatorname{Hom}(GB, G^{n+1}A) \longrightarrow \operatorname{Hom}(GB, G^n A) \longrightarrow \cdots$$
$$\longrightarrow \operatorname{Hom}(GB, GA) \longrightarrow \operatorname{Hom}(GB, A) \longrightarrow 0$$

is contractible.

PROOF. In fact, for $f: GB \longrightarrow G^n A$, let $s_n f = Gf \circ \delta B: GB \longrightarrow G^{n+1}A$. Then

$$\epsilon G^n A \circ s_n f = \epsilon G^n A \circ Gf \circ \delta B = f \circ \epsilon GB \circ \delta B = f$$

while

$$G^{i+1}\epsilon G^{n-i} \circ s_n f = G^{i+1}\epsilon G^{n-i} \circ Gf \circ \delta B = G(G^i \epsilon G^{n-i} \circ f) \circ \delta B$$
$$= s_{n-1}(G^i \epsilon G^{n-i} \circ f)$$

from which it is immediate that $d \circ s_n f = f - s_{n-1} d \circ f$ so that $s_\bullet$ is a contracting homotopy. □

The following proposition can be considered as an early version of acyclic models.

5.5. PROPOSITION. *Suppose $A_\bullet \longrightarrow A_{-1} \longrightarrow 0$ and $B_\bullet \longrightarrow B_{-1} \longrightarrow 0$ are chain complexes in an abelian category. Suppose for all $n \geq 0$, the complex of abelian groups $\operatorname{Hom}(A_n, B_\bullet) \longrightarrow \operatorname{Hom}(A_n, B_{-1}) \longrightarrow 0$ is exact. Then any map $f_{-1}: A_{-1} \longrightarrow B_{-1}$ can be extended to a chain map $f_\bullet: A_\bullet \longrightarrow B_\bullet$ and any two extensions are homotopic.*

PROOF. Since $\operatorname{Hom}(A_0, B_0) \longrightarrow \operatorname{Hom}(A_0, B_{-1}) \longrightarrow 0$ is exact, there is an element $f_0 \in \operatorname{Hom}(A_0, B_0)$ such that $d \circ f_0 = f_{-1} \circ d$. Suppose $f_i: A_i \longrightarrow B_i$ has been constructed for $i < n$. Since $\operatorname{Hom}(A_n, B_n) \longrightarrow \operatorname{Hom}(A_n, B_{n-1}) \longrightarrow \operatorname{Hom}(A_n, B_{n-2})$ is exact and $f_{n-1} \circ d \in \operatorname{Hom}(A_n, B_{n-1})$ is a map such that $d \circ f_{n-1} \circ d = f_{n-2} \circ d \circ d = 0$, there is an arrow $f_n: A_n \longrightarrow B_n$ such that $d \circ f_n = f_{n-1} \circ d$. This proves the existence. Now suppose that $g_\bullet$ is another extension. Let $h_{-1} = 0: A_{-1} \longrightarrow B_0$. Suppose that $h_i: A_i \longrightarrow B_{i+1}$ for $i < n$ such that $d \circ h_i + h_{i-1} \circ d = f_i - g_i$ for $i < n$. Since $\operatorname{Hom}(A_n, B_{n+1}) \longrightarrow \operatorname{Hom}(A_n, B_n) \longrightarrow \operatorname{Hom}(A_n, B_{n-1})$ is exact and

$$\begin{aligned} d \circ (f_n - g_n - h_{n-1} \circ d) &= d \circ f_n - d \circ g_n - d \circ h_{n-1} \circ d \\ &= f_{n-1} \circ d - g_{n-1} \circ g - (h_{n-2} \circ d - f_{n-1} - g_{n-1}) \circ d \\ &= 0 \end{aligned}$$

there is an arrow $h_n: A_n \longrightarrow B_{n+1}$ such that $d \circ h_{n-1} = f_n - g_n - h_{n-1} \circ d$ as required. □

5.6. COROLLARY. *Let $\mathbf{G}$ be a cotriple on the abelian category $\mathscr{A}$ such that ϵA is epic for every object A of $\mathscr{A}$. Then the standard resolution $G^{\bullet+1}A \longrightarrow A \longrightarrow 0$ is exact for each object A of $\mathscr{A}$.*

PROOF. Since there is an exact sequence $GA_\bullet \longrightarrow \mathscr{A} \longrightarrow 0$, the identity map of $A \longrightarrow A$ extends to maps $G^{\bullet+1}A \longrightarrow GA_\bullet$, as well as in the other direction. Each composite extends the identity map of A and is thus homotopic to the identity. Thus the two complexes are homotopic. But the second is exact by construction and hence so is the first. □

We can now complete the argument. If $\mathscr{X}$ is not small, replace it by any small full subcategory that is closed under G and contains the object X. For example, the objects of the form $G^n X$ will do. Thus we can suppose that $\mathscr{X}$ is small. Now let $\widehat{\mathscr{A}}$ be the functor category $\mathrm{Fun}(\mathscr{X}, \mathscr{A})$, which is easily seen to be abelian when $\mathscr{A}$ is. Let $\widehat{\mathbf{G}} = (\widehat{G}, \widehat{\epsilon}, \widehat{\delta})$ be the cotriple on $\widehat{\mathscr{A}}$ defined by $\widehat{G}K_{\bullet} = K_{\bullet}G$ with the obvious $\widehat{\epsilon}$ and $\widehat{\delta}$. Since homotopic complexes have isomorphic homology, it follows that the first is also exact. Thus the chain complex $\widehat{G}^{\bullet+1}K_{\bullet} \longrightarrow \widehat{G}K_{\bullet} \longrightarrow 0$ is exact; but this is exactly $K_{\bullet}G^{\bullet+1} \longrightarrow K_{\bullet} \longrightarrow 0$. □

CHAPTER 6

Cartan–Eilenberg Cohomology

During a four year period more than fifty years ago, a series of papers appeared describing cohomology theories for associative algebras [Hochschild, 1945, 1946], groups [Eilenberg & Mac Lane, 1947], and Lie algebras [Chevalley & Eilenberg, 1948]. Each one described an n-cochain as a function of n variables taking values in a module. There were some differences in that in the case of associative algebra, the cochains were n-linear and in case of Lie algebras they were both n-linear and skew symmetric. But the real differences were in the coboundary operators. Those for associative algebras and groups were essentially the same but the one for Lie algebras was entirely different. The skew symmetric cochains are not closed under any coboundary operator similar to the one used for associative algebras and groups. On the other hand, showing that the square of the Lie coboundary operator is 0 requires not only the skew symmetry of the multiplication, but also the Jacobi identity. In a similar way, when Harrison [1962] created his cohomology theory for commutative algebras, to be discussed in the next chapter, he could not simply take symmetric cochains; there is no obvious coboundary operator that preserves symmetry and also has square 0.

By the time of their 1956 book, Cartan and Eilenberg had found a uniform treatment of the cohomology that included the three examples described above. Suppose $\mathscr{X}$ denotes one of the categories of groups, associative algebras, or Lie algebras. Then for each object X of $\mathscr{X}$, they describe an associative algebra X^e and a canonical X^e-projective resolution

$$\cdots \longrightarrow A_n \longrightarrow A_{n-1} \longrightarrow \cdots \longrightarrow A_1 \longrightarrow A_0 \longrightarrow A \longrightarrow 0$$

of a canonical X^e module A. These had two properties. First, that a left X^e module is the same thing as a coefficient module for the cohomology and second, that for any such module M, the sequence

$$0 \longrightarrow \mathrm{Hom}_{X^e}(A_0, M) \longrightarrow \mathrm{Hom}_{X^e}(A_1, M) \longrightarrow \cdots \longrightarrow \mathrm{Hom}_{X^e}(A_n, M) \longrightarrow \cdots$$

is isomorphic to one of the resolutions used to define the original cohomology. (In the cases of groups and associative algebras, you have to cochains that have been normalized in the sense of the Dold-Puppe theorem as described in the previous chapter.) It follows that in each case the cohomology can be described as $\mathrm{Ext}_{X^e}(A, M)$.

There are two *ad hoc* elements to this description of cohomology. One is the associated algebra X^e. It is always the case that the category of coefficient modules is equivalent to the category of left X^e-modules. This determines X^e, at least up to Morita equivalence. (Two rings are Morita equivalent if and only if one—and hence each—is the ring of endomorphisms of a finitely generated projective generator of the other; essentially, a generalized matrix algebra.) This does not matter much since Ext will not change if a ring is replaced by a Morita equivalent ring.

The second arbitrary element is the choice of the module A. There does not seem to be any such easy characterization of A. However, there is a very interesting observation, due essentially to Jon Beck. In each case, as it happens, $A_0 = X^e$ and the kernel of the map $A_0 \longrightarrow A$ is a module we will call Diff_X, which has the property that $\mathrm{Hom}_{X^e}(\mathrm{Diff}_X, M)$ is isomorphic to the group of *derivations* of X to M. Derivations are defined slightly differently in each of the categories, but we will leave the details to the individual examples. The conclusion is that $\mathrm{Ext}^n_{X^e}(\mathrm{Diff}_X, M) \cong H^{n+1}(X, M)$ for all $n \geq 1$. For $n = 0$, there is some difference, which we will describe later.

Therefore it was of great interest that Jon Beck discovered, in his 1967 thesis, a notion of module over an object in an arbitrary category that also provides an abstract definition of derivation. This would have enabled him to define cohomology theories in a wide class of categories as $\mathrm{Ext}(\mathrm{Diff}_X, -)$. The reason this was not done is that by 1967 other cohomology theories had been defined, none of which followed the Cartan-Eilenberg pattern. One of them, the cohomology theory of commutative algebras, will be discussed in great detail in the next chapter. Our first task here is to describe Beck's notion of module.

1. Beck modules

1.1. The definition of Beck modules. We begin Beck's theory by looking at examples.

If K is a commutative ring, A is an associative K-algebra, and M is a two sided A-module, then the **split singular extension** of A with kernel M is the ring B that is, as an abelian group, just $A \times M$, and whose multiplication is given by $(a, m)(a', m') = (aa', am' + ma')$. If we identify M with $0 \times M$, then M is a 2-sided ideal of B with $M^2 = 0$. Since M is an ideal of B, it is a B-module. Since M annihilates this module, it is a $B/M \cong A$-module. The module structure is evidently the original structure. Beck discovered that the B that arise in this way can be characterized as the abelian group objects in the slice category Alg_K/A, where Alg_K is the category of associative K-algebras.

If $p: C \longrightarrow A$ is an algebra over A, by a *p-derivation*, or simply derivation of C into a M, we mean a linear map $\tau: C \longrightarrow M$ such that $\tau(cc') = p(c)\tau(c') + \tau(c)p(c')$. If p is understood, we often write $\tau(cc') = c\tau(c') + \tau(c)c'$ with the understanding that p induces a C-module structure on any A-module. A particular case is that of a derivation of A into M.

If $p: C \longrightarrow A$ is an algebra homomorphism and $\tau: C \longrightarrow M$ is a function into the A-module M, let B be the split singular extension as described above. Let $q: C \longrightarrow B$ be the function defined by $q(c) = (p(c), \tau(c))$. Then one sees immediately that q is a K-linear function if and only if τ is and from

$$q(c)q(c') = (p(c)p(c'), p(c)\tau(c') + \tau(c)p(c'))$$

that p is an algebra homomorphism if and only if τ is a derivation. Thus we see that $\mathrm{Hom}(C \longrightarrow A, B \longrightarrow A)$ is just the abelian group $\mathrm{Der}(C, M)$ of p-derivations of C to M.

For the second example, suppose that π is a group and M is a left π-module. Let Π denote the group whose underlying set is $\pi \times M$ and whose multiplication is given by $(x, m)(x', m') = (xx', x'^{-1}m + m')$. The identity element is $(1, 0)$ and one can calculate that $(x, m)^{-1} = (x^{-1}, -xm)$. Then M, identified as $1 \times M$, is a commutative normal subgroup of Π. The group Π acts on M by conjugation. Since the action of M on itself is trivial, this action gives an action of $\Pi/M \cong \pi$ on M. This action can be calculated to be the original action. Beck discovered that the algebras that arise in this way are the abelian group objects in the category Grp/π.

For $p: \Phi \longrightarrow \pi$, a function $\tau: \Phi \longrightarrow M$ is a *p-derivation* or simply derivation if $\tau(xy) = p(x)\tau(y) + \tau(x)$. Notice that this reduces to the same formula for the associative algebra case if we take the right action on a module to be the identity. An older name for a derivation of a group into a module is a *crossed homomorphism.* Again it is easy to show that if M is a π-module, then the abelian group $\mathrm{Hom}(\Phi \longrightarrow \pi, \Pi \longrightarrow \pi)$ is isomorphic to the group of p-derivations of $\Phi \longrightarrow M$.

For the third example, let $\mathfrak{g}$ be a K-Lie algebra and M be a $\mathfrak{g}$-module. This is an abelian group that has an action of $\mathfrak{g}$ on M that satisfies $[a, b]m = a(bm) - b(am)$. Given such an action, let $\mathfrak{h}$ be the abelian group $\mathfrak{g} \times M$ with bracket given by $[(a, m), (a', m')] = ([a, a'], am' - a'm)$. This is a Lie algebra and M is an ideal with $[M, M] = 0$ so that M is an $\mathfrak{h}$-module that M acts trivially on and hence is an $\mathfrak{h}/M = \mathfrak{g}$-module. Again, this action can be calculated to be the original. And again, the algebras that arise in this way are just the abelian group objects in the category $\mathsf{Lie}_K/\mathfrak{g}$.

If $p: \mathfrak{f} \longrightarrow \mathfrak{g}$ is Lie algebra homomorphism, then a *p-derivation* or derivation of $\mathfrak{f}$ into M is a linear map $\tau: \mathfrak{h} \longrightarrow M$ such that $\tau[c, c'] = p(c)\tau(c') - p(c')\tau(c)$. This reduces to the same formula as the associative algebra case if we define the left action to be the negative of the left action. As in the other case, the abelian group $\mathrm{Hom}(\mathfrak{f} \longrightarrow \mathfrak{g}, \mathfrak{h} \longrightarrow \mathfrak{g})$ is isomorphic to the group of p-derivations of $\mathfrak{f}$ to M.

Following these examples, Beck defined an *X-module* in any category $\mathscr{X}$ to be an abelian group object in the category $\mathscr{X}/X$. Moreover, he defined, for an object $p: Y \longrightarrow X$ and an abelian group object $Z \longrightarrow X$ of $\mathscr{X}/X$, the abelian group $\mathrm{Hom}(Y \longrightarrow X, Z \longrightarrow X)$ to be $\mathrm{Der}(Y, Z)$, called the group of p-derivations. Amazingly, in all three examples (as well as others, such as commutative algebras), this turned to give exactly the kind of coefficient modules that are used in defining cohomology and the group Der turned out to be exactly the group of derivations. Thus Beck removed all the adhockery from the definition of cohomology.

Of course, although module can be defined in any category, it does not follow that the category of modules is automatically an abelian category. It is not hard to show that when $\mathscr{X}$ is an exact category (in the sense of [Barr, 1971]), so is $\mathscr{X}/X$, as well as the category of abelian group objects in it, which is thereby abelian, see 1.8.12 and 2.2.3.

1.2. Associative algebras: an example. Here is how that works in detail for associative K-algebras. Let A be an associative algebra, M be an A-module, and $B \longrightarrow A$ be the split extension as described above. An abelian group object of a category is determined by certain arrows, namely a zero map $1 \longrightarrow B$, an inverse map $B \longrightarrow B$ and a group multiplication $B \times B \longrightarrow B$. The terminal object of $\mathscr{A}/A$ is id: $A \longrightarrow A$ and the product is the fibered product (pullback) over A, in this case $B \times_A B$. The zero map takes the element a to $(a, 0)$, the inverse map is given by $(a, m) \mapsto (a, -m)$ and the multiplication takes the pair $((a, m), (a, m'))$ in the fiber over a to the element $(a, m + m')$. Observe that each of these operations preserves the first coordinate, as they must, to be arrows in the slice category. This makes B into an abelian group object in Alg_K/A and defines a full and faithful functor $I_A\colon \mathsf{Mod}_A \longrightarrow \mathsf{Alg}_K/A$ whose image is precisely the abelian group objects.

To get a module from an abelian group object, suppose $p\colon B \longrightarrow A$ is an abelian group object in Alg_K/A. The first thing to note is that there has to be a zero section, that is a map $z\colon A \longrightarrow B$ in the category Alg_K/A. This means that z is an algebra homomorphism and that $p \circ z = \mathrm{id}$. Then p is split as a homomorphism of K-algebras. This implies, in particular, that as K-modules, we can write $B = A \times M$, where $M = \ker(p)$. In terms of this splitting, p is the projection on the first coordinate and z is the injection into the first coordinate. Since z is a ring homomorphism, it follows that $(a,0)(a',0) = (aa',0)$. The additive structure takes the form of a commutative multiplication $B \times_A B \longrightarrow B$. Since $B \cong A \times M$, $B \times_A B \cong A \times M \times M$ so that an operation over A takes the form $(a,m) * (a,m') = (a, f(a,m,m'))$. The fact that $*$ is additive implies that $f(a,m,m') = f_1(a) + f_2(m) + f_3(m')$ and the commutativity implies that $f_2 = f_3$. The fact that z is the zero map implies that $(a,m) * (a,0) = (a,m)$, which says, for $m = 0$, that $f_1(a) = 0$ and then for arbitrary m that $f_2(m) = m$. Thus $(a,m) * (a,m') = (a, m+m')$. Now the fact that $z(1) = (1,0)$ is the identity and that $*$ is a ring homomorphism gives

$$\begin{aligned}(1,m)(1,m') &= ((1,m) * (1,0))((1,0) * (1,m')) \\ &= ((1,m)(1,0)) * ((1,0)(1,m')) \\ &= (1,m) * (1,m') = (1, m+m')\end{aligned}$$

Then

$$\begin{aligned}(1, m+m') &= (1,m)(1,m') = ((1,0) + (0,m))((1,0) + (0,m')) \\ &= (1,0)(1,0) + (0,m)(1,0) + (1,0)(0,m') + (0,m)(0,m') \\ &= (1,0) + (0,m) + (0,m') + (0,m)(0,m') \\ &= (1, m+m') + (0,m)(0,m')\end{aligned}$$

which implies that $(0,m)(0,m') = (0,0)$. Since M is the kernel of a homomorphism, it is an ideal so that $(a,0)(0,m) \in M$ and we will denote it by $(0, am)$ and similarly for $(0,m)(a,0)$, which we denote $(0, ma)$. The equations of rings imply readily that M is a two-sided A-module.

$$\begin{aligned}(a,m)(a',m') &= ((a,0) + (0,m))((a',0) + (0,m')) \\ &= (a,0)(a',0) + (0,m)(a',0) + (a,0)(0,m') + (0,m)(0,m') \\ &= (aa',0) + (0,ma') + (0,am') = (aa', ma' + am')\end{aligned}$$

The remaining details are found in [Beck, 1967].

1.3. Differentials. Suppose that $\mathscr{X}$ is one of our categories and X is an object of $\mathscr{X}$. Then the category Mod_X is the category of X-modules and for an object $p\colon Y \longrightarrow X$, there is a functor $\mathrm{Der}(Y, -)\colon \mathsf{Mod}_X \longrightarrow \mathsf{Ab}$. This functor is the composite of the inclusion $I_X\colon \mathsf{Mod}_X \longrightarrow \mathscr{X}/X$ with the homfunctor $\mathrm{Hom}(Y \longrightarrow X, -)$, each of which preserves limits. Thus $\mathrm{Der}(Y,-)$ preserves limits and it is reasonable to suppose that it is represented by an X-module, we call Diff_Y^X or simply Diff_Y, the module of differentials of Y.

The remarkable fact was that in the standard Cartan-Eilenberg cohomology resolution

$$\cdots \longrightarrow A_n \longrightarrow A_{n-1} \longrightarrow \cdots \longrightarrow A_1 \longrightarrow A_0 \longrightarrow A \longrightarrow 0$$

as described above, in every case, $A_0 = X^e$ and the kernel of $A_0 \longrightarrow A$ was exactly Diff_X. Thus

$$\cdots \longrightarrow A_n \longrightarrow A_{n-1} \longrightarrow \cdots \longrightarrow A_1 \longrightarrow \mathrm{Diff}_X \longrightarrow 0$$

is a projective resolution of Diff_X. Thus if we define a cohomology theory, say $\widetilde{H}$ by $\widetilde{H}^n(X,M) = \mathrm{Ext}^n_{X^e}(\mathrm{Diff}_X, M)$ it will be related to the Cartan-Eilenberg cohomology theory by

$$\widetilde{H}^n(X,M) = \begin{cases} \mathrm{Der}(X,M) & \text{if } n = 0 \\ H^{n+1}(X,M) & \text{if } n > 0 \end{cases}$$

For simplicity, we will call this the *dimension-shifted Cartan-Eilenberg cohomology theory*.

This would allow us to define the Cartan-Eilenberg cohomology in wide generality, if we wanted. Unfortunately, the reality is that while Cartan and Eilenberg believed that they had found a uniform approach to algebraic cohomology theories, the three cases they considered turned out to be the only three for which their approach was "right". (Right means that it coincides with the cotriple cohomology theory, which is really the natural one. When it can be described by the Cartan-Eilenberg approach, that is interesting and useful, but that is not the typical case.)

To summarize, here is the Cartan-Eilenberg cohomology. We assume that the category $\mathscr{X}$ has the property that for each X there is given an algebra X^e (which, in fact would have to be determined only up to Morita equivalence) such that the category of X-modules in the sense of Beck is equivalent to that of left X^e-modules. In addition, we suppose that the inclusion $I_X\colon \mathsf{Mod}_X \simeq \mathsf{Mod}_{X^e} \longrightarrow \mathscr{X}/X$ has a left adjoint Diff_X, often denoted Diff. Then the dimension-shifted Eilenberg cohomology is $\mathrm{Ext}^\bullet_{\mathscr{X}^e}(\mathrm{Diff}(X), -)$

1.4. Cotriple cohomology. Given a category $\mathscr{X}$ and a cotriple $\mathbf{G} = (G, \epsilon, \delta)$ on $\mathscr{X}$, there is a natural definition of cohomology of an object A with coefficients in an X-module. This definition is due to [Beck, 1967] and, in fact, was the reason for his introduction of Beck modules. There will also be a homology theory in some cases, but that has been less studied.

Given an object X, there is a simplicial object augmented over X:

$$\cdots \;\vdots\; G^n X \;\vdots\; G^{n-1}X \;\vdots\; \cdots \rightrightarrows GX \longrightarrow X$$

For an X-module M we apply the functor $\mathrm{Der}(-, M)$ and take the alternating sum of the induced homomorphisms to get the cochain complex

$$0 \longrightarrow \mathrm{Der}(GX, M) \longrightarrow \mathrm{Der}(G^2X, M) \longrightarrow \cdots \longrightarrow \mathrm{Der}(G^{n+1}X, M) \longrightarrow \cdots$$

whose cohomology is defined to be $H^\bullet(X,M)$. The group $H^0(X,M)$ can be identified as $\mathrm{Der}(X,M)$ and Beck showed that $H^1(X,M)$ can be identified as the group of equivalence classes of "singular" extensions $Y \longrightarrow X$ with kernel M.

Just for the record, we describe briefly what a singular extension is. If $p\colon M \longrightarrow X$ is an X-module, that is an abelian group object of $\mathscr{X}/X$, and $q\colon Y \longrightarrow X$ is an arbitrary object of $\mathscr{X}/X$, then one can show that $Y \times_X M \longrightarrow Y$

by the projection on the first factor, makes $Y \times_X M$ into a Y-module, assuming that pullback exists. If the pullback $Y \times_X Y$ also exists the first projection $Y \times_X Y \longrightarrow Y$ is an object of $\mathscr{X}/Y$. If $Y \times_X M \cong Y \times_X Y$ as objects of $\mathscr{X}/Y$, then we say that q is a singular extension with kernel M. This turns out to coincide with the usual definitions in the three cases: for associative algebras, a singular extension is a surjection whose kernel has square 0; for groups a singular extension has abelian kernel; and for Lie algebras, where abelian means square 0, it is the same as the other two.

Since it was known (one of the first results shown in the early papers) that the group of singular extensions was classified by H^2 (recall the dimension shift), it was a reasonable conjecture that the dimension-shifted Cartan-Eilenberg cohomology was equivalent to the cotriple cohomology. The early papers on cohomology also had an interpretation of H^3 as certain kinds of obstructions. G. Orzech showed [1972a,b] that this interpretation of the corresponding cotriple H^2 had a similar interpretation as obstructions under certain conditions. Essentially, the objects had to have an underlying group structure and it was required that the annihilator of an ideal be an ideal (in the case of groups, that the centralizer of a normal subgroup be a normal subgroup). This holds for associative and Lie algebras, but not, for example, for Jordan algebras.

1.5. Exercises.

1. Suppose that S is a set. What is an S-module?

2. Show that the category of 2-modules is equivalent to the category of modules over the ring $\mathbf{Z} \times \mathbf{Z}$.

3. This exercise is for people who know about locally presentable categories, which are complete categories that are accessible in the sense of [Makkai & Paré, 1990].

(a) Show that for any category $\mathscr{X}$, the underlying functor $I_A\colon \mathsf{Ab}(\mathscr{X}) \longrightarrow \mathscr{X}$ preserves limits.

(b) Show that if $\mathscr{X}$ is accessible, so is I_A.

(c) Conclude that if $\mathscr{X}$ is locally presentable, then I_A has a left adjoint J_A.

1.6. The standard setting. In order to understand these things in some detail, we describe what we call a *standard Cartan–Eilenberg* or *CE setting.*

We begin with an exact category $\mathscr{X}$. For each object X of $\mathscr{X}$, we denote by $\mathsf{Mod}(X)$ the category $\mathsf{Ab}(\mathscr{X}/X)$ of abelian group objects of $\mathscr{X}/X$. We assume that the inclusion $I_X\colon \mathsf{Mod}(X) \longrightarrow \mathscr{X}/X$ has a left adjoint we denote Diff^X. When $f\colon Y \longrightarrow X$ is an arrow of $\mathscr{X}$, the direct image (or composite with f) determines a functor $f_!\colon \mathscr{X}/Y \longrightarrow \mathscr{X}/X$ that has a right adjoint $f^* = Y \times_X -$ of pulling back along $Y \longrightarrow X$. The right adjoint f^* (but not the direct image $f_!$) restricts to a functor that we will also denote by $f^*\colon \mathsf{Mod}(X) \longrightarrow \mathsf{Mod}(Y)$ and that we will assume has a left adjoint we denote $f_\#$. The diagram is

$$\begin{array}{ccc} \mathscr{X}/X & \underset{I^X}{\overset{\mathrm{Diff}^X}{\rightleftarrows}} & \mathsf{Mod}(X) \\ f_! \uparrow \downarrow f^* & & f_\# \uparrow \downarrow f^* \\ \mathscr{X}/Y & \underset{I^Y}{\overset{\mathrm{Diff}^Y}{\rightleftarrows}} & \mathsf{Mod}(Y) \end{array}$$

The upper and left arrows are left adjoint, respectively to the lower and right arrows and the diagram of the right adjoints commutes, and so, therefore, does the diagram of left adjoints. (In principle, the diagram of left adjoints commute only up to natural equivalence. However the left adjoints are defined only up to natural equivlance and therefore we are free to modify them so that the diagram actually commutes on the nose.) In the familiar cases involving an enveloping algebra X^e, the left adjoint $f_\#$ turns out to be the functor $X^e \otimes_{Y^e} (-)$. That is, X^e becomes a right Y^e-module via f (actually, just the right hand version of f^*) and then that tensor product is an X^e-module.

1.7. PROPOSITION. *If X is an object of the regular category $\mathscr{X}$, the forgetful functor $I_X: \mathsf{Mod}(X) \longrightarrow \mathscr{X}/X$ preserves regular epis.*

PROOF. What we have to show is that if $f: M' \longrightarrow M$ is a regular epimorphism in the category $\mathsf{Mod}(X)$, then it is also a regular epi in $\mathscr{X}/X$. Actually, we will show that if f is strict epic in $\mathsf{Mod}(X)$, then it is regular in $\mathscr{X}$ and hence in $\mathscr{X}/X$.

An object of $\mathsf{Mod}(X)$ is an object $Y \longrightarrow X$ equipped with certain arrows of which the most important is the arrow $m: Y \times_X Y \longrightarrow Y$ that defines the addition. There are also some equations to be satisfied. The argument we give actually works in the generality of the models of a finitary equational theory. So suppose $f: M' \twoheadrightarrow M$ is a strict epimorphism in $\mathsf{Mod}(X)$. If the map $I_X f$ is not strict epi, it can be factored as $Y' = I_X M' \twoheadrightarrow Y'' \rightarrowtail Y = I_X M$ in $\mathscr{X}/X$. Since $\mathscr{X}$, and hence $\mathscr{X}/X$, is regular, the arrow $Y' \times_X Y' \longrightarrow Y'' \times_X Y''$ is also regular epic and we have the commutative diagram

$$\begin{array}{ccccc} Y' \times_X Y' & \twoheadrightarrow & Y'' \times_X Y'' & \rightarrowtail & Y \times_X Y \\ {\scriptstyle m'}\downarrow & & & & \downarrow{\scriptstyle m} \\ Y' & \twoheadrightarrow & Y'' & \rightarrowtail & Y \end{array}$$

The "diagonal fill-in" (here vertical) provides the required arrow $m'': Y'' \times_X Y'' \longrightarrow Y''$ at the same time showing that both of the arrows $Y' \longrightarrow Y'' \longrightarrow Y$ preserve the new operation. A similar argument works for any other finitary operation. As for the equations that have to be satisfied, this follows from the usual argument that shows that subcategories defined by equations are closed under the formation of subobjects. For example, we show that m'' is associative. This requires showing that two arrows $Y'' \times_X Y'' \times_X Y'' \longrightarrow Y''$ are the same. But we have the diagram

$$\begin{array}{ccc} Y'' \times_X Y'' \times_X Y'' & \rightarrowtail & Y \times_X Y \times_X Y \\ \downarrow\downarrow & & \downarrow \\ Y'' & \rightarrowtail & Y \end{array}$$

that commutes with either of the two left hand arrows. With the bottom arrow monic, this means those two arrows are equal. $\square$

2. The main theorem

In this section we prove the main result of this chapter. We start with a "base category" $\mathscr{S}$ and a functor $U: \mathscr{X} \longrightarrow \mathscr{S}$ that preserves regular epics and has a left adjoint F. For group cohomology, $\mathscr{S}$ is the category of sets, while for associative or Lie algebras over the commutative ring K it will be the category of K-modules. In any case, the cohomology will be a K-relative cohomology. Let $\mathbf{G} = (G, \epsilon, \delta)$ denote the cotriple on $\mathscr{X}$ that results from the adjunction $F \dashv U$.

We suppose there is given, for each object X of $\mathscr{X}$, a chain complex functor $C_\bullet^X : \mathscr{X}/X \longrightarrow \mathscr{C}\mathscr{C}\mathsf{Mod}(X)$, the category of chain complexes in $\mathsf{Mod}(X)$. Such a functor assigns to each $Y \longrightarrow X$ a chain complex

$$\cdots \longrightarrow C_n^X(Y) \longrightarrow \cdots \longrightarrow C_0^X(Y) \longrightarrow 0$$

of X-modules. We further suppose that for $f: Y \longrightarrow X$ the diagram

$$\begin{array}{ccc} \mathscr{X}/Y & \xrightarrow{C_\bullet^Y} & \mathscr{C}\mathscr{C}\mathsf{Mod}(Y) \\ {\scriptstyle f_!}\downarrow & & \downarrow{\scriptstyle \mathscr{C}\mathscr{C}f_\#} \\ \mathscr{X}/X & \xrightarrow[C_\bullet^X]{} & \mathscr{C}\mathscr{C}\mathsf{Mod}(X) \end{array}$$

commutes.

Note that all these categories have initial objects. If we take Y to be the initial object, then we get a standard complex for that case and the complex in all the other cases is gotten by applying $i_\#$, where i is the initial morphism. In light of a previous remark, this is just tensoring with X^e.

For the purposes of this theorem, define an object X of $\mathscr{X}$ to be *U-projective* if UX is projective in $\mathscr{S}$ with respect to the class of regular epis. The third hypothesis of this theorem looks a bit mysterious and I will try to explain it. If you look at the complexes $\{C_n\}$ in the Cartan-Eilenberg theory, you will observe that for a fixed n and an object X, $C_n(X)$ does not depend on the group, respectively algebra, structure of X, but only on the underlying set, respectively K-module UX. The full structure of X is used only to define the boundary operator. We formalize this property, which turns out to be important for the analysis here, in the third hypothesis.

2.1. THEOREM. *Suppose that, in the context of a CE setting, when X is U-projective,*

(i) *GX is U-projective;*
(ii) *$C_\bullet^X(X)$ is a projective resolution of $\mathrm{Diff}^X(X)$;*
(iii) *For each $n \geq 0$, there is a functor $\widetilde{C}_n^X : \mathscr{X}/UX \longrightarrow \mathsf{Mod}(X)$ such that the diagram*

$$\begin{array}{ccc} \mathscr{X}/X & \xrightarrow{C_n^X} & \mathsf{Mod}(X) \\ {\scriptstyle U/X}\downarrow & \nearrow{\scriptstyle \widetilde{C}_n^X} & \\ \mathscr{X}/UX & & \end{array}$$

commutes.

Then there is a natural chain transformation $\mathrm{Diff}^X(G^{\bullet+1}-) \longrightarrow C_\bullet^X(-)$ which is a weak homotopy equivalence.

PROOF. We apply the acyclic models theoren with Γ the class of contractible complexes. We have to show that both of the functors are contractible on models and presentable with respect to Γ. However, for $\mathrm{Diff}^X(G^{\bullet+1}-)$, both of these are automatic (Theorem 5.3.6). We turn to these properties for the Cartan–Eilenberg complex.

Presentability: For $n \geq 0$, the complex

$$(*) \qquad \cdots \longrightarrow C_n G^{m+1} \longrightarrow \cdots \longrightarrow C_n G \longrightarrow C_n \longrightarrow 0$$

is equivalent to

$$\cdots \longrightarrow \widetilde{C}_n U G^{m+1} \longrightarrow \cdots \longrightarrow \widetilde{C}_n U G \longrightarrow \widetilde{C}_n U \longrightarrow 0$$

At this point we require,

2.2. LEMMA. *Let the functor $U: \mathscr{X} \longrightarrow \mathscr{S}$ have left adjoint F and let $\mathbf{G}$ be the resultant cotriple on $\mathscr{X}$. Then the simplicial functor*

$$\cdots \underset{\longrightarrow}{\overset{\longrightarrow}{\vdots}} UG^{m+1} \underset{\longrightarrow}{\overset{\longrightarrow}{\vdots}} \cdots \substack{\longrightarrow\\ \longrightarrow\\ \longrightarrow} UG^2 \rightrightarrows UG \longrightarrow U$$

has a natural contraction.

PROOF. We let $s = \eta U G^n : UG^m \longrightarrow UG^{m+1}$. Then

$$Ud^0 \circ s = U\epsilon G^m \circ \eta U G^m = \mathrm{id}$$

while, for $0 < i \leq m$,

$$Ud^i \circ \eta U G^m = UG^i \epsilon G^{m-i} \circ \eta U G^m$$

and the last term equals, by naturality of η,

$$\eta U G^{m-1} \circ UG^{i-1} \epsilon G^{m-i} = \eta U G^{m-1} \circ Ud^{i-1}$$

This shows that s is a contracting homotopy in the simplicial functor. □

If we apply the additive functor $\widetilde{C}_n$ to this contractible complex, we still get a contractible complex, which is $(*)$.

Acyclicity on models: We will show that for each $Y \longrightarrow X$, the complex

$$(**) \qquad \cdots \longrightarrow C_n GY \longrightarrow \cdots \longrightarrow C_1 GY \longrightarrow C_0 GY \longrightarrow \mathrm{Diff}\, GY \longrightarrow 0$$

is contractible.

2.3. LEMMA. *Let P be a regular projective object of $\mathscr{X}$. Then, for any $P \longrightarrow UX$, Diff_{FP} is projective.*

PROOF. When P is projective in $\mathscr{X}$, any $P \longrightarrow X$ is a projective object of $\mathscr{X}/X$. It is immediate that when $L: \mathscr{X} \longrightarrow \mathscr{Y}$ is left adjoint to $R: \mathscr{Y} \longrightarrow \mathscr{X}$, then L takes a projective in $\mathscr{X}$ to a projective in $\mathscr{Y}$ provided R preserves the epimorphic class that defines the projectives. In this case, the right adjoint is the composite UI_X and the class is that of regular epimorphisms. We have assumed that U, and hence U/X, preserves regular epis and Proposition 1.7 says that I_X does.

Since the two complexes reduce to Diff in degree -1, $\mathrm{Diff}\, G^{\bullet+1}$ is contractible with respect to homotopy and C_n is presentable with respect to homotopy, the existence of a natural transformation $\mathrm{Diff}\, G^{\bullet+1} \longrightarrow C_\bullet$ that extends the identity on Diff follows. Since $\mathrm{Diff}\, G^{\bullet+1}$ is presentable with respect to homotopy, hence weak homotopy and $C_\bullet$ is acyclic with respect to weak homotopy, it follows that for each $Y \longrightarrow X$, there is a chain map $C_\bullet(Y) \longrightarrow \operatorname{Diff} G^{\bullet+1}Y$, the conclusion follows. □

With almost the same argument as the proof of 2.1, we can also prove the following.

2.4. THEOREM. *Suppose that, in the context of a CE setting, when X is U-projective,*

(i) *$C_\bullet GX$ has a natural contracting homotopy;*
(ii) *$C_\bullet^X(X)$ is a projective resolution of* $\operatorname{Diff}^X(X)$;
(iii) *For each $n \geq 0$, there is a functor $\widetilde{C}_n^X\colon \mathscr{X}/UX \longrightarrow \mathsf{Mod}(X)$ such that the diagram*

$$\begin{array}{ccc} \mathscr{X}/X & \xrightarrow{C_n^X} & \mathsf{Mod}(X) \\ {\scriptstyle U/X}\downarrow & \nearrow{\scriptstyle \widetilde{C}_n^X} & \\ \mathscr{X}/UX & & \end{array}$$

commutes.

Then there are natural chain transformations $\operatorname{Diff}^X(G^{\bullet+1}-) \longrightarrow C_\bullet^X(-)$ *and* $C_\bullet^X(-) \longrightarrow \operatorname{Diff}^X(G^{\bullet+1}-)$ *which are homotopy inverse to each other.* □

3. Groups

Let Grp be the category of groups and π be a group. The underlying functor $U\colon \mathsf{Grp} \longrightarrow \mathsf{Set}$ evidently satisfies our conditions and the fact that epimorphisms in Set split implies that every group is U-projective. If we fix a group π, the functor $\widetilde{C}_n^\pi\colon \mathsf{Set}/U\pi \longrightarrow \mathsf{Mod}(\pi)$ takes the set $g\colon S \longrightarrow U\pi$ to the free π-module generated by the $(n+1)$st cartesian power S^{n+1}. Now suppose that $g = Up$ for a group homomorphism $p\colon \Pi \longrightarrow \pi$. The value of the boundary operator ∂ on a generator $\langle x_0, x_1, \ldots, x_n\rangle \in \widetilde{C}_n^\pi(Uf\colon U\Pi \longrightarrow U\pi)$ is

$$p(x_0)\langle x_1, \ldots, x_n\rangle + \sum_{i=1}^{n-1}(-1)^i\langle x_0, \ldots, x_{i-1}x_i, \cdots, x_n\rangle + (-1)^n\langle x_0, \ldots, x_{n-1}\rangle$$

which depends on the group structure in Π. This defines the functor $C_\bullet^\pi$ on Grp/π. The standard Cartan–Eilenberg resolution is the special case of this one in which p is the identity $\pi \longrightarrow \pi$. We may denote $C_\bullet^\pi(\mathrm{id}\colon U\pi \longrightarrow U\pi)$ as simply $C_\bullet(\pi)$ (U applied to the identity of π is the identity of $U\pi$).

For any group Π and any Π-module M a **derivation** $\tau\colon \Pi \longrightarrow M$ is a function that satisfies $\tau(xy) = x\tau y + \tau x$. This is sometimes called a crossed homomorphism, since in the case the action is trivial it is just a homomorphism. On the other hand, one can also interpret it as $\tau(xy) = x(\tau y) + (\tau x)y$ with trivial right action.

From this we have that $\tau 1 = \tau(1 \cdot 1) = 1\tau 1 + \tau 1 = 2\tau 1$, which implies that $\tau 1 = 0$. Also, $\tau y = \tau(xx^{-1}y) = x\tau(x^{-1}y) + \tau x$ so that $\tau(x^{-1}y) = x^{-1}\tau y - x^{-1}\tau x$. In particular, $\tau(x^{-1}) = -x^{-1}\tau x$.

3.1. PROPOSITION. *Suppose that Π is free on basis S and M is a Π-module. Then any function $\tau\colon S \longrightarrow M$ extends to a unique derivation $\Pi \longrightarrow M$.*

PROOF. We will extend τ to a function we also call $\tau: \Pi \longrightarrow M$ defined recursively on the word length as follows. First $\tau 1 = 0$ as is required for any derivation from $\tau 1 = \tau(1 \cdot 1) = 1\tau 1 + 1\tau 1 = 2\tau 1$. Assume that τ is defined on all words whose length is less than the length of w. Then either $w = xv$ for some $x \in X$ or $w = x^{-1}v$ for some $x \in X$. In either case, v is a shorter word than w. In the first case, define $\tau w = x\tau v + \tau x$ and in the second, define $\tau w = x^{-1}\tau v - x^{-1}\tau x$. If these give derivations, they are clearly unique, since these definitions are forced, by the remarks above. So suppose that whenever $v = tu$ is shorter than w, it satisfies $\tau(v) = t\tau u + \tau t$. Then for $w = xv = xtu$, we have

$$\tau w = x\tau v + \tau x = xt\tau u + x\tau t + \tau x = xt\tau u + \tau(xt)$$

If, on the other hand, $w = x^{-1}v = x^{-1}tu$, then

$$\tau w = x^{-1}\tau v - x^{-1}\tau x = x^{-1}t\tau u + x^{-1}\tau t - x^{-1}\tau x = x^{-1}t\tau u + \tau(x^{-1}t) \qquad \square$$

This implies that $\mathrm{Diff}^{\pi}(\Pi)$ is the free π-module generated by X.

It is not hard to show that $C^{\pi}_{\bullet}$ is an exact chain complex and hence for any $\Pi \longrightarrow \pi$, $C^{\pi}_{\bullet}(\Pi)$ is a free resolution of $\mathrm{Diff}^{\pi}(\Pi)$. In the case that Π is free, this is then a free resolution of a free module and hence necessarily split. However, we would rather get the extra information available if we know that the splitting is natural, namely that we then get a homotopy equivalence between the two chain complex functors.

Let us simplify notation by dropping the upper index π. We will start by defining a homomorphism $\partial: C_0(\Pi) \longrightarrow \mathrm{Diff}(\Pi)$. There is a function $\tau: X \longrightarrow \mathrm{Diff}(\Pi)$ which is the inclusion of the basis. This extends to a derivation $\tau: \Pi \longrightarrow \mathrm{Diff}(\Pi)$ as above. Since $C_0(\Pi)$ is freely generated by the elements of Π, this derivation τ extends to a π-linear function $\partial: C_0(\Pi) \longrightarrow \mathrm{Diff}(\Pi)$. In accordance with the recipe above, ∂ is defined on elements of Π recursively as follows. We will denote by $\langle w\rangle$ the basis element of $C_0(\Pi)$ corresponding to $w \in \Pi$. As above, either $w = 1$ or $w = xv$ or $w = x^{-1}v$ for some $x \in X$ and some $v \in \Pi$ shorter than w. Then

$$\partial\langle w\rangle = \begin{cases} p(x)\partial\langle v\rangle + x & \text{if } w = xv \\ p(x^{-1})\partial\langle v\rangle - p(x^{-1})x & \text{if } w = x^{-1}v \\ 0 & \text{if } w = 1 \end{cases}$$

Now define $s: \mathrm{Diff}(\Pi) \longrightarrow C_0(\Pi)$ to be the unique π-linear map such that $s(dx) = \langle x\rangle$ for $x \in X$. Since $\mathrm{Diff}(\Pi)$ is freely generated by all dx for $x \in X$, this does define a unique homomorphism. For $x \in X$, we have that $\partial \circ s(dx) = \partial\langle x\rangle = dx$ and so $\partial \circ s = \mathrm{id}$.

For each $n \geq 0$ we define a homomorphism $s: C_n \longrightarrow C_{n+1}$ as follows. We know that C_n is the free π-module generated by Π^{n+1}. We will denote a generator by $\langle w_0, \ldots, w_n\rangle$ where $w_0, \ldots, w_n$ are words in elements of X and their inverses. Then we define $s: C_n \longrightarrow C_{n+1}$ by induction on the length of the first word:

$$s\langle w_0, \ldots, w_n\rangle = \begin{cases} p(x)s\langle w, w_1, \ldots, w_n\rangle - \langle x, w, w_1, \ldots\rangle & \text{if } w_0 = xw \\ p(x^{-1})s\langle w, w_1, \ldots, w_n\rangle + p(x^{-1})\langle x, w_0, w_1, \ldots\rangle & \text{if } w_0 = x^{-1}w \\ \langle 1, 1, w_1, \ldots, w_n\rangle & \text{if } w_0 = 1 \end{cases}$$

3.2. PROPOSITION. *For any word w and any $x \in X$*

$$s\langle xw, w_1, \ldots, w_n\rangle = p(x)s\langle w, w_1, \ldots, w_n\rangle - \langle x, w, w_1, \ldots, w_n\rangle$$
$$s\langle x^{-1}w, w_1, \ldots, w_n\rangle = p(x^{-1})s\langle w, w_1, \ldots, w_n\rangle + \langle x^{-1}, w_0, w_1, \ldots, w_n\rangle$$

PROOF. These are just the recursive definitions unless w begins with x^{-1} for the first equation or with x for the second. Suppose $w = x^{-1}v$. Then from the definition of s,

$$s\langle w, w_1 \ldots, w_n\rangle = x^{-1}s\langle v, w_1, \ldots, w_n\rangle + x^{-1}\langle x, w, w_1, \ldots, w_n\rangle$$

so that

$$\begin{aligned} s\langle xw, w_1, \ldots, w_n\rangle &= s\langle v, w_1, \ldots, w_n\rangle \\ &= xs\langle w, w_1, \ldots, w_n\rangle - \langle x, w, w_1, \ldots, w_n\rangle \end{aligned}$$

The second one is proved similarly. □

Now we can prove that s is a contraction. First we will do this in dimension 0, then, by way of example, in dimension 2; nothing significant changes in any higher dimension. In dimension 0, suppose w is a word and we suppose that for any shorter word v, we have that $s \circ \partial\langle v\rangle + \partial \circ s\langle v\rangle = \langle v\rangle$. If $x = 1$, then

$$s \circ \partial\langle 1\rangle + \partial \circ s\langle 1\rangle = d \circ el\langle 1, 1\rangle = 1\langle 1\rangle - \langle 1\rangle + \langle 1\rangle = \langle 1\rangle$$

If $w = xv$, with $x \in X$, then

$$\begin{aligned} \partial \circ s\langle w\rangle + s \circ \partial\langle w\rangle &= \partial(p(x)s\langle v\rangle - \partial\langle x, v\rangle) + s(dw) \\ &= p(x)\partial \circ s\langle v\rangle - p(x)\langle v\rangle + \langle xv\rangle - \langle x\rangle + s(p(x)\partial(v) + dx) \\ &= \langle w\rangle + p(x)(\partial \circ s + s \circ \partial - 1)\langle v\rangle - \langle x\rangle + \langle x\rangle = \langle w\rangle \end{aligned}$$

A similar argument takes care of the case that $w = x^{-1}v$. In dimension 2, the chain group $C_2(\Pi)$ is freely generated by Π^3. If we denote a generator by $\langle w_0, w_1, w_2\rangle$, we argue by induction on the length of w_0. If $w_0 = 1$, then

$$\begin{aligned} s \circ \partial\langle 1, w_1, w_2\rangle &= s(\langle w_1, w_2\rangle - \langle w_1, w_2\rangle + \langle 1, w_1w_2\rangle - \langle 1, w_1\rangle) \\ &= \langle 1, 1, w_1w_2\rangle - \langle 1, 1, w_1\rangle \end{aligned}$$

while

$$\begin{aligned} \partial \circ s\langle 1, w_1, w_2\rangle &= \partial(\langle 1, 1, w_1, w_2\rangle) \\ &= \langle 1, w_1, w_2\rangle - \langle 1, w_1, w_2\rangle + \langle 1, w_1, w_2\rangle - \langle 1, 1, w_1w_2\rangle + \langle 1, 1, w_1\rangle \end{aligned}$$

and these add up to $\langle 1, w_1, w_2\rangle$. Assume that $(\partial \circ s + s \circ \partial)\langle w\rangle = \langle w\rangle$ when w is shorter than w_0. Then for $w_0 = xw$,

$$\begin{aligned} \partial \circ s\langle xw, w_1, w_2\rangle &= p(x)\partial \circ s\langle w, w_1, w_2\rangle - \partial\langle x, w, w_1, w_2\rangle \\ &= p(x)\partial \circ s\langle w, w_1, w_2\rangle - p(x)\langle w, w_1, w_2\rangle + \langle xw, w_1, w_2\rangle \\ &\quad - \langle x, ww_1, w_2\rangle + \langle x, w, w_1w_2\rangle - \langle x, w, w_1\rangle \end{aligned}$$

while

$$\begin{aligned} &s \circ \partial\langle x, w, w_1, w_2\rangle \\ &\quad = p(xw)s\langle w_1, w_2\rangle - s\langle xww_1, w_2\rangle + s\langle xw, w_1w_2\rangle - s\langle xw, w_1\rangle \\ &\quad = p(xw)s\langle w_1, w_2\rangle - p(x)s\langle ww_1, w_2\rangle + \langle x, ww_1, w_2\rangle \\ &\qquad + p(x)s\langle w, w_1w_2\rangle - \langle x, w, w_1w_2\rangle - p(x)s\langle w, w_1\rangle + \langle x, w, w_1\rangle \\ &\quad = p(x)s \circ \partial\langle w, w_1, w_2\rangle + \langle x, ww_1, w_2\rangle - \langle x, w, w_1w_2\rangle + \langle x, w, w_1\rangle \end{aligned}$$

Then,

$$\begin{aligned}(\partial \circ s + s \circ \partial)\langle xw, w_1, w_2\rangle &= p(x)(\partial \circ s + s \circ \partial)\langle w, w_1, w_2\rangle - p(x)\langle w, w_1, w_2\rangle \\ &+ \langle xw, w_1, w_2\rangle - \langle x, ww_1, w_2\rangle + \langle x, w, w_1w_2\rangle - \langle x, w, w_1\rangle \\ &+ \langle x, ww_1, w_2\rangle - \langle x, w, w_1w_2\rangle + \langle x, w, w_1\rangle\end{aligned}$$

Using the inductive assumption, the first two terms cancel and all the rest cancel in pairs, except for $\langle xw, w_1, w_2\rangle$, which shows that $s \circ \partial + \partial \circ s = 1$ in this case. The second case, that w_0 begins with the inverse of a letter is similar.

This completes the proof of the homotopy equivalence.

4. Associative algebras

The situation with associative algebras is quite similar to that of groups. We begin with a commutative (unitary) ring K. The category $\mathscr{X}$ is the category of K-modules and $\mathscr{A}$ is the category of K-algebras. If A is a K-algebra, the category $\mathsf{Mod}(A)$ is the category of two sided A-modules. The enveloping algebra of A is $A^e = A \otimes_K A^{\mathrm{op}}$ and it is easy to see that two-sided A-modules are the same thing as left A^e-modules. The free algebra generated by a K-module V is the tensor algebra

$$F(V) = K \oplus V \oplus (V \otimes V) \oplus (V \otimes V \otimes V) \oplus \cdots \oplus V^{(n)} \oplus \cdots$$

and it is evident that $F(V)$ is K-projective when V is. Note that we use $V^{(n)}$ to denote the nth tensor power of V. If A is a K-algebra, the functor C_n^A is defined by the formula

$$C_n^A(V \longrightarrow UA) = A \otimes V^{(n+1)} \otimes A^{\mathrm{op}} \cong A^e \otimes V^{(n+1)}$$

for $g\colon V \longrightarrow UA$. The boundary formula is similar to the one for groups. If g has the form $Uf\colon UB \longrightarrow UA$, then

$$\begin{aligned}\partial(a \otimes b_0 \otimes \cdots \otimes b_n \otimes a') &= af(b_0) \otimes b_1 \otimes \cdots \otimes b_n \otimes a' \\ &+ \sum_{i=1}^{n-1}(-1)^i a \otimes b_0 \otimes \cdots \otimes b_{i-1}b_i \otimes \cdots \otimes b_n \otimes a' \qquad (16) \\ &+ (-1)^n a \otimes b_0 \otimes \cdots \otimes b_{n-1} \otimes f(b_n)a'\end{aligned}$$

differing only in the fact that we have operation on the right as well as on the left. The remaining details are essentially similar to those of the group case.

If A is an algebra and M is a two-sided A-module, then a derivation $\tau\colon A \longrightarrow M$ is a linear function such that $\tau(ab) = a(\tau b) + (\tau a)b$.

If $A = F(V)$ is a tensor algebra, then for an A-module M, every linear map $\tau\colon V \longrightarrow M$ extends to a unique derivation defined recursively by the formulas $\tau 1 = 0$ and

$$\tau(v_1 \otimes v_2 \otimes \cdots \otimes v_n) = v_1\tau(v_2 \otimes \cdots \otimes v_n) + \tau v_1$$

If $p\colon A \longrightarrow B$ is an algebra homomorphism, the formula for the contracting homotopy in the Cartan–Eilenberg complex of a tensor algebra is

$$s(a_0 \otimes a_1 \otimes \cdots \otimes a_n) = p(v)s(a \otimes a_1 \otimes \cdots \otimes a_n) + (v \otimes a \otimes a_1 \otimes \cdots \otimes v_n)$$

where $a_0 = v \otimes a$ in the tensor algebra.

5. Lie Algebras

This examples differs from the preceding ones more than just in some details. For one thing, Cartan and Eilenberg limited their treatment to K-Lie algebras that were K-free (instead of K-projective, as they did for associative algebras) and we would like to extend the theory to the projective case. For another, it requires some work to see that the free Lie algebra generated by a projective K-module is still a projective K-module. When the given K-module is free, this fact is buried in an exercise in [Cartan & Eilenberg] (Exercise 8 on page 286), with a long hint that still requires some effort to complete. We give the details, since this argument (nor the fact itself) is certainly not well known.

It is instructive to see why Cartan and Eilenberg limited themselves to K-free Lie algebras. Eilenberg explained it once. They made crucial use of the fact of the Poincaré–Witt theorem that implies that the enveloping algebra of a K-free Lie algebra is K-free. The enveloping algebra functor is not additive and therefore they did not see how to show that the enveloping algebra of a K-projective Lie algebra was K-projective. The problem was that they defined projective as being a direct summand of a free module. In the interim, we have learned, from dealing with non-additive categories, that a more useful definition of projective is as retract of a free. Since the property of being a retract is preserved by all functors, any functor that takes a free object to a free (or even projective) object will also take a projective to a projective. It is interesting to note that at different places in the proof we actually use both definitions of projective.

5.1. The enveloping algebra. There is a functor $K\text{-}\mathsf{Assoc} \longrightarrow K\text{-}\mathsf{Lie}$ that assigns to each associative algebra the Lie algebra with the same underlying K-module and operation

$$[x, y] = xy - yx$$

This functor has a left adjoint given by $\mathfrak{g} \mapsto \mathfrak{g}^e$, which is a quotient of the tensor algebra

$$T(\mathfrak{g}) = K \oplus \mathfrak{g} \oplus (\mathfrak{g} \otimes \mathfrak{g}) \oplus (\mathfrak{g} \otimes \mathfrak{g} \otimes \mathfrak{g}) \oplus \cdots \oplus \mathfrak{g}^{(n)} \oplus \cdots$$

The multiplication is given by

$$(x_1 \otimes \cdots \otimes x_n)(y_1 \otimes \cdots \otimes y_m) = x_1 \otimes \cdots \otimes x_n \otimes y_1 \otimes \cdots \otimes y_m$$

and the quotient is by the 2-sided ideal generated by all terms of the form $x \otimes y - y \otimes x - [x, y]$. Then the Poincaré–Witt theorem says:

5.2. THEOREM. *Suppose that $\mathfrak{g}$ is a K-Lie algebra that is free on the basis $\{x_i \mid i \in I\}$ where I is a totally ordered index set. Then $\mathfrak{g}^e$ is K-free on the basis*

$$\{x_{i_1} \otimes \cdots \otimes x_{i_n} \mid i_1, \ldots, i_n \in I,\ i_1 < \cdots < i_n\}$$

Since every set can be totally ordered, this implies,

5.3. COROLLARY. *If the K-Lie algebra $\mathfrak{g}$ is K-free, then so is $\mathfrak{g}^e$.*

Now we can extend this to K-projectives.

5.4. PROPOSITION. *Let $\mathfrak{g}$ be a K-Lie algebra that is projective as a K-module. Then the enveloping algebra $\mathfrak{g}^e$ is projective as a K-module.*

PROOF. Since $\mathfrak{g}$ is K-projective, we can find a K-module $\mathfrak{g}_0$ such that $\mathfrak{g} \oplus \mathfrak{g}_0$ is a K-free K-module. We can make $\mathfrak{g} \oplus \mathfrak{g}_0$ into a Lie algebra by making $\mathfrak{g}_0$ a central ideal (that is, $[(x, x_0), (y, y_0)] = ([x, y], 0)$ for $x, y \in \mathfrak{g}$ and $x_0, y_0 \in \mathfrak{g}_0$). Then $\mathfrak{g}$ is, as a Lie algebra, a retract of $\mathfrak{g} \oplus \mathfrak{g}_0$. All functors preserve retracts so that $\mathfrak{g}^e$ is a retract of $(\mathfrak{g} \oplus \mathfrak{g}_0)^e$. By the Poincaré–Witt theorem, the latter is K-free, and so $\mathfrak{g}^e$ is K-projective. □

Cartan–Eilenberg made another use of freeness in their development. It was used in the process of showing that if $\mathfrak{h}$ is a Lie subalgebra of the Lie algebra $\mathfrak{g}$, and if $\mathfrak{g}$, $\mathfrak{h}$ and $\mathfrak{g}/\mathfrak{h}$ are K-free, then $\mathfrak{g}^e$ is a free $\mathfrak{h}^e$-module. We will prove this with "free" replaced everywhere by "projective". Of course, if $\mathfrak{g}$ and $\mathfrak{g}/\mathfrak{h}$ are K-projective, so is $\mathfrak{h}$.

5.5. PROPOSITION. *Let* $0 \longrightarrow \mathfrak{h} \longrightarrow \mathfrak{g} \longrightarrow \mathfrak{g}/\mathfrak{h} \longrightarrow 0$ *be an exact sequence of K-projective K-Lie algebras. Then $\mathfrak{g}^e$ is projective as an $\mathfrak{h}^e$-module.*

PROOF. The conclusion is valid when all three of $\mathfrak{g}$, $\mathfrak{h}$ and $\mathfrak{g}/\mathfrak{h}$ are K-free ([Cartan & Eilenberg], Proposition XIII.4.1). For the general case, let $\mathfrak{f} = \mathfrak{g}/\mathfrak{h}$. Since $\mathfrak{h}$ is K-projective, there is a K-module $\mathfrak{h}_0$ such that $\mathfrak{h} \oplus \mathfrak{h}_0$ is K-free. If we give $\mathfrak{h}_0$ the structure of a central ideal, then $\mathfrak{h} \oplus \mathfrak{h}_0$ is a K-free K-Lie algebra. Similarly, choose $\mathfrak{f}_0$ so that $\mathfrak{f} \oplus \mathfrak{f}_0$ is K-free and let $\mathfrak{g}_0 = \mathfrak{f}_0 \oplus \mathfrak{h}_0$. We have a commutative diagram

$$\begin{array}{ccccccccc}
0 & \longrightarrow & \mathfrak{h} & \xrightarrow{\quad f \quad} & \mathfrak{g} & \xrightarrow{\quad g \quad} & \mathfrak{f} & \longrightarrow & 0 \\
 & & \Big\downarrow \begin{pmatrix} 1 \\ 0 \end{pmatrix} & & \Big\downarrow \begin{pmatrix} 1 \\ 0 \\ 0 \end{pmatrix} & & \Big\downarrow \begin{pmatrix} 1 \\ 0 \end{pmatrix} & & \\
0 & \longrightarrow & \mathfrak{h} \oplus \mathfrak{h}_0 & \xrightarrow{\begin{pmatrix} f & 0 \\ 0 & 1 \\ 0 & 0 \end{pmatrix}} & \mathfrak{g} \oplus \mathfrak{h}_0 \oplus \mathfrak{f}_0 & \xrightarrow{\begin{pmatrix} g & 0 & 0 \\ 0 & 0 & 1 \end{pmatrix}} & \mathfrak{f} \oplus \mathfrak{f}_0 & \longrightarrow & 0
\end{array}$$

Since, as K-modules, $\mathfrak{g} \oplus \mathfrak{g}_0 \cong \mathfrak{h} \oplus \mathfrak{h}_0 \oplus \mathfrak{f} \oplus \mathfrak{f}_0$ is a direct sum of free modules, it follows that $\mathfrak{g} \oplus \mathfrak{g}_0$ is free as well. Apply the enveloping algebra functor to the left hand square to get the diagram,

$$\begin{array}{ccc}
\mathfrak{h}^e & \longrightarrow & \mathfrak{g}^e \\
\Big\downarrow & & \Big\downarrow \\
(\mathfrak{h} \oplus \mathfrak{h}_0)^e & \longrightarrow & (\mathfrak{g} \oplus \mathfrak{g}_0)^e
\end{array}$$

According to Proposition XIII.2.1 of [Cartan & Eilenberg], for any two Lie algebras $\mathfrak{g}_1$ and $\mathfrak{g}_2$, there is an isomorphism $(\mathfrak{g}_1 \oplus \mathfrak{g}_2)^e \cong \mathfrak{g}_1^e \otimes \mathfrak{g}_2^e$. This can be proved directly, as Cartan & Eilenberg do, but the easy way is to observe that both sides represent the functor that assigns to an associative algebra A the set of *pointwise commuting* pairs of homomorphisms in $\mathrm{Hom}(\mathfrak{g}_1^e, A) \times \mathrm{Hom}(\mathfrak{g}_2^e, A)$ (see Exercise 1). Moreover, $\mathfrak{h}_0^e$ is K-projective since $\mathfrak{h}_0$ is. If $\mathfrak{h}_0^e$ is K-free, say $\mathfrak{h}_0^e \cong \sum K$, then $\mathfrak{h}^e \otimes \mathfrak{h}_0^e \cong \mathfrak{h}^e \otimes \sum K \cong \sum \mathfrak{h}^e$ is a free $\mathfrak{h}^e$-module. If $\mathfrak{h}_0^e$ is K-projective, then it is a retract of a free K-module and it follows that $\mathfrak{h}^e \otimes \mathfrak{h}_0^e$ is a retract of a free $\mathfrak{h}^e$-module. But $(\mathfrak{g} \oplus \mathfrak{g}_0)^e$ is, as an $(\mathfrak{h} \oplus \mathfrak{h}_0)^e$-module, *a fortiori* as an $\mathfrak{h}^e$-module, isomorphic to a direct sum of copies of $(\mathfrak{h} \oplus \mathfrak{h}_0)^e$ and hence is also $\mathfrak{h}^e$-projective. Finally, $\mathfrak{g}^e$ is a retract as a ring, therefore as a $\mathfrak{g}^e$-module and hence as an $\mathfrak{h}^e$-module, of $(\mathfrak{g} \oplus \mathfrak{g}_0)^e$ and is therefore also $\mathfrak{h}^e$-projective. □

With these two results, the entire chapter XIII of [Cartan & Eilenberg] becomes valid with free replaced by projective.

Now we describe the standard theory from [Cartan & Eilenberg] (with the usual dimension shift). For a K-module M, let $\bigwedge^n(M)$ denote the nth exterior power of M. If $g\colon M \longrightarrow U\mathfrak{g}$ is a module homomorphism, $C_n^{\mathfrak{g}}(M \longrightarrow U\mathfrak{g}) = \mathfrak{g}^e \otimes \bigwedge^{n+1}(M)$. If $g = Uf$ for a Lie algebra homomorphism $f\colon \mathfrak{h} \longrightarrow \mathfrak{g}$, the boundary is described on generators as follows. We adopt here a convention common in algebraic topology in which we denote the omission of a term by putting a $\widehat{\ }$ on it. So, for example, a sequence $x_1, \ldots, \widehat{x_i}, \ldots, x_n$ is shorthand for $x_1, \ldots, x_{i-1}, x_{i+1}, \ldots, x_n$.

$$\partial(x_0 \wedge x_1 \wedge \cdots \wedge x_n) = \sum_{i=0}^{n} (-1)^i f(x_i) \otimes (x_0 \wedge \cdots \wedge \widehat{x_i} \wedge \cdots \wedge x_n) + \sum_{1 \le i < j \le n} (-1)^{i+j} [x_i, x_j] \wedge x_0 \wedge \cdots \wedge \widehat{x_i} \wedge \cdots \wedge \widehat{x_j} \wedge \cdots \wedge x_n$$

In order to apply Theorem 2.1, we must show the following: (See [Cartan & Eilenberg], Exercise 8 on page 286, where this is given as an exercise for the case of a free K-module. The proof for that case follows the long hint given there.)

5.6. PROPOSITION. *Let M be a projective K-module. Then the free Lie algebra FM is also K-projective.*

PROOF. Consider first the case that M is a free K-module. There is a diagram of categories and adjoints, in which $\otimes = \otimes_{\mathbf{Z}}$:

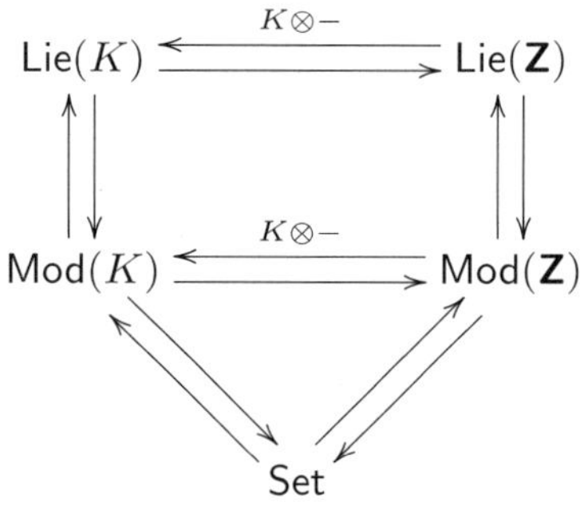

It is clear from this diagram that if we show that the free **Z**-Lie algebra generated by a free **Z**-module (that is, abelian group) is a free abelian group, then by applying the functor $K \otimes -$, it follows that the free K-Lie algebra by a free K-module will be K-free.

So let M be a free abelian group generated by the set X and let $F(M)$ be the free Lie algebra generated by M. By the commutation of adjoints in the diagram

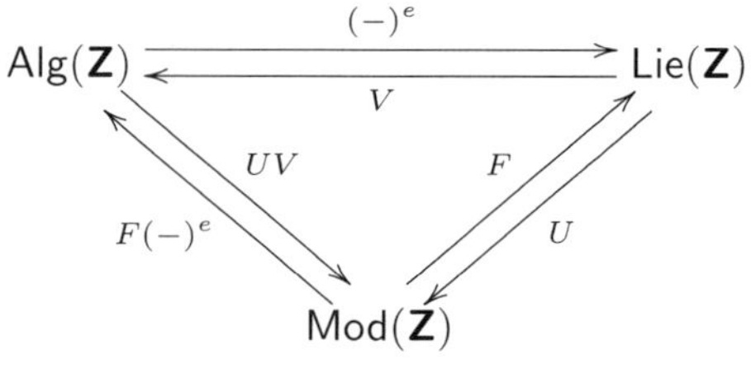

Since $F \dashv U$ and $(-)^e$, it follows that $(F-)^e \dashv UV$. But since UV is just the underlying module functor, it follows that $(F-)^e$ is its left adjoint, which is the tensor algebra functor $T(M) = \mathbf{Z} \oplus M \oplus (M \otimes M) \oplus M^{(3)} \oplus \cdots$ which is $\mathbf{Z}$-free. Then $(FM)^e$ and hence $V(FM)^e$ are free abelian groups. The inner adjunction is a map $e: F(M) \longrightarrow V(F(M)^e)$. If this map can be shown to be injective, then FM is a subgroup of a free abelian group and is therefore free. The map e is certainly injective when K is a field; the Poincaré–Witt theorem gives an explicit linear basis for $\mathfrak{g}^e$, assuming one for $\mathfrak{g}$ and it includes, among other things, the explicit basis of $\mathfrak{g}$.

For a set X, let $F_{\mathrm{ab}}X$ denote the free abelian group generated by X. In particular, we can suppose there is an X with $M = F_{\mathrm{ab}}X$. If e is not injective, then there is a non-zero element $a \in FM$ with $e(a) = 0$. Now, a is made of up a finite sum of a finite number of finitely iterated brackets applied to a finite number of free generators. For each finite integer n let $F_n(M)$ denote the abelian subgroup of FM consisting of the elements that are finite sums of brackets of generators with no more than n brackets. Then there is a finite subset $Y \subseteq X$ and a finite integer n, such that $a \in F_n(F_{\mathrm{ab}}Y)$. Now for any $Y \neq \emptyset$, the inclusion $Y \longrightarrow X$ is a split monic and hence so is $V(F(F_{\mathrm{ab}}Y)^e) \longrightarrow V(FM^e)$. Thus you can see from the diagram

$$\begin{array}{ccc}
F(F_{\mathrm{ab}}Y) & \longrightarrow & FM \\
\downarrow & & \downarrow \\
V(F(F_{\mathrm{ab}}Y)^e) & \rightarrowtail & V(FM^e)
\end{array}$$

that if a is in the kernel of the arrow $FM \longrightarrow V(FM)^e$, it also in the kernel of $F(F_{\mathrm{ab}}Y) \longrightarrow V(F(F_{\mathrm{ab}}Y)^e)$. Now $F_n(F_{\mathrm{ab}}Y)$ is a finitely generated abelian group. At this point we need a lemma.

5.7. Lemma. *Suppose $f: A \longrightarrow B$ is a homomorphism of abelian groups such that f is not injective, A is finitely generated and B is torsion free. Then there is an integer prime p such that the induced map $\mathbf{Z}/p\mathbf{Z} \otimes A \longrightarrow \mathbf{Z}/p\mathbf{Z} \otimes B$ is not injective.*

Proof. We begin by showing that $\mathbf{Z}/p\mathbf{Z} \otimes A \cong A/pA$. In fact, let $g: \mathbf{Z}/p\mathbf{Z} \times A \longrightarrow A/pA$ be defined by $g(n, a) = na + pA$. It is easily seen that this is a bilinear map. If $h: \mathbf{Z}/p\mathbf{Z} \times A \longrightarrow C$ is a bilinear map, let $\widetilde{h}: A/pA \longrightarrow C$ be defined by $\widetilde{h}(a + pA) = h(1, a)$. Since $h(1, pa) = h(p, a) = 0$ this is well defined and evidently a homomorphism such that $\widetilde{h} \circ g = g$. If $l: A/pA \longrightarrow C$ is a homomorphism such that $l \circ g = h$, then $l(a + pA) = l \circ g(1, a) = h(1, a) = \widetilde{h}(a + pA)$ so that $\widetilde{h}$ is unique. Now A is a finitely generated abelian group, hence a direct sum of cyclic groups. Suppose there is an $a \in \ker f$ that is not torsion. Write $a = \sum n_x x$, the sum taken over a chosen set of generators of the cyclic group. There is some x for which $n_x \neq 0$ and x is not torsion. Let p be any prime that does not divide n_x. Then $a \notin pA$ and is in the kernel of the induced map $A/pA \longrightarrow B/pB$. If not, then the kernel of f is the torsion subgroup of $A_t \subseteq A$. Since A_t is a direct sum of cyclic groups, choose a prime that divides one of the orders and then $A_t/pA_t \neq 0$ and the class of any element $a \notin pA_t$ will do. $\square$

This implies that

$$\mathbf{Z}_p \otimes F_n(M) \longrightarrow \mathbf{Z}_p \otimes V_n \longrightarrow \mathbf{Z}_p \otimes U(F(M)^e)$$

is not monic. But $\mathbf{Z}_p$ is a field and both $()^e$ and U commute with $\mathbf{Z}_p \otimes -$, so that reduces the question to the case of a field for which e is monic.

This finishes the case of a free module; projectives are readily handled as retracts of free modules. □

With this, Theorem 2.1 applies and shows that the cotriple resolution is homotopic to the one developed in [Cartan & Eilenberg].

5.8. Exercise.

1. (a) Show that if $\mathfrak{g}_1$, $\mathfrak{g}_2$ and $\mathfrak{g}$ are Lie algebras, there is a one-one correspondence between Lie algebra homomorphisms $f\colon \mathfrak{g}_1 \oplus \mathfrak{g}_2 \longrightarrow \mathfrak{g}$ and pairs of homomorphisms (f_1, f_2) where $f_1\colon \mathfrak{g}_1 \longrightarrow \mathfrak{g}$ and $f_2\colon \mathfrak{g}_2 \longrightarrow \mathfrak{g}$ such that for all $x_1 \in \mathfrak{g}_1$ and $x_2 \in \mathfrak{g}_2$, $[f_1x_1, f_2x_2] = 0$.

(b) Show that if A_1, A_2 and A are associative algebras, there is a one-one correspondence between algebra homomorphisms $f\colon A_1 \otimes A_2 \longrightarrow A$ and pairs of homomorphisms (f_1, f_2) where $f_1\colon A_1 \longrightarrow A$ and $f_2\colon A_2 \longrightarrow A$ such that for all $x_1 \in A_1$ and $x_2 \in A_2$, $f_1x_1f_2x_2 = f_2x_2f_2x_1$.

(c) Conclude that $(\mathfrak{g}_1 \oplus \mathfrak{g}_2)^e \cong \mathfrak{g}_1^e \otimes \mathfrak{g}_2^e$ since both represent the same functor.

CHAPTER 7

Other Applications in Algebra

The previous chapter applied the acyclic models theorems to the homology and cohomology theories from the book of Cartan and Eilenberg. In order that a homology theory fit their patter, it must be a Tor, and therefore the higher homology groups must vanish when the coefficient module is projective (or even flat). Similarly, the higher cohomology groups must vanish when the coefficient module is injective. The theories described in this chapter either do not satisfy that criterion (the Harrison and Shukla theories) or are not (co)homology theories at all (the Eilenberg–Zilber theorem).

We describe the Harrison theory in detail. It has been heavily applied to the theory of commutative algebras. See [André, 1967, 1974] and other references found there. We give only a pointer to the Shukla theory and to the proof that it is a cotriple cohomology. It is extremely complicated and has not, to my knowledge, had any applications. The Eilenberg–Zilber theorem has had important applications. It was also the first occasion, as far as I know, for a proof that uses a method called acyclic models. The connection between that method and the one used here is long and tenuous, but real.

1. Commutative algebras

In 1962, D. K. Harrison defined a cohomology theory for commutative algebras. It turns out that for an algebra over a field of characteristic 0, Harrison's groups are isomorphic to those given by cotriple cohomology. In the process of demonstrating that, we show that the Hochschild cohomology groups of a commutative ring split into a direct sum of the commutative cohomology and a natural complement. This splitting is effected by a series of idempotents in the rational group ring of the symmetric groups, one in each degree. The nth symmetric group acts on the nth chain group by permuting its terms and this allows the idempotents to act as well. Since the series of idempotents commute with the boundary, they also induce splittings on the homology groups and the commutative cohomology turns out to be one of the two summands. M. Gerstenhaber and D. Schack showed that this splitting is just the first step in a splitting of the nth Hochschild cohomology groups into the direct sum of n pieces, one of which is the commutative cohomology. This "Hodge decomposition" has been the subject of intensive study by Gerstenhaber and Schack and others. The "shuffle idempotents" in the rational group ring of the symmetric groups have found other uses, including even in the study of card shuffling (see [Bayer and Diaconis, 1992]).

1.1. The Hochschild cohomology of a commutative algebra. In this chapter, we will, for the time being adhere to the original numbering that shifts the cohomology one dimension from the cotriple cohomology. The reason we do this is that a number of formulas are considerably more natural in that numbering. The shuffle product, described below, is much easier to define and show compatible with the boundary operator.

With the ring K fixed, we write $\otimes$ for $\otimes_K$ and $B^{\otimes n}$ for the nth tensor power of a K-module B. The Hochschild (co)homology groups of a K-algebra homomorphism $p: B \longrightarrow A$ are calculated from a resolution that has $C_n^A(B) = A^e \otimes B^{\otimes n}$ in degree n and the boundary operator described in 6, Equation (16). For a right A^e-module N the homology $H_\bullet(B, N)$ is computed as the homology of the chain complex $C_\bullet^A(B) \otimes_{A^e} N$ and for a left A^e-module M the cohomology $H^\bullet(B, M)$ is the cohomology of the cochain complex $\mathrm{Hom}_{A^e}(C_\bullet^A(B), M)$. Note that

$$C_\bullet^A(B) \otimes_{A^e} M = A^e \otimes B^{\otimes n} \otimes_{A^e} M \cong B^{\otimes n} \otimes M$$

and

$$\mathrm{Hom}_{A^e}(C_\bullet^A(B), M) = \mathrm{Hom}_{A^e}(A^e \otimes B^{\otimes n}, M) \cong \mathrm{Hom}(B^{\otimes n}, M)$$

The module structure is needed only for defining the boundary and coboundary operators. The Harrison (co)homology is defined in the case that A and B are commutative and the coefficient module M has the same action on the left and right. In that case we can replace the Hochschild complex by the complex defined by $C_n^A(B) = A \otimes B^n$ and then compute the homology of $C_n^A(B) \otimes_A M$ and the cohomology of $\mathrm{Hom}_A(C_n^A(B), M)$. The boundary operator will be somewhat different too, being given by

$$\begin{aligned}\partial(a \otimes b_1 \otimes \cdots \otimes b_n) &= ap(b_1) \otimes b_2 \otimes \cdots \otimes b_n \\ &\quad + \sum_{i=1}^{n-1} (-1)^i a \otimes b_1 \otimes \cdots \otimes b_i b_{i+1} \otimes \cdots \otimes b_n \\ &\quad + (-1)^n p(b_n) a \otimes b_1 \otimes \cdots \otimes b_{n-1}\end{aligned}$$

taking advantage of the same action on the right and left.

1.2. Some notation. With $p: B \longrightarrow A$ fixed, we will write C_n for $C_n^A(B)$. We will denote the element $a \otimes b_1 \otimes \cdots \otimes b_n \in C_n$ by $a[b_1, \ldots, b_n]$ or simply $[b_1, \ldots, b_n]$ when $a = 1$. The element $a \in A$ in degree 0 will also be written as $a[\,]$ or simply $[\,]$ when $a = 1$. This notation also makes use of the usual practice of ignoring associations of tensors by making the identification

$$[b_1, \ldots, b_{i-1}, [b_i, \ldots, b_j], b_{j+1}, \ldots, b_n] = [b_1, \ldots, b_{i-1}, b_i, \ldots, b_j, b_{j+1}, \ldots, b_n] \circ$$

If we write $\mathbf{b}$ for $[b_1, \ldots, b_n]$, as we often do, we will let $\mathbf{b}^{@}$, $\mathbf{b}^{@@}$, denote $[b_2, \ldots, b_n]$, $[b_3, \ldots, b_n]$, etc., including $[b_1]^{@} = [\,]$.

The boundary operator can then be described as the as the A-linear map for which

$$\begin{aligned}\partial[b_1, \ldots, b_n] = p(b_1)[b_2, \ldots, b_n] &+ \sum_{i=1}^{n-1} (-1)^i [b_1, \ldots, b_i b_{i+1}, \ldots, b_n] \\ &+ (-1)^n p(b_n)[b_0, \ldots, b_{n-1}]\end{aligned}$$

1.3. Shuffle products. The **shuffle product** $*$ is a graded product defined on C_n as the A-bilinear map defined recursively by $[\,] * \mathbf{b} = \mathbf{b} = \mathbf{b} * [\,]$ and for all $n \geq 1$ and for all $\mathbf{b} \in C_n$ and $\mathbf{c} \in C_m$,

$$\mathbf{b} * \mathbf{c} = [b_1, \mathbf{b}^@ * \mathbf{c}] + (-1)^n [c_1, \mathbf{b} * \mathbf{c}^@]$$

Except for the sign, this is just all the permutations of the terms in $\mathbf{b}$ and in $\mathbf{c}$ that leave those of $\mathbf{b}$ and those of $\mathbf{c}$ in their original relative order, just like the riffle (or dovetail) shuffle. The inductive definition can be understood as follows. In order to shuffle a pack of n cards with one of m, you do nothing if $n = 0$ or $m = 0$. If they are both non-zero, take a card off one of the two piles, shuffle the remaining piles and then replace the missing card on top. The sign is just the ordinary sign of the permutation.

1.4. THEOREM. *For any* $\mathbf{b} = [b_1, \ldots, b_n] \in C_n$ *and* $\mathbf{c} = [c_1, \ldots, c_m] \in C_m$, $\partial(\mathbf{b} * \mathbf{c}) = (\partial \mathbf{b}) * \mathbf{c} + (-1)^n \mathbf{b} * (\partial \mathbf{c})$.

The proof is fairly complicated and uses several lemmas. We introduce some auxiliary operators. We will say that an operator that satisfies this equation is a derivation with respect to $*$ or a $*$-derivation. For $n > 0$ and $m > 0$, define $\mathbf{b} \mathbin{\widetilde{*}} \mathbf{c} = [b_1 c_1, \mathbf{b}^@ * \mathbf{c}^@]$. If either n or m is 0, then $\mathbf{b} \mathbin{\widetilde{*}} \mathbf{c} = 0$. ∂^0 denotes the first face operator given by $\partial^0 \mathbf{b} = p(b_1)\mathbf{b}^@$ and we define $\widetilde{\partial}$ by

$$\widetilde{\partial}\mathbf{b} = \sum_{i=1}^{n-1} (-1)^{i-1} [b_1, \ldots, b_i b_{i+1}, \cdots, b_n] - (-1)^n p(b_n)[b_1, \ldots, b_{n-1}]$$

so that $\partial = \partial^0 - \widetilde{\partial}$.

1.5. LEMMA. $\partial^0(\mathbf{b} * \mathbf{c}) = (\partial^0 \mathbf{b}) * \mathbf{c} + (-1)^n \mathbf{b} * (\partial^0 \mathbf{c})$.

PROOF. We have

$$\begin{aligned} \partial^0(\mathbf{b} * \mathbf{c}) &= \partial^0([b_1, \mathbf{b}^@ * \mathbf{c}]) + (-1)^n \partial^0([c_1, \mathbf{b} * \mathbf{c}^@]) \\ &= p(b_1)(\mathbf{b}^@ * \mathbf{c}) + (-1)^n p(c_1)(\mathbf{b} * \mathbf{c}^@) = (\partial^0 \mathbf{b}) * \mathbf{c} + (-1)^n \mathbf{b} * (\partial^0 \mathbf{c}) \circ \qquad \square \end{aligned}$$

1.6. LEMMA. *Suppose for some* $n \geq 0$ *and* $m \geq 0$, *we know that for all* $\mathbf{b} \in C_n$ *and all* $\mathbf{c} \in C_m$, $\partial(\mathbf{b} * \mathbf{c}) = (\partial \mathbf{b}) * \mathbf{c} + (-1)^n \mathbf{b} * (\partial \mathbf{c})$. *Then* $\widetilde{\partial}(\mathbf{b} * \mathbf{c}) = (\widetilde{\partial} \mathbf{b}) * \mathbf{c} + (-1)^n \mathbf{b} * (\widetilde{\partial} \mathbf{c})$

PROOF. This is immediate since ∂^0 is a $*$-derivation and if ∂ is, so is $\widetilde{\partial} = \partial^0 - \partial$. $\square$

1.7. LEMMA. *Suppose for some* $n \geq 0$ *and* $m \geq 0$, *we know that for all* $\mathbf{b} \in C_n$ *and all* $\mathbf{c} \in C_m$, $\partial(\mathbf{b} * \mathbf{c}) = (\partial \mathbf{b}) * \mathbf{c} + (-1)^n \mathbf{b} * (\partial \mathbf{c})$. *Then for all* $\mathbf{b} \in C_{n+1}$ *and* $\mathbf{c} \in C_{m+1}$, *we have* $\widetilde{\partial}(\mathbf{b} \mathbin{\widetilde{*}} \mathbf{c}) = (\widetilde{\partial} \mathbf{b}) \mathbin{\widetilde{*}} \mathbf{c} + (-1)^n \mathbf{b} \mathbin{\widetilde{*}} (\widetilde{\partial} \mathbf{c})$.

PROOF. Write ∂^1 for the first face operator, namely

$$\partial^1 [b_1, \ldots, b_{n+1}] = [b_1 b_2, \ldots, b_{n+1}] \circ$$

A computation shows that $\widetilde{\partial}\mathbf{b} = \partial^1\mathbf{b} - [b_1, \widetilde{\partial}\mathbf{b}^@]$, where the $\widetilde{\partial}$ on the right hand side is that of one lower dimension. Then

$$\begin{aligned}
\widetilde{\partial}(\mathbf{b} \mathbin{\widetilde{*}} \mathbf{c}) &= \widetilde{\partial}[b_1c_1, \mathbf{b}^@ * \mathbf{c}^@] \\
&= \partial^1 \Big([b_1c_1, \mathbf{b}^@ * \mathbf{c}^@] - [b_1c_1, \widetilde{\partial}(\mathbf{b}^@ * \mathbf{c}^@)]\Big) \\
&= \partial^1 \left([b_1c_1, b_2, \mathbf{b}^{@@} * \mathbf{c}^@] + (-1)^n[b_1c_1, c_2, \mathbf{b}^@ * \mathbf{c}^{@@}]\right) \\
&\quad - \Big([b_1c_1, (\widetilde{\partial}\mathbf{b}^@) * \mathbf{c}^@] + (-1)^n[b_1c_1, \mathbf{b}^@ * (\widetilde{\partial}\mathbf{c}^@)]\Big) \\
&= [b_1b_2c_1, \mathbf{b}^{@@} * \mathbf{c}^@] - [b_1c_1, (\widetilde{\partial}\mathbf{b}^@) * \mathbf{c}^@] \\
&\quad + (-1)^n \Big([b_1c_1c_2, \mathbf{b}^@ * \mathbf{c}^{@@}] - [b_1c_1, \mathbf{b}^@ * (\widetilde{\partial}\mathbf{c}^@)]\Big)
\end{aligned}$$

On the other hand,

$$\begin{aligned}
(\widetilde{\partial}\mathbf{b}) \mathbin{\widetilde{*}} \mathbf{c} &= (\partial^1\mathbf{b} - [b_1, \widetilde{\partial}\mathbf{b}^@]) \mathbin{\widetilde{*}} \mathbf{c} = \partial^1\mathbf{b} \mathbin{\widetilde{*}} \mathbf{c} - [b_1, \widetilde{\partial}\mathbf{b}^@] \mathbin{\widetilde{*}} \mathbf{c} \\
&= [b_1b_2, \mathbf{b}^{@@}] \mathbin{\widetilde{*}} [c_1, c^@] - [b_1, \widetilde{\partial}\mathbf{b}^@] \mathbin{\widetilde{*}} [c_1, \mathbf{c}^@] \\
&= [b_1b_2c_1, \mathbf{b}^{@@} * c^@] - [b_1c_1, (\widetilde{\partial}\mathbf{b}^@) * \mathbf{c}^@]
\end{aligned}$$

and similarly,

$$\mathbf{b} \mathbin{\widetilde{*}} (\widetilde{\partial}\mathbf{c}) = [b_1c_1c_2, \mathbf{b}^@ * \mathbf{c}^{@@}] - [b_1c_1, \mathbf{b}^@ * (\widetilde{\partial}\mathbf{c}^@)] \qquad \square$$

1.8. LEMMA. $\mathbf{b} * \mathbf{c} = \mathbf{b} \mathbin{\widetilde{*}} [1, \mathbf{c}] + (-1)^n[1, \mathbf{b}] \mathbin{\widetilde{*}} \mathbf{c}$.

PROOF. For $\mathbf{b} = [b_1, \ldots, b_n] \in C_n]$ and $\mathbf{c} = [c_1, \ldots, c_m] \in C_m$,

$$\begin{aligned}
\mathbf{b} \mathbin{\widetilde{*}} [1, \mathbf{c}] + (-1)^n[1, \mathbf{b}] \mathbin{\widetilde{*}} \mathbf{c} &= \mathbf{b} \mathbin{\widetilde{*}} [1, \mathbf{c}] + (-1)^n[1, \mathbf{b}] \mathbin{\widetilde{*}} \mathbf{c} \\
&= [b_1, \mathbf{b}^@] \mathbin{\widetilde{*}} [1, \mathbf{b}] + (-1)^n[1, \mathbf{b}] \mathbin{\widetilde{*}} [c_1, \mathbf{c}^@] \\
&= [b_1, \mathbf{b}^@ * \mathbf{c}] + (-1)^n[c_1, \mathbf{b} * \mathbf{c}^@] = \mathbf{b} * \mathbf{c} \qquad \square
\end{aligned}$$

We are now ready to prove the theorem. The case $n = m = 1$ is an immediate computation. We will suppose that the conclusion is valid for all pairs of indices whose sum is smaller than $n + m$. It follows from Lemma 1.7 that if $\mathbf{b} \in C_i$, $\mathbf{c} \in C_j$ and $i + j < m + n + 2$, then

$$\widetilde{\partial}(\mathbf{b} \mathbin{\widetilde{*}} \mathbf{c}) = (\widetilde{\partial}\mathbf{b}) \mathbin{\widetilde{*}} \mathbf{c} + (-1)^i\mathbf{b} \mathbin{\widetilde{*}} (\widetilde{\partial}\mathbf{c})$$

Since ∂^0 is a derivation with respect to $*$, it is sufficient to show that $\widetilde{\partial}$ is. We have, for $\mathbf{b} \in C_i$ and $\mathbf{c} \in C_j$,

$$\begin{aligned}
\widetilde{\partial}(\mathbf{b} * \mathbf{c}) &= \widetilde{\partial}(\mathbf{b} \mathbin{\widetilde{*}} [1, \mathbf{c}]) + (-1)^i\widetilde{\partial}([1, \mathbf{b}] \mathbin{\widetilde{*}} \mathbf{c}) \\
&= (\widetilde{\partial}\mathbf{b}) \mathbin{\widetilde{*}} [1, \mathbf{c}] + (-1)^{i-1}\mathbf{b} \mathbin{\widetilde{*}} (\widetilde{\partial}[1, \mathbf{c}]) \\
&\quad + (-1)^i\widetilde{\partial}[1, \mathbf{b}] \mathbin{\widetilde{*}} \mathbf{c} + [1, \mathbf{b}] \mathbin{\widetilde{*}} (\widetilde{\partial}\mathbf{c}) \\
&= (\widetilde{\partial}\mathbf{b}) \mathbin{\widetilde{*}} [1, \mathbf{c}] + (-1)^{i-1}\mathbf{b} \mathbin{\widetilde{*}} \mathbf{c} + (-1)^i\mathbf{b} \mathbin{\widetilde{*}} [1, \widetilde{\partial}\mathbf{c}] \\
&\quad + (-1)^i\mathbf{b} \mathbin{\widetilde{*}} \mathbf{c} + (-1)^{i+1}[1, \widetilde{\partial}\mathbf{b}] \mathbin{\widetilde{*}} \mathbf{c} - [1, \mathbf{b}] \mathbin{\widetilde{*}} (\widetilde{\partial}\mathbf{c}) \\
&= (\widetilde{\partial}\mathbf{b}) \mathbin{\widetilde{*}} [1, \mathbf{c}] + (-1)^{i-1}[1, \widetilde{\partial}\mathbf{b}] \mathbin{\widetilde{*}} \mathbf{c} \\
&\quad + (-1)^i \Big(\mathbf{b} \mathbin{\widetilde{*}} [1, \widetilde{\partial}\mathbf{c}] + (-1)^i[1, \mathbf{b}] \mathbin{\widetilde{*}} (\widetilde{\partial}\mathbf{c})\Big) \\
&= (\widetilde{\partial}\mathbf{b}) * \mathbf{c} + (-1)^i\mathbf{b} * (\widetilde{\partial}\mathbf{c}) \qquad \square
\end{aligned}$$

1.9. The shuffle idempotent. At this point, we define the Harrison complex of $p\colon B \longrightarrow A$ as follows. Let $S_\bullet^A(B)$ be the subspace of $C_\bullet$ consisting of all shuffles of elements of positive degrees. Then the Harrison complex is $C_\bullet/S_\bullet$. For convenience we will denote this quotient $C_\bullet^{\mathrm{Ha}}(B)$, suppressing mention of A.

1.10. THEOREM. *Suppose K is a field of characteristic 0. Then the projection $\phi\colon C_\bullet(B) \longrightarrow C_\bullet^{\mathrm{Ha}}(B)$ is a split epimorphism and the splitting maps are natural chain transformations.*

PROOF. It is clear that C_n is acted on by the symmetric group S_n by

$$\sigma^{-1}[b_1, \dots, b_n] = [b_{\sigma 1}, \dots, b_{\sigma n}]$$

When K is a field of characteristic 0 (although it would suffice that K be a commutative ring containing $\mathscr{Q}$), C_n becomes a module over the group algebra $\mathscr{Q}[S_n]$. We will be finding appropriate idempotents in the group algebra to give the splitting. If $e = \sum \lambda_\sigma \sigma$ is an element of the group algebra, we let $\operatorname{sgn}(e) = \sum \lambda_\sigma \operatorname{sgn}(\sigma)$, where sgn is the usual signum function. If we denote by ϵ_n the element $\frac{1}{n!}\sum_{\sigma\in S_n} \operatorname{sgn}(\sigma)\sigma$, then it is clear that for any $e \in S_n$, $e\epsilon_n = \epsilon_n e = \operatorname{sgn}(e)\epsilon_n$. The same equation will be true with e replaced by any element of the group. Since, $\operatorname{sgn}(\epsilon_n) = 1$, it also follows that ϵ_n is idempotent.

For the purposes of the next discussion, we will understand the **generic chain** in C_n, or the generic n-chain, to be the chain $[x_1, \dots, x_n]$ in the polynomial ring $B = K[x_1, \dots, x_n]$. An equation is true of the generic chain if and only if it is true for an arbitrary chain in an arbitrary algebra.

1.11. PROPOSITION. *Let $\mathbf{x}$ be the generic chain in C_n. Then $\partial\epsilon_n\mathbf{x} = 0$. If $e \in \mathscr{Q}[S_n]$ has the property that $\partial e\mathbf{x} = 0$, then e is a multiple of ϵ_n, namely $e = \operatorname{sgn}(e)\epsilon_n = e\epsilon_n$. In particular, if $e\epsilon_n = 0$ as well, then $e = 0$.*

PROOF. Let us calculate $n!\partial\epsilon_n\mathbf{x}$ to avoid fractions. For any $\sigma \in S_n$, there are two terms of the form $x_{\sigma 1}[x_{\sigma 2}, \cdots, x_{\sigma n}]$, namely the first term of the boundary of $\sigma^{-1}\mathbf{x}$ and the last term of the boundary of $(\sigma\zeta)^{-1}\mathbf{x}$, where $\zeta = (1 \;\; 2 \;\; \cdots \;\; n)$ is the cyclic permutation. In the first instance, it appears with the coefficient $\operatorname{sgn}(\sigma)$ and in the second with the coefficient $(-1)^n \operatorname{sgn}(\sigma)\operatorname{sgn}(\zeta) = -\operatorname{sgn}(\sigma)$ and so they cancel. In a similar way, the term $[x_{\sigma 1}, \dots, x_{\sigma i}x_{\sigma(i+1)}, \dots, x_{\sigma n}]$ appears twice, but with opposite sign, once as the ith term in the boundary of $\sigma^{-1}\mathbf{x}$ and once as the ith term of the boundary of $((i \;\; i+1)\,\sigma)^{-1}\mathbf{x}$. Thus all terms cancel.

For the converse, let $e = \sum \lambda_\sigma \sigma$. In the polynomial ring $\mathscr{Q}[x_1, \dots, x_n]$ we can carry out the same analysis to $e\mathbf{x}$ to conclude that if $e\mathbf{x} = 0$, then for all transpositions $\tau = (i \;\; i+1)$, we have $\lambda_\sigma = -\lambda_{\sigma\tau}$. Since the adjacent transpositions generate S_n, it follows that $\lambda_\sigma = \operatorname{sgn}(\sigma)\lambda_1$. The remaining statements are now evident. □

1.12. COROLLARY. *Let $\mathbf{x}$ be the generic n-chain and u, $v \in \mathscr{Q}[S_n]$. Then $u = v$ if and only if $\partial u\mathbf{x} = \partial v\mathbf{x}$ and $\epsilon_n u = \epsilon_n v$.*

This corollary is what allows us to apply induction to get the splitting we seek. Let us write s_{ij} for the operator on C_{i+j} defined by

$$s_{ij}\mathbf{x} = [x_1, \dots, x_i] * [x_{i+1}, \dots, x_{i+j}]$$

If we let $\partial_i : C_i \longrightarrow C_{i-1}$ denote the ith boundary operator, then the results of Theorem 1.4 can be summarized by the equation

$$\partial_{i+j} \circ s_{ij} = s_{i-1\,j} \circ (\partial_i \otimes 1) + (-1)^i s_{i\,j-1} \circ (1 \otimes \partial_j)$$

Note that ∂_1 is identically 0 on a symmetric module, so that the first term vanishes when $i = 1$ and the second term does when $j = 1$.

1.13. PROPOSITION. *When* $0 < m < n$, $\partial_n = \partial_{m+1} \otimes 1 + (-1)^m (1 \otimes \partial_{n-m})$.

PROOF. We calculate that

$$\begin{aligned}(\partial_{m+1} \otimes 1)\mathbf{x} &= (\partial[x_1, \dots, x_{m+1}]) \otimes [x_{m+2}, \dots, x_n] \\ &= p(x_1)[x_2, \dots, x_n] + \sum_{i=1}^{m} (-1)^i [x_1, \dots, x_i x_{i+1}, \dots, x_n] \\ &\quad + (-1)^{m+1} p(x_{m+1})[x_1, \dots, x_m, x_{m+2}, \dots, x_n]\end{aligned}$$

while

$$\begin{aligned}(-1)^m (1 \otimes \partial_{n-m})\mathbf{x} &= [x_1, \dots, x_m] \otimes (\partial[x_{m+1}, \dots, x_n]) \\ &= (-1)^m p(x_{m+1})[x_1, \dots, x_m, x_{m+2}, \dots, x_n] \\ &\quad + \sum_{i=m+1}^{n-1} (-1)^i [x_1, \dots, x_i x_{i+1}, \dots, x_n] \\ &\quad + (-1)^n x_n [x_1, \dots, x_{n-1}]\end{aligned}$$

and the sum of these two sums is evidently $\partial\mathbf{x}$. □

We now define $s_n : C_n \longrightarrow C_n$ as $\sum_{i=1}^{n-1} s_{i\,n-i}$.

1.14. PROPOSITION. $\partial \circ s_n = s_{n-1} \circ \partial$.

PROOF. We have

$$\begin{aligned}\partial_n \circ s_n &= \sum_{i=1}^{n-1} \partial_n \circ s_{i\,n-1} \\ &= \sum_{i=2}^{n-1} s_{i-1\,n-i} \circ (\partial_i \otimes 1) + (-1)^i \sum_{i=1}^{n-2} s_{i\,n-1-i} \circ (1 \otimes \partial_{n-i-1}) \\ &= \sum_{i=1}^{n-2} s_{i\,n-1-i} \circ (\partial_{i+1} \otimes 1) + (-1)^i \sum_{i=1}^{n-2} s_{i\,n-1-i} \circ (1 \otimes \partial_{n-i-1}) \\ &= \sum_{i=1}^{n-2} s_{i\,n-1-i} \circ (\partial_{i+1} \otimes 1 + (-1)^i (1 \otimes \partial_{n-i-1})) \\ &= \sum_{i=1}^{n-2} s_{i\,n-1-i} \circ \partial_n = s_{n-1} \circ \partial\end{aligned}$$

□

1.15. THEOREM. *There is a sequence of elements* $e_2 \in \mathcal{Q}[S_2], \dots, e_n \in \mathcal{Q}[S_n], \dots$, *with the following properties*:

(a) e_n *is a polynomial in* s_n *without constant term*;
(b) $\operatorname{sgn} e_n = 1$;
(c) $\partial \circ e_n = e_{n-1} \circ \partial$;
(d) $e_n^2 = e_n$;

(e) $e_n s_{i\,n-i} = s_{i\,n-1}$, *for* $0 < i < n$.

PROOF. One easily proves, using induction, that $\operatorname{sgn}(s_{ij})$ is the binomial coefficient $\binom{i+j}{i}$. It follows that $\operatorname{sgn}(s_n)$ is the sum of all the binomial coefficients except the first and last, so that $\operatorname{sgn}(s_n) = 2^n - 2$. Let $e_2 = \epsilon_2 = \frac{1}{2}s_2$. Assuming that we have found $e_2, e_3, \ldots, e_{n-1}$, satisfying the conditions above, let p be a polynomial such that $p(s_{n-1}) = e_{n-1}$. Define

$$e_n = p(s_n) + (1 - p(s_n))\frac{s_n}{\operatorname{sgn} s_n}$$

It is obvious that (a) is satisfied and (b) is an immediate calculation. Since $\partial \circ s_n = s_{n-1} \circ \partial$, it is immediate that $\partial \circ p(s_n) = p(s_{n-1}) \circ \partial$ and then we have

$$\begin{aligned}
\partial \circ e_n &= \partial \circ p(s_n) + \partial \circ (1 - p(s_n))\frac{s_n}{\operatorname{sgn} s_n} \\
&= p(s_{n-1}) \circ \partial + (1 - p(s_{n-1}))\frac{s_{n-1} \circ \partial}{\operatorname{sgn} s_n} \\
&= e_{n-1} \circ \partial + (1 - e_{n-1})\frac{s_{n-1} \circ \partial}{\operatorname{sgn} s_n} \\
&= e_{n-1} \circ \partial + \frac{(s_{n-1} - e_{n-1}s_{n-1}) \circ \partial}{\operatorname{sgn} s_n} \\
&= e_{n-1} \circ \partial
\end{aligned}$$

From $\partial \circ e_n^2 = e_{n-1}^2 \circ \partial = e_{n-1} \circ \partial = \partial \circ e_n$ together with $\operatorname{sgn}(e_n^2) = 1 = \operatorname{sgn}(e_n)$, we conclude from Proposition 1.11 that e_n is idempotent.

Finally, we calculate that

$$\begin{aligned}
\partial \circ e_n \circ s_{i\,n-i} &= e_{n-1} \circ \partial \circ s_{i\,n-i} \\
&= e_{n-1} \circ (s_{i-1} \circ (\partial_i \otimes 1) + (-1)^i s_{i,n-1-i} \circ (1 \otimes \partial_{n-i})) \\
&= s_{i-1} \circ (\partial_i \otimes 1) + (-1)^i s_{i,n-1-i} \circ (1 \otimes \partial_{n-i}) = s_{n-1} \circ \partial
\end{aligned}$$

In addition, $\epsilon_n e_n s_{i\,n-i} = \epsilon_n s_{i\,n-i}$ so we conclude from Proposition 1.11 that $e_n s_{i\,n-i} = s_{i\,n-i}$. ☐

From (a) it follows that $\operatorname{im} e_n \subseteq \operatorname{im} s_n$, while from (e) and the definition of s_n, we see that $\operatorname{im} s_{i\,n-i} \subseteq \operatorname{im} e_n$ for $i = 1, \ldots, n-1$, so that

$$\operatorname{im} e_n \subseteq \operatorname{im} s_n \subseteq \sum_{i=1}^{n-1} \operatorname{im} s_{i\,n-i} \subseteq \operatorname{im} e_n$$

and thus the image of e_n is exactly the shuffles. Since e_n is idempotent, this image along with the quotient modulo it is split and, from (c), the splitting is compatible with the boundary operator and so we have split the chain (as well as cochain) complexes, as seen in the following.

1.16. COROLLARY. *In characteristic 0, the Harrison chain complex is a direct summand of the Hochschild chain complex, being the chains that are annihilated by* e_n. ☐

We call e_n the nth **shuffle idempotent** since it characterizes sums of shuffles as though permutations it preserves. Gerstenhaber and Schack have made remarkable use of it as the first idempotent of their "Hodge decomposition" of the Hochschild cohomology groups of a commutative algebra, which we briefly describe.

We begin with the observation that $e_2 = \epsilon_2$, which we call e_{22}. If we let $e_{21} = 1 - e_2$, we have $1 = e_{21} + e_{22}$ is a sum of orthogonal idempotents, one of which is the shuffle idempotent. The idempotent e_3 can be written as $e_3 = e_{32} + e_{33}$, where $e_{32} = e_3 - \epsilon_3$ and $e_{33} = \epsilon_3$ are orthogonal idempotents. If we let $e_{31} = 1 - e_3$, then we have $1 = e_{31} + e_{32} + e_{33}$ and $e_3 = e_{32} + e_{33}$. Moreover, $\partial \circ e_{31} = e_{21} \circ \partial$, $\partial \circ e_{32} = e_{22} \circ \partial$ and $\partial \circ e_{33} = 0$. For general n, they make clever use of Corollary 1.12 to show inductively that the characteristic polynomial of s_n acting on the subring of $\mathscr{Q}[s_n] \subseteq \mathscr{Q}[S_n]$ generated by s_n is

$$\mu_n(t) = t(t-2)(t-6)\cdots(t-(2^n-2)) = (t-(2^n-2))\mu_{n-1}(t)$$

Since the eigenvalues are distinct, it follows that $\mathscr{Q}[s_n]$ is a direct sum of one dimensional ideals generated by idempotents e_{ni}, which generates the kernel of $s_n - (2^i - 2)$. The idempotent corresponding to $2^n - 2$ is ϵ_n and it is not hard to prove, using 1.12 once more, that $\partial \circ e_{ni} = e_{n-1\,i}$ for $i < n$ and $\partial \circ e_{nn} = 0$. It is also not hard to show that

$$e_n = 1 - e_{n1} = e_{n2} + \cdots + e_{nn}$$

These idempotents divides the nth Hochschild chain group into a direct sum of n pieces and, being compatible with the boundary, do the same for the homology and cohomology groups. The commutative cohomology is the first piece. The remaining pieces are not well understood, but the whole process is reminiscent of the Hodge decomposition of the (co-)homology of a manifold.

1.17. Harrison cohomology of polynomial rings. Free commutative algebras are polynomial algebras. We want to use the splitting of the Hochschild cohomology to show that the Harrison cohomology of polynomial algebras vanishes in characteristic 0, which is a necessary condition for the Harrison cohomology being isomorphic to the cohomology derived from the free algebra resolution. In light of the acyclic models theorem, it is close to being sufficient. The basic idea is to begin by analyzing the Hochschild cohomology and showing that all the cohomology classes are represented by cochains that are in the image of s_n. We will do this for polynomial rings in a finite number of indeterminates by a counting argument and then use a direct limit construction for the general case. Let $B = K[x_1, \ldots, x_n]$ be a polynomial ring in n variables. We will be treating K as a B-module in which the variables annihilate K. We begin with a preliminary result. In the statement, we form the tensor product (over K) of the ring B with D-modules M and N. It is understood that the resultant objects are $B \otimes D$-modules in the obvious way, which is to say that $(b \otimes d)(b' \otimes m) = bb' \otimes dm$.

1.18. PROPOSITION. *Let B and D be algebras over the commutative ring K with B being K-flat. Then for any right D-module M and left D-module N, we have* $\mathrm{Tor}_\bullet^{B\otimes D}(B \otimes M, B \otimes N) \cong B \otimes \mathrm{Tor}_\bullet^D(M, N)$.

PROOF. Let $P_\bullet$ be an D-projective resolution of N. Then $B \otimes P_\bullet$ is an $B \otimes D$ projective resolution of $B \otimes N$. It is exact because B is flat. If we apply $(B \otimes M) \otimes_{B\otimes D} -$, we get

$$(B \otimes M) \otimes_{B\otimes D} (B \otimes P_\bullet) \cong B \otimes M \otimes_D P_\bullet$$

Since B is K-flat, the functor $B \otimes -$ is exact, which means it commutes with homology, so that the homology of the right hand side is $B \otimes \mathrm{Tor}_\bullet^D(M, N)$. The homology of the left hand side of that isomorphism is, of course, $\mathrm{Tor}_\bullet^{B\otimes D}(B \otimes M, B \otimes N)$. □

1.19. COROLLARY. $\mathrm{Tor}^{B\otimes B}(B\otimes K, B\otimes K)\cong B\otimes \mathrm{Tor}^B(K,K)$

Of course, $B\otimes K\cong B$ as an abelian group (and K-module) but not as an B-module. The reason is that on $B\otimes K$ the variables in the second copy of B act trivially, which is not true in B. We have, however, the following.

1.20. PROPOSITION. *In the category of all modules (see 1.11), $(B\otimes B, B)\cong (B\otimes B, B\otimes K)$.*

PROOF. The ring $B\otimes B$ is a polynomial ring in $2n$ indeterminates $x_1\otimes 1,\dots,x_n\otimes 1, 1\otimes x_1,\dots,1\otimes x_n$. The action on B is that both $x_i\otimes 1$ and $1\otimes x_i$ act as multiplication by x_i. The action on $B\otimes K\cong B$ is that $x_i\otimes 1$ acts as multiplication by x_i, but $1\otimes x_i$ acts trivially. We now define $(\phi, f)\colon (B\otimes B, B)\longrightarrow (B\otimes B, B\otimes K)$ by $\phi(x_i\otimes 1)=x_i\otimes 1$, $\phi(1\otimes x_i)=x_i\otimes 1-1\otimes x_i$ and $f(r)=r\otimes 1$. Then

$$\phi(x_i\otimes 1)f(r)=(x_i\otimes 1)(r\otimes 1)=x_ir\otimes 1=f(x_ir)=f((x_i\otimes 1)r)$$

and

$$\phi(1\otimes x_i)f(r)=(x_i\otimes 1-1\otimes x_i)(r\otimes 1)=x_ir\otimes 1=f(x_ir)=f((1\otimes x_i)r)$$

which shows that (ϕ, f) is a morphism in the category of modules. The inverse is given by (γ, g) where $\gamma(x_i\otimes 1)=x_i\otimes 1$, $\gamma(1\otimes x_i)=x_i\otimes 1+1\otimes x_i$ and $g(r\otimes a=ar)$. These maps are readily seen to be inverse to each other. □

1.21. PROPOSITION. $\mathrm{Tor}^{B\otimes B}(B,B)\cong B\otimes \mathrm{Tor}^B(K,K)$ *as K-modules.*

PROOF. If I is a set, denote by $I\cdot B$ the direct sum of I copies of B. There is a free resolution of K of the form

$$\cdots\longrightarrow I_n\cdot B\longrightarrow I_{n-1}\cdot B\longrightarrow\cdots\longrightarrow I_1\cdot B\longrightarrow I_0\cdot B$$

By using 1.11.1, we have a long sequence augmented over $B\otimes K$ using the same (or isomorphic) objects. Exactness is a property of sequences of abelian groups so the isomorphic sequence is also exact. The result will follow from the next proposition.

1.22. PROPOSITION. *Tensor product is defined as a functor*

$$B\text{–}\mathsf{Rmod}\times_{\mathsf{Ring}} B\text{–}\mathsf{Lmod}\longrightarrow \mathsf{Ab}$$

PROOF. The domain category has as objects all pairs $((B,M),(B,M'))$ where M is a right B-module and M' is a left B-module. An arrow is a 3-tuple

$$(\phi, f, f')\colon ((B,M),(B,M'))\longrightarrow ((D,N),(D,N'))$$

such that (ϕ, f) is an arrow of B-Rmod and (ϕ, f') is an arrow of Lmod. On objects we define $(B,M)\otimes(B,M')=M\otimes_B M'$. Suppose that

$$(\phi, f, f')\colon ((B,M),(B,M'))\longrightarrow ((D,N),(D,N'))$$

is an arrow. Then the composite $g\colon M\times M'\xrightarrow{f\times f'} N\times N'\longrightarrow N\otimes_D N'$ has to be shown to be middle bilinear. It is evidently biadditive since f and f' are additive. For the middle exchange, we have

$$\begin{aligned} g(mr,m') &= f(mr)\otimes f'(m')=f(m)\phi(r)\otimes f'(m')\\ &= f(m)\otimes\phi(r)f'(m')=f(m)\otimes f'(rm')=g(m,rm') \end{aligned}$$ □

1.23. PROPOSITION. *Let B be as above. Then $\mathrm{Tor}_m^B(K,K)$ is a free B-module of dimension $\binom{n}{m}$.*

PROOF. We will prove this by induction on the number of variables. When $n = 0$, it is obvious. From the inductive definition of the binomial coefficients, it will follow immediately from,

1.24. PROPOSITION. *For any ring B, right B-module M and left B-module N, we have $M \otimes_{B[x]} N \cong M \otimes_B N$ and for $n > 0$, $\mathrm{Tor}_n^{B[x]}(M, N) \cong \mathrm{Tor}_n^B(M, N) \oplus \mathrm{Tor}_{n-1}^B(M, N)$.*

PROOF. There is an exact sequence of $B[x]$-modules

$$0 \longrightarrow B[x] \xrightarrow{x} B[x] \longrightarrow B \longrightarrow 0$$

whose first three terms constitute an $B[x]$-projective resolution of B. Let $P_\bullet$ be an B-projective resolution of N. Then in the double complex

$$\begin{array}{ccccccccc}
 & & 0 & & 0 & & & & 0 \\
 & & \downarrow & & \downarrow & & & & \downarrow \\
\cdots & \longrightarrow & B[x] \otimes P_n & \to & B[x] \otimes P_{n-1} & \longrightarrow & \cdots & \longrightarrow & B[x] \otimes P_0 \to 0 \\
 & & \downarrow & & \downarrow & & & & \downarrow \\
\cdots & \longrightarrow & B[x] \otimes P_n & \to & B[x] \otimes P_{n-1} & \longrightarrow & \cdots & \longrightarrow & B[x] \otimes P_0 \to 0 \\
 & & \downarrow & & \downarrow & & & & \downarrow \\
\cdots & \longrightarrow & P_n & \longrightarrow & P_{n-1} & \longrightarrow & \cdots & \longrightarrow & P_0 \longrightarrow 0 \\
 & & \downarrow & & \downarrow & & & & \downarrow \\
 & & 0 & & 0 & & & & 0
\end{array}$$

The double complex as a whole is acyclic, from Theorem 3.6.3, since each column is. It follows that the subcomplex consisting of the top two non-zero rows is homologous to the bottom row, whose homology is M concentrated in degree 0. Since each term in that subcomplex is $B[x]$ projective, it follows that those top two rows are an $B[x]$-projective resolution of M. Condensing this into a single complex, we have an $B[x]$ projective resolution of M

$$\cdots \longrightarrow (B[x] \otimes P_n) \oplus (B[x] \otimes P_{n-1}) \longrightarrow (B[x] \otimes P_{n-1}) \oplus (B[x] \otimes P_{n-2})$$
$$\longrightarrow \cdots \longrightarrow B[x] \otimes P_0 \longrightarrow 0$$

The boundary operator has matrix

$$\begin{pmatrix} B[x] \otimes d & x \otimes \mathrm{id} \\ 0 & B[x] \otimes d \end{pmatrix}$$

After applying $M \otimes_{B[x]} -$ we get the complex

$$\cdots \longrightarrow (M \otimes P_n) \oplus (M \otimes P_{n-1}) \longrightarrow (M \otimes P_{n-1}) \oplus (M \otimes P_{n-2})$$
$$\longrightarrow \cdots \longrightarrow M \otimes P_0 \longrightarrow 0$$

However, since x is the 0 operator on M, the boundary is now

$$\begin{pmatrix} M \otimes d & 0 \\ 0 & M \otimes d \end{pmatrix}$$

Now the conclusion is evident. □

1.25. PROPOSITION. *Let B be a polynomial ring over K and M be a submodule of a free B-module such that $K \otimes_B M = 0$. Then $M = 0$.*

PROOF. Let J be the ideal generated by the variables. Since

$$0 \longrightarrow J \longrightarrow B \longrightarrow K \longrightarrow 0$$

is exact, so is

$$J \otimes_B M \longrightarrow B \otimes_B M \cong M \longrightarrow K \otimes_B M \longrightarrow 0$$

so that $K \otimes_B M = 0$ implies that $JM = M$. Therefore $M = \bigcap_i J^i M$, which is impossible for a free module or any non-zero submodule since $\bigcap_{i=1}^{\infty} J^i M \subseteq \bigcap_{i=1}^{\infty} J^i F = 0$. □

1.26. PROPOSITION. *In $C_m(B) \otimes_B K$, there is a K-subspace of dimension (nm) consisting of cocycles, that is independent modulo boundaries and on which s_m acts as the identity.*

PROOF. For each subset $i_1 < i_2 < \cdots < i_m$ of $\{1, \ldots, n\}$, let $\langle x_{i_1}, \ldots, x_{i_m} \rangle$ denote $\epsilon_m[x_{i_1}, \ldots, x_{i_m}]$. These are clearly linearly independent and there exactly (nm) of them. It follows from $d \circ \epsilon_m = 0$ that these are cycles and from $s_m \circ \epsilon_m = \epsilon_m$ that s_m acts as the identity.

Let $B_m(B) \otimes_B K$ denote the group of boundaries. Now $C_m(B) \otimes_B K$ is free on all forms $\mathbf{b} = [b_1, \ldots, b_m]$, where $b_1, \ldots, b_m$ are monomials, so that $B_m(B)$ is generated by all $d[b_1, \ldots, b_m]$. If for some i, $b_i = 1$ while $b_{i-1} \neq 1 \neq b_{i+1}$, then $d^{i-1}\mathbf{b} = d^i\mathbf{b}$ so that those terms cancel. If there are two or more consecutive 1s, then every term of $d\mathbf{b}$ has a 1. If $b_1 \neq 1$, then $d^0\mathbf{b} = 0$ because all variables annihilate K and similarly for d^m. Finally, if $b_{i-1} \neq 1 \neq b_i$, then the $i-1$st term of $d^i\mathbf{b}$ has degree greater than 1. Thus no term of $d\mathbf{b}$ has a single term consisting of monomials of degree exactly 1. But every $\langle x_{i_1}, \ldots, x_{i_m} \rangle$ has exactly that form. □

1.27. COROLLARY. *$C_\bullet^{\mathrm{Ha}}(B) \otimes_B K = 0$ is exact.*

PROOF. For we have the exact sequence

$$0 \longrightarrow C_\bullet^{\mathrm{Ha}}(B) \otimes B_K \longrightarrow C_\bullet(B) \otimes B_K \longrightarrow s_n C_\bullet(B) \otimes_B K \longrightarrow 0$$

and we have just seen that the second map induces an isomorphism of homology. □

1.28. PROPOSITION. *$H(C_\bullet^{\mathrm{Ha}}(B) \otimes B_K) \cong H(C_\bullet^{\mathrm{Ha}}(B)) \otimes B_K$.*

PROOF. We know from 1.23 that the homology of the Hochschild complex consists of B projectives. These facts are still true for the Harrison subcomplex, because that is a retract. Thus it is sufficient to show,

1.29. LEMMA. *Suppose $C_\bullet \longrightarrow 0$ is a chain complex of B-projectives whose homology also consists of B projectives. Then for any B-module M, $H(C_\bullet \otimes_B M) \cong H(C_\bullet) \otimes_B M$.*

PROOF. Suppose the starting degree of $C_\bullet$ is 0, that is that $C_n = 0$ for $n < 0$. Let B_i, Z_i and H_i denote the ith boundary, cycle and homology groups. We have exact sequences

$$0 \longrightarrow Z_i \longrightarrow C_i \longrightarrow B_{i-1} \longrightarrow 0$$
$$0 \longrightarrow B_i \longrightarrow Z_i \longrightarrow H_i \longrightarrow 0$$

Since $B_{-1} = 0$, it follows that $Z_0 = C_0$ is projective. But then Z_0 and H_0 are projective and therefore B_0 is. Continuing in this way, we see inductively that each Z_i and B_i is projective. But then the sequences

$$0 \longrightarrow Z_i \otimes_B M \longrightarrow C_i \otimes_B M \longrightarrow B_{i-1} \otimes_B M \longrightarrow 0$$

$$0 \longrightarrow B_i \otimes_B M \longrightarrow Z_i \otimes_B M \longrightarrow H_i \otimes_B M \longrightarrow 0$$

are exact from which it follows that the homology of $C_\bullet \otimes_B M$ is $H_\bullet \otimes_B M$.

1.30. COROLLARY. *A polynomial ring in a finite number of indeterminates has trivial Harrison homology and cohomology.*

PROOF. Just put together the preceding with Corollary 1.27 and Proposition 1.25. □

1.31. THEOREM. *The Harrison homology and cohomology of a polynomial ring is trivial.*

PROOF. For a polynomial ring in finitely many indeterminates, we have seen this already. Any cycle in the Harrison chain complex of an arbitrary polynomial ring involves only a finitely many indeterminates and so is a cycle in the chain group of a finite polynomial algebra and hence a boundary. Thus the chain complex is exact. When B is a polynomial ring, the module of differentials is B-projective and so the chain complex augmented over Diff(B) is contractible, whence the cohomology is also trivial. □

1.32. COROLLARY. *For algebras over a field of characteristic 0, the Harrison homology and cohomology theories are weakly equivalent to the cotriple homology and cohomology.*

PROOF. On each ring there is a map of the complexes that is a homotopy equivalence. Thus we can use the acyclic models by taking Γ to be the weakly contractible complexes. This gives us weak homotopy equivalences between the Harrison and cotriple cohomology theories. We have not constructed, nor do we know how to construct, a natural homotopy inverse. □

1.33. An example in finite characteristic. Let K be a field of characteristic $p \neq 0$. We will show that there is a non-trivial Harrison cohomology class in degrees $2p^m$ for any $m > 0$. In particular, $\mathscr{H}^4_{\mathrm{Ha}}(B, B) \neq 0$ when K has characteristic 2.

We begin by counting the number of even less the number of odd permutations in s_{ij}. Call this number q_{ij}. It is apparent from the inductive definition of s_{ij} that

$$q_{ij} = q_{ji}$$

$$q_{i1} = \begin{cases} 1 & \text{if } i \text{ is even} \\ 0 & \text{if } i \text{ is odd} \end{cases}$$

$$q_{ij} = \begin{cases} q_{i-1\,j} + q_{i\,j-1} & \text{if } i \text{ is even} \\ q_{i-1\,j} - q_{i\,j-1} & \text{if } i \text{ is odd} \end{cases}$$

from which we can show by induction that

$$q_{ij} = \begin{cases} 0 & \text{if } i \text{ and } j \text{ are both odd} \\ (\,[i/2] + [j/2][i/2]\,) & \text{otherwise} \end{cases}$$

From this and standard properties of binomial coefficients, we see that when $i+j = n = 2p^m$, then for all $0 < i < n$, p divides n. Now define a cochain f of degree n on $K[x]$ with coefficients in K by

$$f[x^{i_1}, \dots, x^{i_n}] = \begin{cases} 1 & \text{if } i_1 = \dots = i_n = 1 \\ 0 & \text{otherwise} \end{cases}$$

Then $fs_{i\,n-i}[x, \dots, x] = q_{i,n-i}f[x, \dots, x] = 0$ for $0 < i < n$ from which it follows that $fs_n[x, \dots, x] = 0$ and since e_n is a polynomial without constant in s_n, it follows that $fe_n[x, \dots, x] = 0$ and, since f vanishes on all other terms, that $fe_n = 0$ and therefore f is a Harrison cochain. On the other hand, it is trivially seen to be a cocycle and, from the previous analysis, the linear space generated by this cycle does not meet the coboundaries except in 0.

1.34. $n!$ **suffices.** Although there is no known application for this fact, it is at least of minor interest that in the construction of the idempotent $e_n \in \mathcal{Q}[S_n]$, the denominator is only $n!$ instead of the $\prod_{i=1}^{n}(2^i - 2)$ that appears in the construction. Thus, for example, e_4, e_5 and e_6 are definable over a field of characteristic 7, even though $7|2^4 - 2$.

Let $B = \mathcal{Q}[x_1, x_2, \dots]$ be the polynomial ring in countably many variables and let $\mathbf{x}$ be the chain $[x_1, x_2, \dots, x_n] \in C_n(B)$. Also write C_n for $C_n(B)$.

A most important observation is that if $c \in \mathcal{Q}[S_n]$ is such that $\partial c\mathbf{x} \in \mathbf{Z}[S_{n-1}C_{n-1}]$, then there is an $a \in \mathcal{Q}$ such that $c - a\epsilon_n \in \mathbf{Z}[S_n]$. The proof of this assertion is essentially the same as the proof that if $\partial c\mathbf{x} = 0$, then c is a multiple of ϵ_n. In the computation of $\partial c\mathbf{x}$ each term appears twice one from a term of the form $r_\sigma \sigma\mathbf{x}$ and once from $r_\tau \tau\mathbf{x}$ where $\sigma = (i \quad i+1)\tau$ and the only way that they can contribute an integer to the sum is if $r_\sigma - r_\tau$ is an integer. Let b_σ be the fractional part of r_σ. Then $b_\sigma - b_\tau$ is an integer which must be 0. This is true whenever σ and τ differ by an adjacent transposition. But the adjacent transpositions generate S_n.

From this we conclude that if $\partial c\mathbf{x} \in \mathbf{Z}[S_{n-1}]C_{n-1}$ and $c\epsilon_n$ is an integer multiple of ϵ_n, then $c \in \mathbf{Z}[S_n]$. In fact, if we write $c = c' + a\epsilon_n$ with $c' \in \mathbf{Z}[S_n]$, then $c\epsilon_n = c'\epsilon_n + a\epsilon_n$. Since c' has integer coefficients, $c'\epsilon_n$ clearly is an integer multiple of ϵ_n and if $c\epsilon_n$ is too, then a must be an integer.

Now suppose that $p \in \mathcal{Q}[t]$ is a polynomial such that $p(s_{n-1})$ has integer coefficients. (We do not assume that p has integer coefficients.) Then $\partial p(s_n)\mathbf{x} = p(s_{n-1})\partial\mathbf{x}$ has integer coefficients. Moreover, $p(s_n)\epsilon_n = p(s_n\epsilon_n) = p((2^n - 2)\epsilon_n) = p(2^n - 2)\epsilon_n$.

We apply this to the polynomial $p(t) = (n-1)!p_{n-1,i}(t)$ where

$$p_{n-1\,i}(t) = \frac{(t - \lambda_1) \cdots (t - \lambda_{i-1})(t - \lambda_{i+1}) \cdots (t - \lambda_{n-1})}{(\lambda_i - \lambda_1) \cdots (\lambda_i - \lambda_{i-1})(\lambda_i - \lambda_{i+1}) \cdots (\lambda_i - \lambda_{n-1})}$$

Here $\lambda_i = 2^i - 2$ à la Gerstenhaber and Schack. I claim that

$$p_{n-1\,i}(\lambda_n) = \frac{(\lambda_n - \lambda_1) \cdots (\lambda_n - \lambda_{i-1})(\lambda_n - \lambda_{i+1}) \cdots (\lambda_n - \lambda_{n-1})}{(\lambda_i - \lambda_1) \cdots (\lambda_i - \lambda_{i-1})(\lambda_i - \lambda_{i+1}) \cdots (\lambda_i - \lambda_{n-1})}$$

is an integer (even without the $(n-1)!$ factor). We break it up into two factors. The second is

$$\frac{(2^n - 2^{i+1}) \cdots (2^n - 2^{n-1})}{(2^i - 2^{i+1}) \cdots (2^i - 2^{n-1})}$$

which after reversing the numerator is, up to a sign

$$\frac{(2^n - 2^{i+1}) \cdots (2^n - 2^{n-1})}{(2^{n-1} - 2^i) \cdots (2^{i+1} - 2^i)} = 2 \cdot 4 \cdots 2^{n-i-1}$$

(Note that $i \leq n-1$.) The first factor is

$$\frac{(2^n - 2) \cdots (2^n - 2^{i-1})}{(2^i - 2) \cdots (2^i - 2^{i-1})}$$

This numerator and denominator in this fraction are clearly divisible by the same factor of 2. For the rest, we work modulo powers of two in the multiplicative group of positive rationals. We have

$$\begin{aligned}\frac{(2^n-2)\cdots(2^n-2^{i-1})}{(2^i-2)\cdots(2^i-2^{i-1})} &= \frac{(2^{n-1}-1)\cdots(2^{n-1}-2^{i-2})}{(2^{i-1}-1)\cdots(2^{i-1}-2^{i-2})}\\ &= \frac{(2^{n-1}-1)\cdots(2^{n-1}-2^{i-2})(2^{n-1}-2^{i-1})(2^{n-1}-2^{n-2})}{(2^{i-1}-1)\cdots(2^{i-1}-2^{i-2})(2^{n-1}-2^{i-1})(2^{n-1}-2^{n-2})}\\ &= \frac{(2^{n-1}-1)\cdots(2^{n-1}-2^{i-2})(2^{n-1}-2^{i-1})\cdots(2^{n-1}-2^{n-2})}{(2^{i-1}-1)\cdots(2^{i-1}-2^{i-2})(2^{n-i}-1)\cdots(2^{n-i}-2^{n-i-1})}\\ &= \frac{f(n-1)}{f(i-1)f(n-i)}\end{aligned}$$

where $f(k) = (2^k-1)(2^k-2)\cdots(2^k-2^{k-1})$. One way of seeing that $\frac{f(k+l)}{f(k)f(l)}$ is always an integer, is to begin by showing that $f(k)$ is the order of $\mathrm{GL}_k(2)$. For in choosing an invertible kxk matrix we must first choose a non-zero vector in k dimensional space, of which there are exactly 2^k-1 choices. Having chosen one such vector, the second row is any vector not a multiple of the first row, of which there are 2^k-2 choices. The first two rows span a space of four vectors and any vector not in that space is an admissible third row, thus there are 2^k-4 choices and so on. Now we can embed $\mathrm{GL}_k(2)$ and $\mathrm{GL}_l(2)$ into $\mathrm{GL}_{k+l}(2)$ as the automorphisms of some k dimensional subspace and some complementary l dimensional subspace of an m dimensional space. These subgroups are disjoint and commute pointwise so that their product is a subgroup of order $f(k)f(l)$ and the index is the number we seek.

The conclusion is that $(n-1)!p_{n-1\,i}(s_n)$ is an integer. Now suppose that $p(t)$ is a polynomial such that $p(s_{n-1})$ is idempotent and such that $p(s_n)\epsilon_n = a\epsilon_n$. Then I claim that a is the unique number such that $p(s_n)-a\epsilon_n$ is an idempotent orthogonal to ϵ_n. In fact, from the fact that $p(s_{n-1})$ is idempotent, it is immediate that $p(s_n)^2 - p(s_n)$ is a multiple of ϵ_n and clearly the coefficient is a^2-a. Then $(p(s_n)-a\epsilon_n)^2 = p(s_n)^2 - 2ap(s_n)\epsilon_n + a^2\epsilon_n = p(s_n) + (a^2-a)\epsilon_n - 2a^2\epsilon_n + a^2\epsilon_n = p(s_n) - a\epsilon_n$. The uniqueness is clear. It follows from the preceding analysis that $p_{ni} = p_{n-1\,i} - p_{n-1\,i}(2^n-2)\epsilon_n$ and the first term uses only $(n-1)!$ in the denominator, while the second uses $n!$.

1.35. Exercise.

1. The purpose of this exercise is to sketch another proof that when $A = K[x_1,\dots,x_n]$, then $\mathrm{Tor}_m^A(K,K)$ is a vector space of dimension $\binom{n}{m}$. Let A_m denote the free A-module generated by all sequences $[x_{j_1},\dots,x_{j_m}]$ such that $j_1 < j_2 < \cdots < x_{j_m}$. The module $A_0 = A$ is thought of as generated by an empty bracket. Define $\partial\colon A_m \longrightarrow A_{m-1}$ to be the A-linear map such that

$$\partial[x_{j_1},\dots,x_{j_m}] = \sum_{i=1}^{m}(-1)^i x_{j_i}[x_{j_1},\dots,\widehat{x}_{j_i},\dots,x_{j_m}]$$

Also define $\partial\colon A_0 \longrightarrow A$ by $\partial[\,] = 1$.

(a) Show that this is a chain complex.

(b) Define a K-linear map $\sigma\colon A_m \longrightarrow A_{m+1}$ as follows. There is a K-basis of A_m consisting of all $\mu[x_{j_1},\dots,x_{j_m}]$ where μ is a monomial, possibly 1. If $\mu \neq 1$, let j be the smallest index of a variable in μ. Then define

$$s(\mu[x_{j_1},\dots,x_{j_m}]) = \begin{cases}(\mu/x_j)[x_j,x_{j_1},\dots,x_{j_m}] & \text{if } j<j_1\\ 0 & \text{otherwise}\end{cases}$$

In particular $s[x_{j_1}, \ldots, x_{j_m}] = 0$. In degree -1, $\sigma 1 = [\,]$. Show that $\sigma \circ \partial + \partial \circ \sigma = \mathrm{id}$.

(c) Show that $K \otimes_A A_m$ is the vector space generated by all $[x_{j_1}, \ldots, x_{j_m}$ for $j_1 < \cdots < j_m$ and that $K \otimes_A \partial = 0$.

(d) Conclude that $\mathrm{Tor}^A_m(K, K)$ is a vector space of dimension $(\,nm\,)$.

1.36. Historical comment. In my 1962 dissertation, I gave explicit proofs of the splitting of the Hochschild complex in dimensions 3 and 4 over fields of characteristic not 2 or 3. (The remaining dimensions were settled in [Barr, 1968b]). The argument above, from 1.17 to here, was sketched out *instantly* by Harrison as soon as I had told him about the splitting. Although we never discussed it (and his memory for this part of his career is now virtually nil), it seems clear in retrospect that he was already aware that the splitting of the Hochschild complex would lead to this proof of the vanishing of the cohomology of polynomial rings. The argument he gave in [Harrison, 1962] (limited to dimension 2 and 3) was based on giving an explicit proof for a polynomial ring in one generator and then proving that the cohomology of a tensor product of commutative algebras is the direct product of the cohomology groups of the factors. My guess is that he had used this argument in dimension 2 (where the splitting is obvious) then did not push it through in dimension 3, but instead found the different argument using tensor products. André proves the general tensor product theorem in his 1967 notes, but his definition of the cohomology is rather obviously equivalent to the one derived from the cotriple resolution.

2. More on cohomology of commutative cohomology

2.1. André cohomology. Michel André [1967, 1974] developed the commutative cohomology into a powerful tool for studying commutative algebras. He used a definition of cohomology via polynomial algebra resolutions that was equivalent to the polynomial algebra cotriple cohomology.

Let K be a commutative ring and A be a commutative K-algebra. A polynomial resolution of A is a simplicial K-algebra

$$\cdots \;\underset{d^{n+1}}{\overset{d^0}{\vdots}}\; A_n \;\underset{d^{n}}{\overset{d^0}{\vdots}}\; A_{n-1} \;\underset{d^{n-1}}{\overset{d^0}{\vdots}}\; \cdots \underset{d^1}{\overset{d^0}{\rightrightarrows}} A_0 \xrightarrow{d} A$$

for which each A_n, $n \geq 0$ is a polynomial algebra and for which the associated chain complex

$$\cdots \xrightarrow{\sum(-1)^{d^i}} A_n \xrightarrow{\sum(-1)^{d^i}} A_{n-1} \longrightarrow \cdots \xrightarrow{d^0 - d^1} A_0 \xrightarrow{d} A \longrightarrow 0$$

is exact. Then the cohomology with coefficients in the A-module M is defined to be that of the cochain complex

$$0 \longrightarrow \mathrm{Der}(A_0, M) \longrightarrow \cdots \longrightarrow \mathrm{Der}(A_n, M) \longrightarrow \mathrm{Der}(A_{n+1}, M) \longrightarrow \cdots$$

using, in degree n, the map $\sum_{i=0}^{n+1}(-1)^i d^i$. Since the polynomial algebra cotriples complex satisfies André's conditions, his cohomology theory is just the cotriple cohomology.

2.2. The long exact sequence. We wish to show that the homology and cohomology of pairs introduced in 3.8 takes a particularly useful form in the case of commutative algebras. A commutative algebra homomorphism $A \longrightarrow B$ defines B as an A-algebra. Thus for a B-module M we have both $H^\bullet_A(B,M)$ and $H^\bullet_K(B,A,M)$. What we want to do in this section is to show that the two groups are naturally isomorphic. The facts for homology are similar, but we give proofs mainly for cohomology. We begin with,

2.3. THEOREM. *Suppose K is a commutative ring and A and B are K-algebras such that $\mathrm{Tor}^K_n(A,B) = 0$ for all $n > 0$. Then for any A, B-bimodule M, the natural map $H^\bullet(A \otimes B, M) \longrightarrow H^\bullet(A,M) \oplus H^\bullet(B,M)$ is an isomorphism.*

PROOF. Let $A_\bullet \longrightarrow A$ and $B_\bullet \longrightarrow B$ be simplicial resolution of A and B, respectively by polynomial algebras. Then I claim that the homology associated to the simplicial set $C_\bullet$ defined by $C_n = A_n \otimes B_n$ with each face and degeneracy operator being the tensor product of the corresponding face and degeneracy operators, is $\mathrm{Tor}(A,B)$. Consider the double chain complex associated to the double simplicial object whose m,nth term is $A_m \otimes B_n$. The mth row is

$$\cdots \longrightarrow A_m \otimes B_n \longrightarrow A_m \otimes B_{n-1} \longrightarrow \cdots \longrightarrow A_m \otimes B_0$$

which is the tensor product of A_m with a K-projective resolution of B, whose homology is $\mathrm{Tor}(A_m, B)$. But A_m is K-projective and hence the homology is $A_m \otimes B$ concentrated in degree 0. Thus the complex augmented by $A_m \otimes B$ is acyclic and the homology of the double complex is the same as the homology of the augmentation row, which is

$$\cdots \longrightarrow A_m \otimes B \longrightarrow A_{m-1} \otimes B \longrightarrow \cdots \longrightarrow A_0 \otimes B$$

which is the homology of a K-projective resolution of A tensored with B, whose homology is thereby $\mathrm{Tor}(A,B)$. If all the higher Tor groups vanish, then this homology is $A \otimes B$ concentrated in degree 0, so that the double complex is a polynomial algebra resolution of $A \otimes B$. The Eilenberg-Zilber theorem, which we will be taking up in Section 4 below, states that the total chain complex of the double chain complex associated to a double simplicial object is homotopic to the chain complex associated to the diagonal simplicial object. In this case the diagonal simplicial algebra is

$$\cdots \ \substack{\longrightarrow \\ \vdots \\ \longrightarrow} \ A_n \otimes B_n \ \substack{\longrightarrow \\ \vdots \\ \longrightarrow} \ A_{n-1} \otimes B_{n-1} \ \substack{\longrightarrow \\ \vdots \\ \longrightarrow} \ \cdots \rightrightarrows A_0$$

and so it follows that $A_n \otimes B_n$ is a resolution of $A \otimes B$. It is easy to prove that $\mathrm{Der}(A_n \otimes B_n, M) \longrightarrow \mathrm{Der}(A_n, M) \oplus \mathrm{Der}(B_n, M)$ is an isomorphism, from which the conclusion follows. □

2.4. COROLLARY. *Suppose X is a set $B = A[X]$ is the algebra of polynomials in X over A. Then for any B-module M*

$$H^n(B,M) \cong \begin{cases} H^n(A,M) & \text{if } n > 0 \\ \mathrm{Der}(A,M) \oplus M^X & \text{if } n = 0 \end{cases}$$

PROOF. This is an immediate consequence of the fact that $B \cong A \otimes K[X]$, the preceding theorem and the homology of a polynomial algebra. □

2.5. THEOREM. *Suppose that $f\colon A \longrightarrow B$ is a homomorphism of commutative K-algebras. Then for any B-module M, there is a natural equivalence $H^\bullet_A B, M \cong H^\bullet(B,A,M)$.*

PROOF. Let $\mathbf{G} = (G, \epsilon, \delta)$ denote the free commutative K-algebra cotriple on the category of A-algebras and $\mathbf{G}_A = (G_A, \epsilon_A, \delta_A)$ denote the free commutative A-algebra cotriple on the same category. The groups $H^\bullet_A(B, M)$ are the cohomology of the cochain complex

$$0 \longrightarrow \mathrm{Der}_A(G_A B, M) \longrightarrow \cdots \longrightarrow \mathrm{Der}_A(G^n_A B, M) \longrightarrow \mathrm{Der}_A(G^{n+1}_A B, M) \longrightarrow \cdots$$

with boundary operator given by

$$\sum_{i=0}^{n} (-1)^i \,\mathrm{Der}(G^i_A \epsilon_A G^{n-i}_A, M)$$

while the groups $H^\bullet(B, A, M)$ are the cohomology of the cochain complex

$$\begin{aligned} &0 \longrightarrow \mathrm{Der}(GB, M) \longrightarrow \mathrm{Der}(G^2 B, M) \oplus \mathrm{Der}(GA, M) \longrightarrow \cdots \\ &\mathrm{Der}(G^n B, M) \oplus \mathrm{Der}(G^{n-1} A, M) \longrightarrow \mathrm{Der}(G^{n+1}, M) \oplus \mathrm{Der}(G^n A, M) \longrightarrow \cdots \end{aligned}$$

with boundary operator given by

$$\begin{pmatrix} \sum_{i=0}^{n} (-1)^i \,\mathrm{Der}(G^i \epsilon G^{n-i}) & 0 \\ \mathrm{Der}(G^n f, M) & \sum_{i=0}^{n-1} (-1)^i \,\mathrm{Der}(G^i \epsilon G^{n-i-1} A, M) \end{pmatrix}$$

We will use acyclic models to compare these two cochain complexes. Consider the case that $B = A[X]$ is a polynomial algebra over A with variable set X. Then

$$H^n_A(B, M) = \begin{cases} M^X & \text{if } n = 0 \\ 0 & \text{if } n > 0 \end{cases}$$

From the fact that $B \cong A \otimes k[X]$ we have just seen that

$$H^n(B, M) = \begin{cases} H^0(A, M) \oplus M^X & \text{if } n = 0 \\ H^n(A, M) & \text{if } n > 0 \end{cases}$$

However, the inclusion $A \longrightarrow B$ is split monic in the category of k-algebras so that the map $H^n(B, M) \longrightarrow H^n(A, M)$ is split epic for all $n \geq 0$ and so the long exact mapping cone sequence breaks up into a series of short exact sequences

$$0 \longrightarrow H^0(B, A, M) \longrightarrow \mathrm{Der}(B, M) \longrightarrow \mathrm{Der}(A, M) \longrightarrow 0$$

for $n = 0$ and

$$0 \longrightarrow H^n(B, A, M) \longrightarrow H^n(B, M) \longrightarrow H^n(A, M) \longrightarrow 0$$

for $n > 0$. Thus $H^0(B, A, M)$ is the kernel of $\mathrm{Der}(B, M) \longrightarrow \mathrm{Der}(A, M)$, which is easily seen to be $\mathrm{Der}_A(B, M) = M^X$ so that the augmented complex $C^\bullet(B, A, M) \longrightarrow M^X \longrightarrow 0$ is acyclic. Similarly, the augmented complex $C^\bullet_A(B, M) \longrightarrow M^X$ is acyclic. Thus the two complexes are acyclic on models and have the same 0 dimensional group. Finally, for $n > 0$, we need maps $C^n_A(G_A B, M) \longrightarrow C^n_A(B, M)$ that splits $C^n_A(\epsilon_A, M)$ and similarly for $C^n(B, A, M)$. For the first, take

$$\mathrm{Der}_A(\delta_A G^n_A B, M)\colon \mathrm{Der}_A(G^{n+2}_A B, M) \longrightarrow \mathrm{Der}_A(G^{n+1}_A B, M)$$

and for the second,

$$\begin{aligned} \mathrm{Der}(\delta G^n B, M) \oplus \mathrm{id}\colon \mathrm{Der}(G^{n+2} B, M) &\oplus \mathrm{Der}(G^{n+1} A, M) \\ &\longrightarrow \mathrm{Der}(G^{n+2} B, M) \oplus \mathrm{Der}(G^n A, M) \end{aligned}$$

Then the two theories are equivalent. □

3. Shukla cohomology

Hochschild's original cohomology theory for associative algebras, [1945] was for algebras over a field. In the Cartan–Eilenberg version, the ground ring was allowed to be an arbitrary commutative ring. However, the theory was relative to that ring. For example, the second cohomology group (in the original numbering) classified the singular extensions of the ring with the coefficient module as kernel. The Cartan–Eilenberg version classified only those extensions that split as modules over the ground ring. Such a theory is called a **relative cohomology theory**. Actually, the same was true for Harrison's theory. His original paper, [1962], included an appendix that computed the absolute H^2 of a commutative ring, written by me. The referee insisted on an appendectomy, since the results had no obvious application. The referee was probably right and the same comment could be made for the Shukla cohomology groups we discuss briefly below. This is why we give no details, but refer to the original papers for the proofs.

In is dissertation, published as [1961], Shukla produced a cohomology theory for associative algebras that takes into account both the additive and multiplicative structure. In dimension k, there were k different kinds of chain groups that had to be sum to produce the k-dimensional chain groups. Each kind had its own coboundary operator. In each dimension, one of the kinds concerned purely the multiplicative structure and one the linear structure and the others were a mixture. The crucial step in showing that this theory was equivalent to the cohomology defined by the (absolutely) free algebra functor was, as usual, to show that it vanished on free algebras. The details are found in [Barr, 1967]. A crucial part of the argument was the use of distributive laws among cotriples, another important idea of Jon Beck's (but unpublished by him).

4. The Eilenberg–Zilber theorem

The Eilenberg–Zilber theorem states that the two functors K and TL, described in 3.7.3, from the category of double simplicial objects to the category of chain complexes are homotopic. The classical proof is geometric, based on a triangulation of a product of two simplexes. But a straightforward acyclic models proof is quite easy. For more details on the statement of the theorem, see 7.3 of Chapter 3

4.1. THEOREM. *Let $\mathscr{A}$ be an abelian category, let $\mathscr{B}$ be the category of augmented double simplicial objects over $\mathscr{A}$, and let $\mathscr{C}$ be the category of chain complexes over $\mathscr{A}$. Let $C\Delta: \mathscr{B} \longrightarrow \mathscr{C}$ assign to each augmented double simplicial object the chain complex associated to the diagonal complex and TL assign to each double simplicial object the total complex of the double chain complex associated to it. Then $C\Delta$ is homotopic to TL.*

PROOF. We use the cotriple $\mathbf{G}_T$ from 3.3 of Chapter 4. In order to apply it to ordinary double simplicial objects, we consider every double simplicial object an augmented double simplicial object with 0 objects in all terms with at least one negative index. An object of the form $G_T\mathscr{A}$ has contractible rows and columns and hence so does the double chain complex associated. Either the row or column contractions suffice to give a contraction on the total complex. Hence TL is acyclic on models. Since the rows and columns are contractible, so is the diagonal and hence so is $C\Delta$ and hence that functor is also acyclic on models. For the G_T-presentability, we first observe that the adjunction arrow $G_T \longrightarrow \mathrm{Id}$ is induced by

d^0 and ∂^0. On $C\Delta$ it is just $d^0\partial^0$ and has a right inverse $\sigma^0 s^0$. These faces and degeneracies are just the ones that are dropped from the simplicial objects when G_T is applied. The case of TL is a little more complicated. In degree n,

$$TL\epsilon\mathscr{A}\colon A_{n+1\,0} \oplus A_{n1} \oplus \cdots \oplus A_{0\,n+1} \longrightarrow A_{n0} \oplus A_{n-1\,0} \oplus \cdots \oplus A_{0n}$$

has an $n{+}1 \times n$ matrix

$$\begin{pmatrix} d^0 & \partial^0 & 0 & \cdots & 0 \\ 0 & d^0 & \partial^0 & \cdots & 0 \\ \cdot & \cdot & \cdot & \cdots & \cdot \\ 0 & 0 & 0 & \cdots & d^0 \end{pmatrix}$$

and has a right inverse given by the $n \times n{+}1$ matrix

$$\begin{pmatrix} s^0 & -s^0\partial^0 s^0 & s^0\partial^0 s^0\partial^0 s^0 & \cdots & (-1)^{n+1}s^0(\partial^0 s^0)^n \\ 0 & s^0 & -s^0\partial^0 s^0 & \cdots & (-1)^n s^0(\partial^0 s^0)^{n-1} \\ \cdot & \cdot & \cdot & \cdots & \cdot \\ 0 & 0 & 0 & \cdots & s^0 \\ 0 & 0 & 0 & \cdots & 0 \end{pmatrix}$$

as can be verified by direct calculation.

Finally, we have to show that the two chain complex functors give naturally equivalent 0 dimensional homology. This means showing that the coequalizer of

$$A_{10} \oplus A_{01} \xrightarrow{(\,d^0 - d^1 \;\; \partial^0 - \partial^1\,)} A_{00}$$

is naturally equivalent to the coequalizer of

$$A_{11} \xrightarrow{d^0\partial^0 - d^1\partial^1} A_{00}$$

To see this, we show that both squares commute in

$$\begin{array}{ccc} A_{10} \oplus A_{01} & \xrightarrow{(\,d^0 - d^1 \;\; \partial^0 - \partial^1\,)} & A_{00} \\ {\scriptstyle (\,\sigma^0 \;\; s^0\,)}\big\downarrow & & \big\downarrow{\scriptstyle \mathrm{id}} \\ A_{11} & \xrightarrow{d^0\partial^0 - d^1\partial^1} & A_{00} \\ {\scriptstyle \binom{\partial^0}{d^1}}\big\downarrow & & \big\downarrow{\scriptstyle \mathrm{id}} \\ A_{10} \oplus A_{01} & \xrightarrow[(\,d^0 - d^1 \;\; \partial^0 - \partial^1\,)]{} & A_{00} \end{array}$$

In fact,

$$(\,d^0 - d^1 \quad \partial^0 - \partial^1\,)\begin{pmatrix} \partial^0 \\ d^1 \end{pmatrix} = \partial^0 d^0 - \partial^0 d^1 + d^1\partial^0 - \partial^1 d^1 = \partial^0 d^0 - \partial^1 d^1$$

and

$$\begin{aligned} (\partial^0 d^0 - \partial^1 d^1)(\,\sigma^0 \quad s^0\,) &= (\,(\partial^0 d^0 - \partial^1 d^1)\sigma^0 \quad (\partial^0 d^0 - \partial^1 d^1)s^0\,) \\ &= (\,d^0 - d^1 \quad \partial^0 - \partial^1\,) \end{aligned}$$

That this implies that the induced map between the coequalizers is the identity follows from:

4.2. Proposition. *Suppose both squares of*

$$\begin{array}{ccc} A_1 & \xrightarrow{d} & A_0 \\ {\scriptstyle f_1}\downarrow & & \downarrow{\scriptstyle f_0} \\ A_1' & \xrightarrow{d'} & A_0' \\ {\scriptstyle g_1}\downarrow & & \downarrow{\scriptstyle g_0} \\ A_1 & \xrightarrow{d} & A_0 \end{array}$$

commute and that f_0 and g_0 are inverse isomorphisms. Then the induced $\operatorname{coker} d \longrightarrow \operatorname{coker} d'$ *is an isomorphism.*

Proof. The composite square

$$\begin{array}{ccc} A_1 & \xrightarrow{d} & A_0 \\ {\scriptstyle g_1 f_1}\downarrow & & \downarrow{\scriptstyle g_0 f_0 = \mathrm{id}} \\ A_1 & \xrightarrow[d]{} & A_0 \end{array}$$

and the identity induces $\mathrm{id}\colon \operatorname{coker} d \longrightarrow \operatorname{coker} d$. The same is true of the map induced by $f_0 g_0$ and the conclusion follows. □

CHAPTER 8

Applications in Topology

1. Singular homology

1.1. Singular chains. This is the time to read 1.6. Although singular chain groups can, and usually are, defined on the category Top, we will find it helpful to define them as additive functors on the additive category **Z**Top. To emphasize this, we will write **Z** $\mathrm{Hom}(X, Y)$ for the set of morphisms between two spaces in **Z**Top.

We denote by Δ_n the subset of $(n+1)$-dimensional euclidean space consisting of all $(t_0, \ldots, t_n)$ for which $t_i \geq 0$, $i = 0, \ldots, n$ and $t_0 + \cdots + t_n = 1$. This set is, in fact, the convex hull of the basis vectors in $(n+1)$-dimensional euclidean space. Geometrically, it is an n-dimensional simplex. There is a map $\partial^i \colon \Delta_{n-1} \longrightarrow \Delta_n$ defined for $i = 0, \ldots, n$ by $\partial^i(t_0, \ldots, t_{n-1}) = (t_0, \ldots, t_{i-1}, 0, t_i, \ldots, t_n)$. Then $\partial = \sum_{i=0}^{n}(-1)^i \partial^i \colon \Delta_{n-1} \longrightarrow \Delta_n$ is an arrow of **Z**Top. One easily shows that $\partial \circ \partial = 0$. In fact, assuming that $i < j$ the term in which $t_i = t_j = 0$ appears in the sum, one in which the coefficient is $(-1)^{i+j}$ because first t_j and then t_i was set to 0 and second with coefficient $(-1)^{i+j-1}$ since first t_i was set to 0 and second the $(j-1)$st entry, now t_j was.

The nth singular chain group, $C_n(X)$ of a space X is defined as the free abelian group generated by the set $\mathrm{Hom}(\Delta_n, X)$ of continuous functions of the n-simplex into X. An element of $\mathrm{Hom}(\Delta_n, X)$ is called an n-simplex (or singular n-simplex) in X, while an element of **Z** $\mathrm{Hom}(\Delta_n, X)$ is called an n-chain. We define $d \colon C_n(X) \longrightarrow C_{n-1}(X)$ by $dc = c \circ \partial$. From $\partial \circ \partial = 0$, it immediately follows that $d \circ d = 0$.

1.2. Cone construction. If U is a convex open subset of a euclidean space, b is an element of U and $\sigma \colon \Delta_n \longrightarrow U$ is a singular n-simplex, the cone $b \cdot \sigma$ is a singular $n+1$-simplex. There is a standard definition, but later on we will want to vary it, so we use a somewhat less obvious definition.

Let $r(t)$ be any continuous bijective function $I \longrightarrow I$ such that $r(0) = 0$ and $r(1) = 1$. The identity function obviously qualifies, but we will have occasion to use a different one. Fix a choice of r and define

$$b \cdot \sigma(t_0, t_1, \ldots, t_{n+1}) = \begin{cases} r(t_0)b + (1 - r(t_0))\sigma\left(\frac{t_1}{1-t_0}, \ldots, \frac{t_{n+1}}{1-t_0}\right) & \text{if } t_0 \neq 1 \\ b & \text{if } t_0 = 1 \end{cases}$$

This is continuous since $1 - r(t_0) \longrightarrow 0$ as $t_0 \longrightarrow 1$, while the second term is bounded, since it is a continuous function on a compact set.

1.3. Proposition. *For a singular n-simplex σ in a convex set U in euclidean space,*

$$(b \cdot \sigma) \circ d = \begin{cases} \sigma - b \cdot (\sigma \circ \partial) & \text{if } n > 0 \\ \sigma - [b] & \text{if } n = 0 \end{cases}$$

PROOF. We will show that $(b \cdot \sigma) \circ \partial^0 = \sigma$ and $(b \cdot \sigma) \circ \partial^i = b \cdot (\sigma \circ \partial^{i-1})$ for $i > 0$. For the first, we have

$$\begin{aligned}(b \cdot \sigma) \circ \partial^0(t_0, \dots, t_n) &= b \cdot \sigma(0, t_0, \dots, t_n) \\ &= r(0)b + (1 - r(0))\sigma(t_0, \dots, t_n) \\ &= \sigma(t_0, \dots, t_n)\end{aligned}$$

For $i > 0$ and $t \neq 1$, we have

$$\begin{aligned}(b{\cdot}\sigma) \circ \partial^i(t_0, \dots, t_n) &= b \cdot \sigma(t_0, \dots, t_{i-1}, 0, t_i, \dots, t_n) \\ = r(t)b + (1 - r(t))\sigma &\left(\frac{t_1}{1-t_0}, \dots, \frac{t_{i-1}}{1-t_0}, 0, \frac{t_i}{1-t_0}, \dots, \frac{t_n}{1-t_0}\right)\end{aligned}$$

while

$$\begin{aligned}b{\cdot}(\sigma \circ \partial^{i-1})(t_0, \dots, t_n) &= r(t)b + (1 - r(t))\sigma \circ \partial^{i-1}\left(\frac{t_1}{1-t_0}, \dots, \frac{t_n}{1-t_0}\right) \\ = r(t)b + (1 - r(t))\sigma &\left(\frac{t_1}{1-t_0}, \dots, \frac{t_{i-1}}{1-t_0}, 0, \frac{t_i}{1-t_0}, \dots, \frac{t_n}{1-t_0}\right)\end{aligned}$$

The argument when $t_0 = 1$ is trivial. □

When you see a special case in a formula as above, you might wonder if there is something behind it. There is good reason for thinking that there ought to be an empty, dimension -1-simplex that is the boundary of every 0 dimensional simplex and whose boundary is 0. If you are familiar with homology theory, you will see that that would give reduced homology instead of ordinary. The formula for the boundary of $b \cdot \sigma = \sigma - b \cdot \partial\sigma$ would then work without restriction, including dimension 0. This is not a very important point, however.

1.4. COROLLARY. *The singular chain complex of a convex subset of euclidean space is contractible.*

PROOF. Let U be a convex subset of euclidean space and let $b \in U$ be arbitrary. Then the maps $s_n: C_n(U) \longrightarrow C_{n+1}$ defined by $s\sigma = b \cdot \sigma$ satisfy

$$\begin{aligned}ds_n\sigma + s_{n-1}d\sigma &= (b \cdot \sigma) \circ \partial + b \cdot (\sigma \circ \partial) \\ &= \sigma - b \cdot (\sigma \circ \partial) + b \cdot (\sigma \circ \partial) = \sigma\end{aligned}$$

so that $ds_n + s_{n-1}d = \text{id}$. □

1.5. Barycentric subdivision. Barycentric subdivision is a way of dividing simplexes into smaller pieces and is crucial for proving such things as that homology with small simplexes (see 2.1 below) is the same as the full homology. The basic idea is not really hard; however the formal description is complicated. Basically, the standard simplex is divided into a number of smaller pieces and a singular simplex is equivalent, up to homotopy with the sum of singular simplexes on the pieces. We illustrate with the subdivision of a triangle as the sum of the six individual triangles shown below.

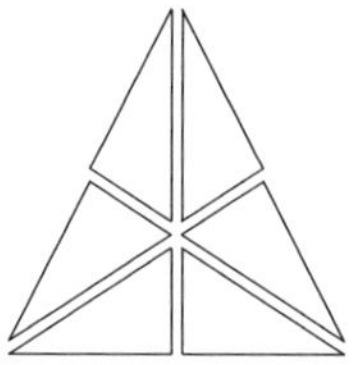

We will show that a simplex is homotopic to its subdivision. This fact is a good illustration of the fact that the algebraic operation of formal addition of simplexes turns into the topological union, at least in a certain sense. It is the same sense in which algebraic topology turns topology into algebra.

We begin by inductively defining an arrow $\xi_n : \Delta_n \longrightarrow \Delta_n$ in **Z**Top. Let b_n denote the point $(1/(n+1), 1/(n+1), \dots, 1/(n+1))$ of the n-simplex. This point is called the **barycenter**. Then ξ_0 is the identity and for $n > 0$,

$$\xi_n = b_n \cdot (\partial \circ \xi_{n-1})$$

1.6. PROPOSITION. *For $n > 0$, $\xi_n \circ \partial = \partial \circ \xi_{n-1}$.*

PROOF. For $n = 1$,

$$\begin{aligned} \xi_1 \circ \partial &= (b_1 \cdot (\partial \circ \xi_0)) \circ \partial \\ &= \partial \circ \xi_0 - [b_1] \circ \partial = \partial \circ \xi_0 \end{aligned}$$

For $n > 1$, assume the result inductively for $n - 1$. Then

$$\begin{aligned} \xi_n \circ \partial &= (b_n \cdot (\partial \circ \xi_{n-1})) \circ \partial = \partial \circ \xi_{n-1} - b_n \cdot (\partial \circ \xi_{n-1} \circ \partial) \\ &= \partial \circ \xi_{n-1} - b_n \cdot (\partial \circ \partial \circ \xi_{n-2}) = \partial \circ \xi_{n-1} \end{aligned}$$

For a space X define $\mathrm{Sd}_n X : C_n(X) \longrightarrow C_n(X)$ on simplexes by $\mathrm{Sd}_n \sigma = \sigma \circ \xi_n$.

1.7. PROPOSITION. *The transformation $\mathrm{Sd}_n(X)$ is the component at X of a natural chain transformation.*

PROOF. For a map $f : X \longrightarrow Y$ and $\sigma \in C_n(X)$

$$\begin{aligned} C_n(f) \circ \mathrm{Sd}_n(X)(\sigma) &= C_n(f)(\sigma \circ \xi_n) = f \circ \sigma \circ \xi_n \\ &= \mathrm{Sd}_n(Y)(f \circ \sigma) = \mathrm{Sd}_n(Y) \circ C_n(f)(\sigma) \\ d \circ \mathrm{Sd}_n(X)(\sigma) &= d(\sigma \circ \xi_n) = \sigma \circ xi_n \circ \partial \\ &= \sigma \circ \partial \circ \xi_{n-1} = d(\sigma) \circ \xi_{n-1} = \mathrm{Sd}_{n-1} \circ d(\sigma) \end{aligned}$$ □

Let U be a convex subset of euclidean space and $\sigma : \Delta_n \longrightarrow U$ be a singular simplex. We will say that σ is **totally convex** if for any face $\phi : \Delta_m \longrightarrow \Delta_n$ the image of $\sigma \circ \phi$ is a convex subset of U. It follows, among other things, that the only extreme points of $\sigma(\Delta_n)$ occur at the vertices.

1.8. PROPOSITION. *Suppose U is a subset of euclidean space, $b \in U$ and σ is a totally convex n-simplex in U. Then $b \cdot \sigma$ is totally convex.*

PROOF. If $b \notin \Delta_n$, the equation

$$b \cdot \sigma(t_0, \dots, t_{n+1}) = r(t_0)b + (1 - r(t_0))\sigma\left(\frac{t_1}{1 - t_0}, \dots, \frac{t_{n+1}}{1 - t_0}\right)$$

implies that the image of $b\cdot\sigma$ is convex. The faces for $t_0 = 0$ are the faces of σ and are convex by assumption. The remaining faces are $b\cdot\tau$ where τ is a face of σ and it is convex by the first remark. □

1.9. PROPOSITION. *The diameter of any simplex of the barycentric subdivision of the standard n-simplex does not exceed $\frac{n}{n+1}$ times the diameter of the original simplex.*

PROOF. Let us suppose that U is a subset of euclidean space and that $\sigma\colon \Delta_n \longrightarrow U$ is an affine simplex. Suppose the vertices of σ are $v_0, \ldots, v_n$. Let

$$b(\sigma) = \frac{v_0 + v_1 + \cdots + v_n}{n+1}$$

The subdivision formula on such a simplex is

$$\mathrm{Sd}(\sigma) = b(\sigma)\cdot(\sigma\circ\partial)$$

and a simplex of the subdivision is a simplex of $b(\sigma)\cdot(\sigma\circ\partial^i)$ for $0 \le i \le n$. By induction, a simplex $\mathrm{Sd}_n(\sigma)$ is a sum of simplexes of the form

$$b(\sigma)\cdot b(\sigma\circ\partial^i)\cdot b(\sigma\circ\partial^i\circ\partial^j)\cdot\cdots\cdot b(\sigma\circ\partial^i\circ\partial^j\circ\cdots\circ\partial^k)$$

and these factors are the vertices. Thus the diameter is the maximum distance between any two of these barycenters. After renumbering, we can assume that the maximum is assumed between the barycenters $(v_0 + \cdots + v_i)/(i+1)$ and $(v_0 + \cdots + v_j)/j+1$ with $i < j$. The distance between them is

$$\begin{aligned}
&\left\|\frac{v_0 + \ldots + v_i}{i+1} - \frac{v_0 + \cdots + v_j}{j+1}\right\| \\
&\qquad = \left\|\frac{v_0 + \cdots + v_i}{i+1} - \frac{v_0 + \cdots + v_i}{j+1} - \frac{v_{i+1} + \cdots + v_j}{j+1}\right\| \\
&\qquad = \left\|(v_0 + \cdots + v_i)\left(\frac{1}{i+1} - \frac{1}{j+1}\right) - \frac{v_{i+1} + \cdots + v_j}{j+1}\right\| \\
&\qquad = \left\|(v_0 + \cdots + v_i)\frac{j-i}{(i+1)(j+1)} - \frac{v_{i+1} + \cdots + v_j}{j+1}\right\| \\
&\qquad = \frac{\|(v_0 + \cdots + v_i)(j-i) - (i+1)(v_{i+1} + \cdots + v_j)\|}{(i+1)(j+1)}
\end{aligned}$$

What is inside the distance sign is the difference of two expressions, each of which is the sum of exactly $(i+1)(j-i)$ vertices of the simplex. No matter how these are arranged, none of those differences can exceed r and so the whole sum is at most

$$\frac{(i+1)(j-i)}{(i+1)(j+1)}r = \frac{j-i}{j+1}r \le \frac{j}{j+1}r \le \frac{n}{n+1}r$$ □

1.10. LEMMA (Hausdorff covering lemma.). *Let X be a compact metric space and suppose $\mathscr{U}$ is an open cover of X. Then there is an $r > 0$ such that every set of diameter less than r is in some element of $\mathscr{U}$.*

PROOF. For $x \in X$, let $N_\epsilon(x)$ denote the ϵ sphere around x. For each $\epsilon > 0$, let V_ϵ consist of all points $x \in X$ for which there is a $\delta > \epsilon$ such that N_δ is included in some set in $\mathscr{U}$. It is clear that $\epsilon_1 < \epsilon_2$ implies that $V_{\epsilon_1} \supseteq V_{\epsilon_2}$. If $x \in V_\epsilon$ and $\delta > \epsilon$ is such that $N_\delta(x)$ is in some member of $\mathscr{U}$, then for any $y \in N_{(\delta-\epsilon)/2}(x)$, it is immediate that $N_{(\delta+\epsilon)/2}(y)$ is in the same member of $\mathscr{U}$ and thus V_ϵ is open.

Since the set of all V_ϵ covers X, a finite subset does. Since they are nested, a single one does. □

1.11. COROLLARY. *Let Δ be a simplex and $\mathscr{U}$ a cover of Δ. Then there is an integer k such that every simplex of $\mathrm{Sd}^k(\Delta)$ is contained in some single member of $\mathscr{U}$.* □

1.12. A cotriple. We will use a model-induced cotriple on the category of topological spaces that is gotten by taking the simplexes as models (see 2). This means that

$$GX = \sum_n \sum_{\sigma:\Delta_n \to X} \Delta_n$$

For $\sigma\colon \Delta_n \longrightarrow X$, we denote by $\langle\sigma\rangle\colon \Delta_n \longrightarrow GX$ the inclusion into the sum. Then $\epsilon X\colon GX \longrightarrow X$ is defined by $\epsilon X \circ \langle\sigma\rangle = \sigma$ and $\delta X\colon GX$ by $\delta X \circ \langle\sigma\rangle = \langle\langle\sigma\rangle\rangle$. Then $\mathbf{G} = (G, \epsilon, \delta)$ is a cotriple.

1.13. PROPOSITION. *The singular chain complex functor $\mathscr{C}_\bullet$ is $\mathbf{G}$-presentable and $\mathbf{G}$-acyclic on models with respect to the class of homotopy equivalences.*

PROOF. If $\sigma\colon \Delta_n \longrightarrow X$ is a singular n-simplex in X, then $\langle\sigma\rangle\colon \Delta_n \longrightarrow GX$ is a singular n-simplex in GX for which $\epsilon\circ\langle\sigma\rangle = \sigma$. If $f\colon X \longrightarrow Y$ is a continuous map, then, by definition, $Gf \circ \langle\sigma\rangle = \langle f \circ \sigma\rangle$, which shows that $\langle -\rangle\colon C_n(X) \longrightarrow C_n(GX)$ is natural. This is the $\mathbf{G}$-presentability. As for the G acyclicity, GX is a disjoint union of simplexes and the chain complex of every simplex is contractible. □

1.14. COROLLARY. *Suppose $\alpha_\bullet\colon C_\bullet \longrightarrow C_\bullet$ is an endomorphism of the singular chain complex functor which induces the identity arrow on H_0. Then $\alpha_\bullet$ is homotopic to the identity.* □

PROOF. From the commutative

$$\begin{array}{ccc} C_0 & \xrightarrow{\ d\ } & H_0 \\ {\scriptstyle \mathrm{id}}\downarrow & & \downarrow{\scriptstyle \alpha_{-1}} \\ C_0 & \xrightarrow[\ d\ ]{} & H_0 \end{array}$$

we conclude that α_{-1} is the identity. □

1.15. COROLLARY. *The chain map $\mathrm{Sd}_\bullet\colon C_\bullet \longrightarrow C_\bullet$ induces a natural homotopy equivalence.* □

1.16. An explicit formula. It will be useful to have an explicit formula for the homotopy. Define $\eta_{n+1}\colon \Delta_{n+1} \longrightarrow \Delta_n$ in $\mathbf{ZTop}$ for $n \geq 0$ by $\eta_1 = 0$ and $\eta_{n+1} = b_n \cdot (1 - \xi_n - \partial \circ \eta_n)$. Assume inductively that $\eta_n \partial = 1 - \xi_{n-1} - \partial \circ \xi_{n-1}$. Then

$$\begin{aligned}
\eta_{n+1} \circ \partial &= (b_n \cdot (1 - \xi_n - \partial \circ \eta_n) \circ \partial) \\
&= 1 - \xi_n - \partial \circ \eta_n + b_n \cdot (\partial - \xi_n \circ \partial - \partial \circ \eta_n \circ \partial) \\
&= 1 - \xi_n - \partial \circ \eta_n + b_n \cdot (\partial - \partial \circ \xi_{n-1} - \partial \circ (1 - \xi_{n-1} - \partial \circ \xi_{n-1})) \\
&= 1 - \xi_n - \partial \circ \eta_n
\end{aligned}$$

so that $\eta_{n+1} \circ \partial + \partial \circ \eta_n = 1 - \xi_n$. It follows immediately that if we define $h_n(X): C_n(X) \longrightarrow C_{n+1}(X)$ by $h_n(\sigma) = \sigma \circ \eta_{n+1}$, then $d \circ h_n(X) + h_{n-1}(X) = 1 - \mathrm{Sd}_n(X)$. The naturality of h_n is also clear.

1.17. Exercise.

1. Show that the diameter of an affine simplex is the largest distance between any two vertices. One way is to first show that if this maximum is M, then the distance from any point in the simplex to any vertex is at most M and then use that to show that the distance of any point in the simplex to any other is at most M.

2. Covered spaces

Some of what we do in algebraic topology is best dealt with by considering the category Cov of covered topological spaces. An object of Cov is a pair $(X, \mathscr{U})$ where X is a topological space and $\mathscr{U}$ is an open cover of X. An arrow $f: (X, \mathscr{U}) \longrightarrow (Y, \mathscr{V})$ in this category consists of a continuous function f such that for $U \in \mathscr{U}$, there is a $V \in \mathscr{V}$ such that $fU \subseteq V$. Since this is equivalent to $U \subseteq f^{-1}V$, we can also describe this by saying that $\mathscr{U}$ refines $f^{-1}\mathscr{V}$.

2.1. Small simplexes. If $(X, \mathscr{U})$ is a covered space, we denote by $\widehat{C}_\bullet(X, \mathscr{U})$ the chain complex defined by $\widehat{C}_n(X, \mathscr{U}) = \mathbf{Z}\,\mathrm{Hom}_{\mathsf{Cov}}((\Delta_n, \{\Delta_n\}), (X, \mathscr{U}))$. This means that Δ_n is given the singleton cover and an n-simplex is a singular simplex whose image is entirely contained in some member of $\mathscr{U}$. Such a simplex will be called a **$\mathscr{U}$-small simplex** or simply small simplex, if no confusion can result. Since any face of a small simplex is small, it follows that $C_\bullet(X, \mathscr{U})$ is a subcomplex of $C_\bullet(X)$. Moreover, $C_\bullet(X) = C_\bullet(X, \{X\})$. We will also write $C_\bullet(X, \mathscr{U}) = C_\bullet(X)$ so that we can compare $C_\bullet$ with $\widehat{C}_\bullet$ as functors on the same category.

2.2. Two cotriples. There are two ways of extending the cotriple $\mathbf{G}$ to Cov. The first is not actually model induced, although the same formulas apply. We continue to call it $\mathbf{G}$ and it is defined at the covered space $(X, \mathscr{U})$ by

$$G(X, \mathscr{U}) = \sum_{n \geq 0} \sum_{\sigma: \Delta_n \longrightarrow X} (\Delta_n, \sigma^{-1}\mathscr{U})$$

Note that we do not use only small simplexes, which is why this is not model induced. On the other hand, the cover of X survives to become the covers of the components of $G(X, \mathscr{U})$, which is why ϵ is an arrow of Cov.

We use the same notation as for the model induced cotriple, that is for $\sigma: \Delta_n \longrightarrow X$, we denote by $\langle\sigma\rangle: (\Delta_n, \sigma^{-1}\mathscr{U}) \longrightarrow G(X, \mathscr{U})$ the inclusion into the sum. Then $\epsilon(X, \mathscr{U}): G(X, \mathscr{U}) \longrightarrow (X, \mathscr{U})$ is defined by $\epsilon(X, \mathscr{U}) \circ \langle\sigma\rangle = \sigma$ and $\delta(X, \mathscr{U}): G(X, \mathscr{U})$ by $\delta(X, \mathscr{U}) \circ \langle\sigma\rangle = \langle\langle\sigma\rangle\rangle$. The thing to note is that these are maps in the category of covered spaces, even though the original simplexes did not respect the covers. Then $\mathbf{G} = (G, \epsilon, \delta)$ is a cotriple. As with model induced cotriples, if $f: (X, \mathscr{U}) \longrightarrow (Y, \mathscr{V})$ is an arrow in the category of covered spaces, then Gf is defined by $Gf \circ \langle\sigma\rangle = \langle f \circ \sigma\rangle$. If $\sigma: \Delta_n \longrightarrow X$, then the cover on Δ_n is $\sigma^{-1}\mathscr{U}$. In order to show that Gf is an arrow in the category of covered spaces, we have to show that the identity map on Δ_n is an arrow $(\Delta_n, \sigma^{-1}\mathscr{U}) \longrightarrow (\Delta_n, (f \circ \sigma)^{-1}\mathscr{V})$. This is the same thing as saying that the cover by $\sigma^{-1}\mathscr{U}$ refines that of $(f \circ \sigma)^{-1}\mathscr{V}$. Now if $U \in \mathscr{U}$, there is some $V \in \mathscr{V}$ such that $f(U) \subseteq V$ or $V \subseteq f^{-1}V$. But then $\sigma^{-1}U \subseteq \sigma^{-1} \circ f^{-1}V = (f \circ \sigma)^{-1}V$. We note that this cotriple also makes sense on the category of ordinary spaces, thought of as having a single element cover. In that case it is model induced.

The second cotriple, $\widehat{\mathbf{G}} = (\widehat{G}, \widehat{\epsilon}, \widehat{\delta})$ is a model induced cotriple, induced by the simplexes Δ_n, covered by themselves. So a map $\Delta_n \longrightarrow (X, \mathscr{U})$ is, by definition, a small simplex. The cotriples are similar, but each has a role to play in the development.

2.3. Singular homology on Cov. We will not repeat the arguments, but the same development that we have just carried out with $\mathbf{G}$ and $C_\bullet$ can be repeated, *mutatis mutandis*, with $\widehat{\mathbf{G}}$ and $\widehat{C}_\bullet$. We conclude that,

2.4. THEOREM. *Any chain map $\alpha: \widehat{C}_\bullet \longrightarrow \widehat{C}_\bullet$ that induces an isomorphism of 0 dimensional homology and, in particular any for which α_0 is the identity, is homotopic to the identity.*

2.5. COROLLARY. *The chain map $\mathrm{Sd}_\bullet: \widehat{C}_\bullet(X, \mathscr{U}) \longrightarrow \widehat{C}_\bullet(X, \mathscr{U})$ induces a homotopy equivalence.*

2.6. PROPOSITION. *If $(\Delta_n, \mathscr{U})$ is a covered simplex, the complex $\widehat{C}_\bullet(\Delta_n, \mathscr{U})$ is contractible.*

PROOF. Since it is a complex of free abelian groups, it suffices to show it is acyclic. Given an m-cycle c, it is an m-cycle in the complex $C_m(\Delta_n)$. That complex is certainly contractible, so there is a $c' \in C_{m+1}(\Delta_n)$ such that $\partial c' = c$. Of course, c' is not necessarily a sum of small simplexes, but it is the sum of a finitely many simplexes, so there is a k such that each simplex of $\beta^k c'$ is small. Then $\partial \beta^k c' = \beta^k \partial c' = \beta^k c$ and we know that in $C_n(\Delta, \mathscr{U})$, $\beta^k c \sim c$. Thus $c - \beta^k c$ and $\beta^k c$ are both boundaries and hence so is c. □

Notice that no naturality is claimed here. It is hard to see how there could be.

2.7. THEOREM. *The inclusion $\widehat{C}_\bullet \longrightarrow C_\bullet$ is a weak homotopy equivalence.*

PROOF. We use the acyclic models theorem with weak contractions as the acyclic class and $\mathbf{G}$ as the cotriple. If $\sigma: \Delta_n \longrightarrow (X, \mathscr{U})$ is a small simplex, then there is a corresponding summand $\langle \sigma \rangle: \Delta_n \longrightarrow G(X, \mathscr{U})$. Moreover, since σ is small, the summand $\langle \sigma \rangle$ is covered by itself. That is, the cover will generally contain many, even infinitely many sets, but one of them will be the whole simplex. Then we can think of $\langle \sigma \rangle$ as defining a simplex of $\widehat{C}_n(G(X, \mathscr{U}))$. Clearly, $\widehat{\epsilon}(X, \mathscr{U}) \circ \langle \sigma \rangle = \sigma$. If $\sigma: \Delta_n \longrightarrow X$ is a simplex, then $\langle \sigma \rangle: \Delta_n \longrightarrow C_n(GX)$ is also a simplex and clearly $\epsilon X \circ \langle \sigma \rangle = \sigma$. Thus both functors are G-presentable. Now $G(X, \mathscr{U})$ is simply a sum of simplexes and so the standard argument that the augmented chain complex of a contractible space carries over to a disjoint sum of them and shows that $C_\bullet$ is $\mathbf{G}$-contractible on models. Finally Proposition 2.6 shows that $\widehat{C}_\bullet$ is as well. □

2.8. Homology of the nerve of a cover. Let $(X, \mathscr{U})$ be a covered space. Define a chain complex as follows: An n-simplex is a string $[U_0, U_1, \ldots, U_n]$ with each $U_i \in \mathscr{U}$ and such that $U_0 \cap U_1 \cap \cdots \cap U_n \neq \emptyset$. If $[U_0, U_1, \ldots, U_n]$ is an n-simplex, then for $i = 0, \ldots, n$, we define

$$d^i[U_0, U_1, \ldots, U_n] = [U_0, U_1, \ldots, \widehat{U_i}, \ldots, U_n]$$

and

$$s^i[U_0, U_1, \ldots, U_n] = [U_0, U_1, \ldots, U_i, U_i, \ldots, U_n]$$

We let $K_n(\mathscr{U})$ denote the free abelian group generated by the n-simplexes of the cover with the boundary operator defined as usual for the simplicial abelian group defined by the d^i and s^i. The simplicial set is called the **nerve** of the cover and its homology is called the **homology of the nerve** of the cover.

2.9. THEOREM. *Let the topological space X be contractible to a point. Then the singular homology complex of X is contractible.*

PROOF. There is a continuous map $H: X \times [0,1] \longrightarrow X$ such that $H(-,0)$ is the identity and $H(-,1)$ is constant at a point $*$. Define $s: C_n(X) \longrightarrow C_{n+1}$ by the formula

$$sf(x_0, \ldots, x_{n+1}) = \begin{cases} * & \text{if } x_0 = 1 \\ H\left(f\left(\frac{x_1}{1-x_0}, \ldots, \frac{x_{n+1}}{1-x_0}\right), x_0\right) & \text{if } x_0 \neq 1 \end{cases}$$

This is obviously continuous for $x_0 \neq 1$ and the continuity at the remaining point follows readily from the fact that $H(-,1)$ is constant. Now we calculate (assuming $x_0 \neq 1$; the remaining case is similar), writing $y_0 = 1 - x_0$

$$\begin{aligned}
&(d \circ s)f(x_0, \ldots, x_n) \\
&= sf(0, x_0, \ldots, x_n) + \sum_{i=1}^{n+1}(-1)^i sf(x_0, \ldots, x_{i-1}, 0, x_i, \ldots, x_n) \\
&= H(f(x_0, \ldots, x_n), 0) + \sum_{i=1}^{n+1}(-1)^i H\left(f\left(\frac{x_1}{y_0}, \ldots, \frac{x_{i-1}}{y_0}, 0, \frac{x_i}{y_0}, \ldots, \frac{x_n}{y_0}\right), x_0\right) \\
&= f(x_0, \ldots, x_n) + \sum_{i=1}^{n+1}(-1)^i H\left(f\left(\frac{x_1}{y_0}, \ldots, \frac{x_{i-1}}{y_0}, 0, \frac{x_i}{y_0}, \ldots, \frac{x_n}{y_0}\right), x_0\right) \\
&= f(x_0, \ldots, x_n) + \sum_{i=0}^{n}(-1)^{i+1} H\left(f\left(\frac{x_1}{y_0}, \ldots, \frac{x_i}{y_0}, 0, \frac{x_{i+1}}{y_0}, \ldots, \frac{x_n}{y_0}\right), x_0\right)
\end{aligned}$$

and

$$\begin{aligned}
(s \circ d)f(x_0, \ldots, x_n) &= \sum_{i=0}^{n}(-1)^i (s \circ d^i) f(x_0, \ldots, x_n) \\
&= \sum_{i=0}^{n}(-1)^i H\left(d^i f\left(\frac{x_1}{y_0}, \ldots, \frac{x_n}{y_0}\right), x_0\right) \\
&= \sum_{i=0}^{n}(-1)^i H\left(f\left(\frac{x_1}{y_0}, \ldots, \frac{x_i}{y_0}, 0, \frac{x_{i+1}}{y_0}, \ldots, \frac{x_n}{y_0}\right), x_0\right)
\end{aligned}$$

and adding these we see that

$$s \circ d + d \circ s = \mathrm{id}$$

□

2.10. Simple covers. A cover $\mathscr{U}$ of a space X is called **simple** if for every finite set $U_0, \ldots, U_n$ of sets in $\mathscr{U}$, either $U_0 \cap \cdots \cap U_n$ is empty or it is contractible to a point. In the rest of this section, $\mathscr{U}$ will always be a simple cover.

Now, let Y be the disjoint union of the members of $\mathscr{U}$. There is an obvious map $Y \longrightarrow X$ which includes each $U \in \mathscr{U}$ into X. It is clear that the image of $C_n(Y) \longrightarrow C_n(X)$ is exactly $C_n(Y, \mathscr{U})$.

An element of Y can be denoted (x, U) where $x \in U \in \mathscr{U}$. Let Y_X^n be the nth fiber power of $Y \longrightarrow X$, which we can describe as a subset of the nth cartesian power Y^n. An element of Y^n is an n-tuple $((x_1, U_1), \ldots, (x_n, U_n))$. This element is in Y_X^n if and only if $x_1 = \cdots = x_n$. The element x must then be in $U_1 \cap \cdots \cap U_n$. Thus Y_X^n is a disjoint union of all the non-empty sets $U_1 \cap \cdots \cap U_n$, each of which is connected. These intersections are thus the connected components of the fiber power and each is, by assumption, contractible to a point.

A point to note is that the cover induced on each component of Y_X^n by $\mathscr{U}$ is refined by the whole component. Thus the cover induced on Y_X^n is that of the connected components. Since simplexes are connected, there is no difference between $C_\bullet$ and $\widehat{C}_\bullet$ on Y_X^n.

There are projections $\partial^i : Y_X^{n+1} \longrightarrow Y_X^n$ for $i = 0, \ldots, n$ given by

$$\partial^i((x, U_0), \ldots, (x, U_n)) = ((x, U_0), \ldots, \widehat{(x, U_i)}, \ldots, (x_n, U_n))$$

We now form the double complex $C_{\bullet\bullet}(X, \mathscr{U})$ as follows:

$$C_{mn} = \begin{cases} C_n(Y_X^{m+1}), & n \geq 0 \text{ and } m \geq 0 \\ K_n(\mathscr{U}), & n \geq 0 \text{ and } m = -1 \\ \widehat{C}_m(X, \mathscr{U}), & m \geq 0 \text{ and } n = -1 \\ 0, & \text{otherwise} \end{cases}$$

The operators d^i and ∂^j induce the horizontal and vertical boundary operators. They naturally commute, but if we negate the boundary in every other row, they will anticommute.

2.11. PROPOSITION. *The rows and columns of this double complex are contractible, except for the bottom row and right hand column.*

PROOF. The columns of this double complex are the singular complex of a space that is the disjoint union of spaces contractible to a point, augmented over the free abelian group generated by its components. To prove this contractible, it is sufficient to show that if the space X is contractible, then the augmented singular complex $C_\bullet(X) \longrightarrow \mathbf{Z}$ is contractible, which is the content of Theorem 2.9. As for the rows, we must find a contraction in the complex

$$\cdots \longrightarrow C_n(Y_X^m) \longrightarrow \cdots \longrightarrow C_n(Y) \longrightarrow C_n(X, \mathscr{U}) \longrightarrow 0$$

Our search will be greatly helped by good notation. An n-simplex in Y_X^m consists of a component $\langle U_1, \ldots, U_m \rangle$ of the latter, together with an n-simplex $\sigma : \Delta_n \longrightarrow X$ whose image is included in $U_1 \cap \cdots \cap U_m$. Let us denote this generator of $C_n(Y_X^m)$ by $\langle \sigma; U_1, \ldots, U_m \rangle$. Choose a function $f : C_n(X, \mathscr{U}) \longrightarrow \mathscr{U}$ such that $\sigma(\Delta_n) \subseteq f(\sigma)$. Then define

$$s\langle \sigma; U_1, \ldots, U_m \rangle = \langle \sigma; f(\sigma), U_1, \ldots, U_m \rangle$$

In degree 0, this is interpreted to mean that $s\sigma = \langle \sigma; f(\sigma) \rangle$. Then one sees immediately that $d^0 \circ s = \text{id}$ and $d^{i+1} \circ s = s \circ d^i$ so that s is a contracting homotopy in the row. □

2.12. COROLLARY. *The homology of the nerve of a simple cover is equivalent to that of the singular homology with small simplexes.* □

PROOF. This follows immediately from 3.6.1. □

Here is another application.

2.13. THEOREM. *Suppose X is a topological space and A and B are two open subsets such that $X = A \cup B$. Then there is an exact homology sequence*

$$\cdots \longrightarrow H_n(A \cap B) \longrightarrow H_n(A) \oplus H_n(B) \longrightarrow H_n(X) \longrightarrow H_{n-1}(A \cap B) \longrightarrow \cdots$$

PROOF. Let $\mathscr{U} = \{A, B\}$ be the 2 element cover of X. Let $i: A \longrightarrow X$, $j: B \longrightarrow X$, $u: A \cap B \longrightarrow A$, and $v: A \cap B \longrightarrow B$ be the inclusions. Then we have an exact sequence

$$0 \to C_n(A \cap B) \xrightarrow{\begin{pmatrix} C_n(u) \\ -C_n(v) \end{pmatrix}} C_n(A) \oplus C_n(B) \xrightarrow{(C_n(i)\ C_n(j))} C_n(X, \mathscr{U}) \to 0$$

from which the conclusion follows. □

This is, essentially, the Mayer-Vietoris exact sequence which is proved under hypotheses on A and B that guarantee that they are retracts of open neighborhoods so that this theorem can be applied to other than open sets.

3. Simplicial homology

In this section, we describe the simplicial homology on the category of triangulated spaces using so-called oriented simplexes. In addition, we will show that for a triangulated space, the simplicial homology and singular homology coincide.

3.1. Triangulated spaces. By a **triangulated space** we mean a space that is a union of simplexes subject to two conditions

1. A set is open if and only if its intersection with each simplex is open.
2. The intersection of two simplexes is a face of each.

The first condition says the space is a quotient of a disjoint union of simplexes and the second is a restriction on the nature of the kernel pair of that quotient mapping.

3.2. The simplicial category. Let T be a triangulated space. Let $V(T)$ denote the set of all the vertices of the simplexes of T. We denote by Simp the category whose objects are triangulated spaces and whose morphisms are functions that take simplexes to simplexes and are linear on the interiors. If T is a triangulated space, we let $|T|$ denote the underlying topological space.

3.3. Chains. Let T be a triangulated space. We define a chain complexes, called $C_\bullet(T)$. An n-simplex, denoted $[v_0, \ldots, v_n]$, is a string of vertices of T, not necessarily distinct, such that $\{v_0, \ldots, v_n\}$ is the set of vertices of some simplex. Then $C_n(T)$ is the free abelian group generated by the set of n-simplexes. For $i = 0 \ldots n$ we define face operators $d^i: C_n(T) \longrightarrow C_{n-1}(T)$ by $d^i[v_0, \ldots, v_n] = [v_0, \ldots, \widehat{v_i}, \ldots, v_n]$ and degeneracy operators $s^i: C_n(T) \longrightarrow C_{n-1}(T)$ by $s^i[v_0, \ldots, v_n] = [v_0, \ldots, v_i, v_i, \ldots, v_n]$.

We define a cotriple on Simp that takes a triangulated space T to the space GT defined as the disjoint union of all the simplexes of T, each triangulated in the standard way. So an element of GT is a pair (x, σ), where $x \in \sigma$ and σ is a simplex of T. A vertex of GT is then a pair (v, σ) where σ is a simplex of T and v a vertex of σ. We define $\epsilon T: GT \longrightarrow T$ by $\epsilon T(x, \sigma) = x$. A simplex of GT is a pair (σ, τ)

where τ is a simplex of T and σ is a face of τ. Thus an element of G^2T is a triplet (x, σ, τ) where τ is a simplex of T, σ is a face of τ and $x \in \sigma$. Then we define $\delta T: GT \longrightarrow G^2T$ by $\delta T(x, \sigma) = (x, \sigma, \sigma)$. If $f: T \longrightarrow T'$ is an arrow in Simp, then $Gf(x, \sigma) = (fx, f\sigma)$, which makes sense since f takes simplexes to simplexes.

3.4. PROPOSITION. *The complex $C_\bullet(T)$ is* **G***-presentable.*

PROOF. Let $[v_0, \ldots, v_n]$ be an n-simplex of $C_n(T)$. Let σ be the simplex of T whose vertices are the set $\{v_0, \ldots, v_n\}$. Then we let $\theta[v_0, \ldots, v_n] = [(v_0, \sigma), \ldots, (v_n, \sigma)]$. It is clear that $\epsilon T \circ \theta = \text{id}$. To show naturality, suppose $f: T \longrightarrow T'$ is a simplicial map. Then $f(\sigma)$ is the unique simplex whose vertices are $fv_0, \ldots, fv_n$. Then

$$\begin{aligned}\theta \circ C_n(T)f[v_0, \ldots, v_n] &= \theta[fv_0, \ldots, fv_n] \\ &= [(fv_0, f\sigma), \ldots, (fv_n, f\sigma)] = C_n(GT)f[(v_0, \sigma), \ldots, (v_n, \sigma)] \\ &= C_n(GT)f \circ \theta[v_0, \ldots, v_n]\end{aligned}$$

3.5. PROPOSITION. *The complex $C_\bullet(T)$ is* **G***-acyclic.*

PROOF. Since GT is a disjoint union of simplexes and both chain groups take disjoint unions to direct sums, it is sufficient to show this for simplexes. So let Δ_n denote an n-simplex, with a total order on its vertices, say $v_0 < \cdots < v_n$. An m-simplex in $C_m(\Delta_n)$ is an m-tuple $[v_{i_0}, \ldots, v_{i_m}]$. Now let $s[v_{i_0}, \ldots, v_{i_m}] = [v_0, v_{i_0}, \ldots, v_{i_m}]$. It is clear that $d^0 \circ s = \text{id}$ and $d^i \circ s = s \circ d^{i-1}$ for $0 < i \leq m$, so that s is a contracting homotopy on the complex $C_\bullet(\Delta_n)$. □

3.6. COROLLARY. *Suppose $\alpha_\bullet: C_\bullet \longrightarrow C_\bullet$ is an endomorphism of the simplicial chain complex functor which induces the identity arrow on H_0. Then $\alpha_\bullet$ is homotopic to the identity.* □

4. Singular homology of triangulated spaces

4.1. Barycentric coordinates. There is a special coordinate system induced by a triangulation of a space that is quite useful. We begin by observing that each point of a triangulated space is in the interior of a unique simplex. For each point of the space is in at least one simplex and each point of a simplex is interior to some face of the simplex (a vertex is interior to itself). Second, since two simplexes can intersect only at a common face and the points in the face cannot be interior to both, no point can be interior to more than one simplex.

If $\{v_0, \ldots, v_n\}$ are the set of vertices of an n-simplex, then a point of that simplex can be written uniquely as $t_0v_0 + \cdots t_nv_n$ where each t_i is a non-negative real number and $t_0 + \cdots t_n = 1$. The point is interior to the simplex if every $t_i > 0$.

Now we think of the n-simplex as being the set of points $(t_0, t_1, \ldots, t_n) \in \mathscr{R}^n$ such that $\sum_{i=0}^n t_i = 1$ and each $t_i \geq 0$. It is interior if and only if each $t_i > 0$. These coordinates are called the **barycentric coordinates** of the point. (Question: Why are they called the "barycentric coordinates"? The barycenter is the center of gravity. "Affine coordinates" would make a lot more sense.)

We can make use of this in the following way. Let V be the set of vertices of the triangulated space. Let $\{t_v \mid v \in V\}$ be a set of real numbers, finitely non-zero, all non-negative with $\sum_{v \in V} t_v = 1$. Let $v_0, \ldots, v_n$ denote the set of vertices for which $t_{v_i} \neq 0$. If there is a simplex σ whose vertices are $v_0, \ldots, v_n$ then there is a unique point $x \in \sigma$ whose barycentric coordinates are $(t_{v_1}, \ldots, t_{v_n})$. If there is no such simplex, then there is no point with those barycentric coordinates. However, each point gives rise to a unique set $\{t_v \mid v \in V\}$ which will be called its barycentric coordinates.

4.2. Open star cover. If σ is a simplex of T, the **open star** of σ, denoted $\operatorname{st}\sigma$, is the union of the interiors of all the simplexes of which σ is a face. In particular, if v is a vertex, $\operatorname{st} v$ is the union the interiors of all the simplexes that v is a vertex of and the interior of a 0-simplex is itself. Clearly every point of $X = |T|$ is an interior point of at least one simplex. Thus the open stars of the vertices are a cover of X called the **open star cover**.

4.3. PROPOSITION. *Let $v_0, \ldots, v_n$ be vertices of simplexes in X. Then*

$$\operatorname{st} v_0 \cap \cdots \cap \operatorname{st} v_n = \begin{cases} \operatorname{st}\sigma & \text{if } v_0, \ldots, v_n \text{ are the vertices of } \sigma \\ \emptyset & \text{if they are not the vertices of any simplex} \end{cases}$$

PROOF. Suppose that $v_0, \ldots, v_n$ are the vertices of the simplex σ. Then each v_i is a face of any simplex that σ is a face of, so that any point in the interior of one of those simplexes in the star of each v_i and hence in their intersection. On the other hand, any simplex of which σ is not a face cannot have every v_i as a vertex and hence any interior point of such a simplex is not in the star of at least one v_i. The same argument applies in the case that there is no simplex that $v_0, \ldots, v_n$ are all members of. □

4.4. PROPOSITION. *For any simplex σ, $\operatorname{st}\sigma$ is contractible.*

PROOF. Define $H: \operatorname{st}\sigma \times [0,1] \longrightarrow \sigma$ as follows, using the barycentric coordinates. Fix a vertex v_0 of σ. If p is a point of $\operatorname{st}\sigma$ with barycentric coordinates $\{p_v \mid v \in V\}$, define $H(p,t)$ to be the point whose barycentric coordinates are

$$\begin{cases} (1-t)p_v + t, & \text{if } v = v_0 \\ (1-t)p_v, & \text{otherwise} \end{cases}$$

Since p is in the open star of σ, it has non-zero coordinates corresponding to each vertex of σ, in particular for v_0. It follows that for $0 \le t < 1$, p has the same non-zero coordinates as those of $H(p,t)$, which implies that the latter point actually exists. Evidently $H(p,0) = p$, while $H(p,1) = v_0$. □

Put the two preceding propositions together to conclude:

4.5. THEOREM. *The open star cover of a simplicial complex is simple.* □

This fact is a direct consequence of the requirement that in a triangulation any two simplexes, if they meet, do so in a common face.

4.6. THEOREM. *The homology of the nerve of the open star cover of a simplicial complex is isomorphic to the simplicial homology.*

PROOF. If Y is the disjoint union of the stars of the vertices, then Y_X^n is the disjoint unions of the all the non-empty sets

$$\operatorname{st} v_0 \cap \cdots \cap \operatorname{st} v_n$$

This set is $\operatorname{st}\sigma$ if the vertices of σ are $v_0, \ldots, v_n$. Note that these vertices may be repeated, so that σ is not necessarily an n-simplex. Thus we get one copy of $\operatorname{st} v$ for each ordering of the vertices of σ. But this is the definition of the chain complex of a simplicial complex. □

5. Homology with ordered simplexes

The original definitions of homology went to some effort to avoid simplexes with repeated vertices and having simplexes that differed only in the order of the vertices. I do not know whether this was merely for computational efficiency or that it was not realized at first that two such simplexes were necessarily in the same homology class. The same thing happened for the early definitions of singular homology. In this section, we will use acyclic models to show that these older definitions give the same homology and cohomology.

5.1. Ordered simplicial homology. Let T be a triangulated space and suppose a total order is chosen on the set V of vertices. Then an **ordered n-simplex** $\langle v_0, v_1, \dots, v_n\rangle$ consists of a string of vertices such that

1. $v_0,\ v_1,\ \dots,\ v_n$ are the vertices of some simplex of T; and
2. $v_0 < v_1 < \cdots < v_n$.

Note that this means, among other things, that the v_i are all distinct. We let $C_n^{\mathrm{ord}}(T)$ denote the free abelian group generated by the ordered n-simplexes. Although there are no degeneracy operators, the same formula works to give a boundary operator. Namely, define $d^i\colon C_n^{\mathrm{ord}}(T) \longrightarrow C_{n-1}^{\mathrm{ord}}(T)$ for $i = 0, \dots, n$ by

$$d^i\langle v_0, v_1, \dots, v_n\rangle = \langle v_0, v_1, \dots, v_{i-1}, v_{i+1}, \dots, v_n\rangle$$

The right hand side of that formula is usually denoted $\langle v_0, v_1, \dots, \widehat{v_i}, \dots, v_n\rangle$. Then

$$d = \sum_{i=0}^{n}(-1)^i d^i\colon C_n^{\mathrm{ord}}(T) \longrightarrow C_{n-1}^{\mathrm{ord}}(T)$$

One readily shows that $d \circ d = 0$ and then there are homology groups $H_\bullet^{\mathrm{ord}}(T)$.

There is an obvious natural inclusion $\alpha_\bullet\colon C_\bullet^{\mathrm{ord}} \longrightarrow C_\bullet$ that commutes with the boundary operator. It is less obvious, but this inclusion has a natural splitting. To see this, define an action of the symmetric group S_{n+1} on $C_n(T)$ by letting, for $\pi \in S_{n+1}$ and $\sigma = [v_0, v_1, \dots, v_n]$,

$$\pi^{-1}\sigma = \operatorname{sgn}(\pi)[v_{\pi 0}, v_{\pi 1}, \dots, v_{\pi n}]$$

Define $\beta_n\colon C_n \longrightarrow C_n^{\mathrm{ord}}$ as follows. Suppose that T is a triangulated space with a total order on its vertices. and $\sigma = [v_0, \dots, v_n] \in C_n(T)$ is a simplex. If σ has a repeated vertex, let $\beta_n(\sigma) = 0$. Otherwise, there is a unique permutation $\pi \in S_{n+1}$ such that $\pi\sigma$ is $\pm$ an ordered simplex and we let $\beta_n(\sigma) = \pi\sigma$. This definition on simplexes extends to a unique homomorphism $C_n(T) \longrightarrow C_n^{\mathrm{ord}}(T)$. Clearly $\beta_n \circ \alpha_n = \mathrm{id}$.

5.2. THEOREM. *For any triangulated space T, we have $d \circ \beta_n = \beta_{n-1} \circ d$.*

PROOF. We begin the proof by letting $K_n = \ker(\beta_n)$. I claim that $d(K_n) \subseteq K_{n-1}$. For an ordered simplex σ, let $K_n(\sigma)$ denote the subgroup of K_n spanned by all $\sigma - \pi\sigma$ for $\pi \in S_{n+1}$. It is clear that $K_n = D_n \oplus \sum K_n(\sigma)$, the sum taken over all the ordered simplexes. Hence it is sufficient to show that d takes every degenerate simplex as well as every simplex of the form $\sigma - \pi\sigma$ into K_{n-1}. If $\sigma = \langle v_0, \dots, v_n\rangle$ is multiply degenerate, that is two vertices are repeated or one vertex is repeated more than once, then every $d^i\sigma$ is degenerate and hence $d\sigma \in K_{n-1}$. Thus we can suppose that $v_i = v_j$, $i < j$ and there is no other degeneracy. In that case, all but two terms of $d\sigma$ are degenerate and those terms are

$$(-1)^i \langle v_0, \ldots, \widehat{v}_i, \ldots, v_j, \ldots, v_n \rangle + (-1)^j \langle v_0, \ldots, v_i, \ldots, \widehat{v}_j, \ldots, v_n \rangle$$

and these two terms differ only by $j - i - 1$ transpositions and hence add up to an element of K_{n-1}.

We now consider the case of $\sigma - \pi\sigma$, where σ is an ordered simplex and $\pi \in S_{n+1}$, by induction on the number of adjacent transpositions necessary to express π. If this number is 0, π is the identity and there is nothing to prove. Write $\pi = \theta\phi$, where θ is an adjacent transposition and ϕ is expressible as a composite of fewer adjacent transpositions than π. Then $\sigma - \pi\sigma = (\sigma - \phi\sigma) + (\phi\sigma - \theta\phi\sigma)$. We assume that $d(\sigma - \phi\sigma) \in K_{n-1}$. We will let $\tau = \pm\phi\sigma$, the sign chosen so that τ is a simplex and will show that $d(\tau - \theta\tau) \in K_{n-1}$.

It is clear that when σ is an ordered simplex, so is $d\sigma$ and so $d \circ \beta_n(\sigma) = d\sigma = \beta_{n-1} \circ d\sigma$. For general simplexes, it will be sufficient to show that $d(K_n) \subseteq K_{n-1}$. For then we see that if σ is degenerate, then

$$\beta_{n-1} \circ d\sigma = 0 = d \circ \beta_n \sigma$$

Otherwise we infer from $\pi\sigma - \sigma \in K_n$, that modulo K_{n-1}

$$\begin{aligned}
\beta_{n-1} \circ d\pi\sigma &= \beta_{n-1} \circ d\sigma + \beta_{n-1} \circ d(\pi\sigma - \sigma) \\
&= d \circ \beta_n \sigma \\
&\equiv d \circ \beta_n \sigma + d \circ \beta_n(\pi\sigma - \sigma) \\
&= d \circ \beta_n \circ \pi\sigma
\end{aligned}$$

Since every non-degenerate simplex is $\pm\pi\sigma$ for some ordered simplex, the conclusion follows. □

5.3. THEOREM. *The inclusion* $C_\bullet^{\mathrm{ord}} \longrightarrow C_\bullet$ *is a homotopy equivalence on the category of triangulated spaces.*

PROOF. If we identify $C_\bullet^{\mathrm{ord}}$ with $C_\bullet/D_\bullet$, then we have that the composite $C_\bullet^{\mathrm{ord}} \longrightarrow C_\bullet \longrightarrow C_\bullet^{\mathrm{ord}}$ is the identity. Since $C_0^{\mathrm{ord}} = C_0$, the other composite is the identity in degree 0 and the conclusion now follows from Corollary 3.6. □

5.4. Ordered singular chains. Now we consider the category of spaces. There is no analog of the subgroup of ordered singular chains, but there is an analog of the $C_\bullet/D_\bullet$ construction. If $\sigma\colon \Delta_n \longrightarrow X$ is a singular n-simplex and $\pi \in S_{n+1}$, define $\pi\sigma$ by $(\pi\sigma)(t_0, \ldots, t_n) = \mathrm{sgn}(\pi)\sigma(t_{\pi 0}, \ldots, t_{\pi n})$. This extends to a unique additive operation on singular chains. Let $D_n(X)$ consist of those singular chains c for which $\pi c = -c$.

5.5. PROPOSITION. *Let* θ *be an adjacent transposition. Then* $d(c - \theta c) \in D_{n-1}$.

PROOF. Let us denote by θ^j the transposition that interchanges j and $j+1$. As usual $d^i\colon C_n(X) \longrightarrow C_{n-1}(X)$ is defined by

$$d^i\sigma(t_0, \ldots, t_{n-1}) = \sigma(t_0, \ldots, t_{i-1}, 0, t_i, \ldots, t_{n-1})\circ$$

Then we claim that

$$d^i \circ \theta^j = \begin{cases} \theta^{j-1} \circ d^i, & \text{if } i < j \\ d^{i+1}, & \text{if } i = j \\ d^{i-1}, & \text{if } i = j+1 \\ \theta^j \circ d^i, & \text{if } i > j+1 \end{cases}$$

We prove, for example, the first one. The others are proved similarly. We have, for $i < j$,

$$\begin{aligned} d^i \circ \theta^j \sigma(t_0, \ldots, t_{n-1}) &= \theta^j \sigma(t_0, \ldots, t_{i-1}, 0, t_i, \ldots, t_{j-1}, t_j, \ldots, t_{n-1}) \\ &= \sigma(t_0, \ldots, t_{i-1}, 0, t_i, \ldots, t_j, t_{j-1}, \ldots, t_{n-1}) \\ &= d^i \sigma(t_0, \ldots, t_{i-1}, t_i, \ldots, t_j, t_{j-1}, \ldots, t_{n-1}) \\ &= \theta^{j-1} d^i \sigma(t_0, \ldots, t_{i-1}, t_i, \ldots, t_{j-1}, t_j, \ldots, t_{n-1}) \end{aligned}$$

From this we calculate that

$$\begin{aligned} d \circ (1 - \theta^j) &= \sum_{i=0}^{n} (-1)^i d^i \circ (1 - \theta^j) \\ &= \sum_{i=0}^{j-1} (-1)^i d^i \circ (1 - \theta^j) + (-1)^j d^j (1 - \theta^j) \\ &\qquad d + (-1)^{j+1} d^{j+1} (1 - \theta^j) + \sum_{i=j+2}^{n} (-1)^i d^i \circ (1 - \theta^j) \\ &= \sum_{i=0}^{j-1} (-1)^i (1 - \theta^{j-1}) \circ d^i + (-1)^j (d^j + d^{j+1}) \\ &\qquad d + (-1)^{j+1} (d^{j+1} + d^j) + \sum_{i=j+2}^{n} (-1)^i (1 - \theta^j) \circ d^i \\ &= (1 - \theta^{j-1}) \circ \sum_{i=0}^{j-1} (-1)^i d^i + (1 - \theta^j) \sum_{i=j+2}^{n} (-1)^i \circ d^i \end{aligned}$$ □

Now if $\theta^j \sigma = \sigma$, then $(1 - \theta^j)\sigma = 2\sigma$ and then

$$\begin{aligned} d(2\sigma) &= d(1 - \theta^j)\sigma \\ &= (1 - \theta^{j-1}) \circ \sum_{i=0}^{j-1} (-1)^i d^i \sigma + (1 - \theta^j) \sum_{i=j+2}^{n} (-1)^i \circ d^i \sigma \\ &\subseteq D_{n-1} \end{aligned}$$

This implies that there is a natural homomorphism $C_\bullet \longrightarrow C_\bullet^{\mathrm{ord}}$. We would now like to show it is a homotopy equivalence. Unlike the case of simplicial complexes, there does not seem to be any way of making it natural, but rather will show that it is a weak homotopy equivalence. In order to do this, we make use of a cotriple that was first used by Kleisli [1974] for similar purposes. $I = [0, 1]$ is the unit interval of real numbers.

For a space X and element $x \in X$, let $I \underset{x}{\multimap} X$ denote the space of continuous functions (called **paths**) $p: I \longrightarrow X$ such that $p(0) = x$, topologized with the compact/open topology. This means that for a compact subset $K \subseteq I$ and an open subset $U \subseteq X$, we let $N(K,U)$ denote the set $\{p: I \longrightarrow X \mid p(K) \subseteq U\}$. Then the compact/open topology is the one that takes the set of all $N(K,U)$ as a basis for the topology.

Define $GX = \sum_{x\in X} I \underset{x}{\multimap} X$. Of course, the point set of GX is just the set of paths in X, but the topology is not that of the path space, since paths starting at distinct points are in different components. We define $\epsilon X: GX \longrightarrow X$ as evaluation at 1. We could also define $\delta: G \longrightarrow G^2$ so as to make (G, ϵ, δ) a cotriple, but this part of the structure is not needed.

In order to interpret the next theorem, we will think of Σ_{n+1} as embedded as the subgroup of Σ_{n+2} consisting of those permutations of $\{0, \ldots, n+1\}$ that fix $n+1$.

5.6. PROPOSITION. *There is a natural chain contraction s in the augmented chain complex functor $C_\bullet G \longrightarrow C_{-1}G \longrightarrow 0$ such that for an n-simplex σ and $\pi \in \Sigma_{n+1}$, $s(\pi\sigma) = \pi s(\sigma)$.*

PROOF. Let X be any space and suppose that $\sigma: \Delta_n \longrightarrow GX$ is a singular n-simplex. Since σ is a continuous function from a connected space into a disjoint sum of spaces, it necessarily factors through one of the summands. Thus we can think of σ as being a continuous function $\Delta_n \longrightarrow I \underset{x}{\multimap} X$ for a uniquely determined $x \in X$. A function $\sigma: \Delta_n \longrightarrow I \underset{x}{\multimap} X$ transposes to a function that will also denote $\sigma: \Delta \times I \longrightarrow X$ and we will denote its value at the point $\mathbf{t} = (t_0, \ldots, t_n)$ such that $t_0 + \cdots + t_n = 1$ and $u \in I$ by $\sigma(\mathbf{t}; u)$. First we consider the case of dimensions -1 and 0. Any path in $I \underset{x}{\multimap} X$ is homotopic to the constant path at x which we will denote p_x. In fact, the map $H: I \times I \longrightarrow X$ given by $H(u_0, u_1) = p(u_0 u_1)$ gives the path p when $u_0 = 1$ and p_x when $u_0 = 0$. On the other hand, two distinct elements of x correspond to distinct components of GX. Thus $C_{-1}(GX) = H_0(GX)$ is the free abelian group generated by the elements of X. We now let $s: C_{-1}(GX) \longrightarrow C_0(GX)$ given by $sx = p_x$. The group $C_0(GX)$ is the free abelian group generated by the paths in GX. For $n \geq 0$, define $s: C_n(GX) \longrightarrow C_{n+1}(GX)$ by

$$(s\sigma)(\mathbf{t}, t_{n+1}; u) = (-1)^{n+1} \begin{cases} \sigma\left(\frac{t_0}{1-t_{n+1}}, \ldots, \frac{t_n}{1-t_{n+1}}; (1-t_{n+1})u\right), & \text{if } t_{n+1} \neq 1 \\ x, & \text{if } t_{n+1} = 1 \end{cases}$$

This is obviously continuous for $t_{n+1} < 1$. We will defer to later the proof that it is continuous at $t_{n+1} = 1$.

The first thing we want to do is calculate $sd + ds$ in degree 0. A 0-path in GX is just an element of GX, that is a path in X. To conform with the notation in higher degrees, we denote it $p(1; u)$. The 1 stands for the single element of Δ_0 and $u \in I$. By definition, $dp = p(1; 0)$, the starting point, which we will call x. Then $sdp = p_x$, which we can write as $sdp(1; u) = p(1; 0)$. We have,

$$dsp(1; u) = sp(0, 1; u) - sp(1, 0; u) = -p(1; 0) + p(1; u) = p(1; u) - sd(1; u)$$

as required. For $n > 0$, we calculate, assuming $t_n \neq 1$,

$$
\begin{aligned}
ds\sigma(t_0,\ldots,t_n;u) &= \sum_{i=0}^{n+1}(-1)^i s\sigma\,(t_0,\ldots,t_{i-1},0,t_i,\ldots,t_n;u)\\
&= (-1)^{n+1}\sum_{i=0}^{n}(-1)^i s\sigma\,(t_0,\ldots,t_{i-1},0,t_i,\ldots,t_n;u)\\
&\quad + (-1)^{n+1}s\sigma\,(t_0,\ldots,t_n,0;u)\\
&= (-1)^{n+1}\sum_{i=0}^{n}(-1)^i\sigma\left(\frac{t_0}{1-t_n},\ldots,\frac{t_{i-1}}{1-t_n},0,\frac{t_i}{1-t_n},\ldots,\frac{t_{n-1}}{1-t_n};(1-t_n)u\right)\\
&\quad + (-1)^{n+1}(-1)^{n+1}\sigma\,(t_0,\ldots,t_n;u)\\
&= (-1)^{n+1}d\sigma\left(\frac{t_0}{1-t_n},\ldots,\frac{t_{n-1}}{1-t_n};(1-t_n)u\right)+\sigma\,(t_0,\ldots,t_n;u)\\
&= -sd\sigma\,(t_0,\ldots,t_n;u)+\sigma\,(t_0,\ldots,t_n;u)
\end{aligned}
$$

so that $ds = -sd + 1$. The case that $t_n = 1$ can be handled similarly (or use continuity).

Next we show that s is continuous at $t_{n+1} = 1$. Let U be a neighborhood of x. Since σ is continuous and $\sigma(\mathbf{t};1) = x$ for all $\mathbf{t} = (t_0,\ldots,t_n) \in \Delta_n$, there is a neighborhood $V_{\mathbf{t}}$ of $\mathbf{t}$ and a number $\epsilon_{\mathbf{t}} > 0$ such that $\mathbf{t}' \in V_{\mathbf{t}}$ and $0 < u < \epsilon_{\mathbf{t}}$ implies that $\sigma(\mathbf{t},u) \in U$. Since Δ_n is compact, it is covered by a finite number of these $V_{\mathbf{t}}$. Taking ϵ as the minimum of the $\epsilon_{\mathbf{t}}$ corresponding to those finitely many $V_{\mathbf{t}}$, we see that when $0 < u < \epsilon$, then $\sigma(\mathbf{t},u) \in U$ for all $\mathbf{t} \in \Delta_n$ and hence

$$\lim_{t_{n+1}\longrightarrow 1}\sigma\left(\frac{t_0}{1-t_{n+1}},\ldots,\frac{t_n}{1-t_{n+1}};(1-t_{n+1})u\right)=x$$

The fact that $\pi s = s\pi$ for $\pi \in \Sigma_{n+1}$ is obvious. □

5.7. PROPOSITION. *For all $n \geq 0$, there is a natural transformation $\theta_n\colon C_n \longrightarrow C_nG$ such that $C_n\epsilon \circ \theta_n = \mathrm{id}$ and such that for each $\pi \in \Sigma_{n+1}$, $\theta_n(\pi\sigma) = \pi\theta_n(\sigma)$.*

PROOF. Let X be a topological space. Define $\theta_nX\colon C_nX \longrightarrow C_nGX$ by

$$\theta_n(\sigma)(t_0,\ldots,t_n)(u) = \sigma\left(ut_0+\frac{1-u}{n+1},\ldots,ut_n+\frac{1-u}{n+1}\right)$$

which is a simplex in the component of GX based at $\sigma(\frac{1-u}{n+1},\ldots,\frac{1-u}{n+1})$. It is clear that $\theta_n(\sigma)(t_0,\ldots,t_n)(1) = \sigma(t_0,\ldots,t_n)$ and that $\theta(\sigma\circ p) = \theta(\sigma)\circ p$, from which the second claim follows. □

Now we can apply acyclic models to $C_\bullet$ as well as $C_\bullet^{\mathrm{ord}}$. We have shown that $C_\bullet$ is ϵ-presentable and that $C_\bullet \longrightarrow C_{-1} \longrightarrow 0$ is G-contractible and both natural transformations commute with the action of the symmetric groups. This implies that $s(D_n) \subseteq D_{n+1}$, and similarly that $\theta_n(D_n) \subseteq D_nG$ and so $C_\bullet^{\mathrm{ord}}$ is also ϵ-presentable and that $C_\bullet^{\mathrm{ord}} \longrightarrow C_{-1}^{\mathrm{ord}} \longrightarrow 0$ is G-contractible. Since $C_0 = C_0^{\mathrm{ord}}$, it follows that

5.8. THEOREM. *$C_\bullet \longrightarrow C_\bullet^{\mathrm{ord}}$ is a homotopy equivalence.* □

6. Application to homology on manifolds

Consider a differentiable manifold M of class C^q, for $0 \leq q \leq \infty$. For $p \leq q$, we can form the group $C_n^p(M)$ of singular n-simplexes in M that are p times continuously differentiable. Intuitively, we feel that the resultant chain complex should not depend, up to homology, on q. We would expect a process analogous to simplicial approximation to allow us to smooth a simplex of class C^p to obtain a homologous simplex of class C^q. The case $q = \infty$ and $p = 0$ is well known, in connection with de Rham cohomology see [Bredon, 1993]. Here we deal with the general case and show homotopy equivalence directly. The basic argument is a modification of the one used in Bredon. We have previously given an argument that is based even more directly on our acyclic models theorem, but it depends on the existence of a simple cover and on paracompactness, which this one does not. See [Barr, 1996].

A simplex is not a manifold, so that one has to define what it means for a singular simplex in a manifold to be smooth. One possibility would be to extend the category to manifolds with boundary and thus to include the simplexes. Then smoothness would have to be defined in terms of one-sided derivative. I know no reason that this would not work, but simplexes have lots of corners and this leads us into uncharted (for me!) waters that I would rather avoid. Thus I will follow Bredon and define a smooth simplex as one that has a smooth extension to some neighborhood of Δ in the space defined by $t_0+\cdots+t_n = 1$, which is, essentially, $\mathscr{R}^n$.

The main consequence of this decision for us is that the various cotriples that are used to prove, for example, the subdivision and Mayer-Vietoris theorems are no longer available. However, the explicit formula of 1.16 will be valid as soon as we can describe a "smooth cone" construction that takes smooth cones to smooth cones.

Recall that in 1.2 we defined, for a convex subset U of a euclidean space, a point $b \in U$ and a singular simplex $\sigma\colon \Delta_n \longrightarrow U$ a cone $b \cdot \sigma\colon \Delta_{n+1} \longrightarrow U$ by

$$b \cdot \sigma(t_0, t_1, \ldots, t_{n+1}) = \begin{cases} r(t_0)b + (1 - r(t_0))\sigma\left(\frac{t_1}{1-t_0}, \ldots, \frac{t_{n+1}}{1-t_0}\right) & \text{if } t_0 \neq 1 \\ b & \text{if } t_0 = 1 \end{cases}$$

The function $r\colon I \longrightarrow I$ is any continuous bijective function such that $r(0) = 0$ and $r(1) = 1$. The problem is the $1 - t_0$ in the denominator that does not interfere with continuity since σ is bounded on the compact set Δ_n, but does interfere with smoothness. We will show that for the choice of

$$r(t) = \begin{cases} 1 - e^{1-1/(1-t)} & \text{if } 0 \leq t \leq 1 \\ 1 & \text{if } t = 1 \end{cases}$$

In order to extend this to a neighborhood, we extend the definition of $r(t)$ to the entire line by

$$r(t) = \begin{cases} 1 - e^{1-1/(1-t)} & \text{if } t \leq 1 \\ 1 & \text{if } t \geq 1 \end{cases}$$

Since U is open, b is at positive distance from the complement and hence there will be some $\epsilon > 0$ such that for all $t_0 \in [-\epsilon, 1 + \epsilon]$, the extension defined by

$$\begin{aligned} b \cdot \sigma(t_0, &t_1, \ldots, t_{n+1}) \\ &= \begin{cases} r(t_0)b + (1 - r(t_0))\sigma\left(\frac{t_1}{1-t_0}, \ldots, \frac{t_{n+1}}{1-t_0}\right) & \text{if } -\epsilon < t_0 < 1 \\ b & \text{if } 1 \leq t_0 \leq 1 + \epsilon \end{cases} \end{aligned}$$

still takes values in U.

6.1. PROPOSITION. *If σ is p times continuous differential, then so is $b \cdot \sigma$.*

PROOF. There is obviously no problem except when $t_0 = 1$. As $t_0 \longrightarrow 1-$, the exponent goes to $-\infty$ and hence $\lim_{t_0 \longrightarrow 1-} r(t) = 1$, which establishes continuity. The differentiability follows from the following lemma. □

6.2. LEMMA. *Suppose we write $\tau = 1/(1-t_0)$ and that k and l are positive integers, s is a polynomial in $t_0, \dots, t_n$ and ϕ is at least k times continuously differentiable on a neighborhood of Δ_n. Let f be the function defined on a neighborhood of Δ_n to be 0 for $t_0 \geq 1$ for $t_0 < 1$ by*

$$(*) \qquad e^{-\tau}\tau^{-l}s(t_0, \dots, t_n)\phi(\tau t_0, \dots, \tau t_n)$$

has a continuous partial derivative that is a finite sum of terms of the same form involving the functions and their first partial derivatives of those appearing in $(*)$ and is, therefore, at least $k-1$ times continuously differentiable.

PROOF. Let ϕ_i denote the partial derivative of ϕ with respect to its ith variable. Then the partial derivative of $(*)$ with respect to t_0 is

$$\begin{aligned}
&- e^{-\tau}\frac{\partial \tau}{\partial t_0}\tau^{-l}s(t_0, \dots, t_n)\phi(\tau t_0, \dots, \tau t_n) \\
&- 2ke^{-\tau}\tau^{-l-1}s(t_0, \dots, t_n)\phi(\tau t_0, \dots, \tau t_n) \\
&+ e^{-\tau}\tau^{-l}\frac{\partial s(t_0, \dots, t_n)}{\partial t_0}\phi(\tau t_0, \dots, \tau t_n) \\
&+ e^{-\tau}\tau^{-l}s(t_0, \dots, t_n)\phi_0(\tau t_0, \dots, \tau t_n)\tau \\
&+ e^{-\tau}\tau^{-l}s(t_0, \dots, t_n)\sum_{i=0}^{n}\phi_i(\tau t_0, \dots, \tau t_n)t_i\frac{\partial \tau}{\partial t_0}
\end{aligned}$$

Since $\partial\tau/\partial t_0 = -1/\tau^2$ each term has the required form and is at least $k-1$ times differentiable. Since $\lim_{\tau \longrightarrow \infty} e^{-\tau}\tau^{-l} = 0$, each term is continuous at the boundary. □

Let M be a manifold of class C^q as above, $p \leq q$ and suppose that $j(M)\colon C^p_\bullet(M) \longrightarrow C_\bullet(M)$ is the inclusion of the group of q-smooth singular chains into the group of all chains. It is clear that $j(M)$ is the M component of a natural transformation between these functors. We will be showing that it is a quasi-homotopy equivalence, that is a homotopy equivalence at each object, without necessarily having a natural homotopy inverse. We note that with the smooth cone, the proof of 1.4 remains valid and we can conclude that

6.3. LEMMA. *Let U be a non-empty convex open subset of $\mathscr{R}^n$. Then the simplicial set of smooth simplexes on U is contractible.*

Similarly the construction of the simplicial subdivision, as well as the proof, using the explicit formula of 1.16 that it is homotopic to the identity, remain unchanged.

If M is a manifold of class C^p, $1 \leq p \leq \infty$, let $j_\bullet(M)\colon C^p_\bullet(M) \longrightarrow C_\bullet(M)$ denote the inclusion of the subgroup of p times differentiable chains into the group of all singular chains. We wish to show that this is, for each M, a homotopy equivalence. To this end, we let $J_\bullet$ denote the mapping cone of $j_\bullet$. In accordance with 2.9, to prove that $j_\bullet$ is a homotopy equivalence, it suffices to show that $J_\bullet$ is contractible. Since the terms are all projective, it suffices to show that it is exact.

6.4. PROPOSITION. *The complex $J_\bullet$ satisfies the Mayer-Vietoris theorem.*

PROOF. Suppose U and V are open subsets of the manifold M. Let $C^p_\bullet = C^p_\bullet(U \cup V)$, $C_\bullet = C_\bullet(U \cup V)$ and $C^p_\bullet = C^p_\bullet(U \cup V)$. Let $\widetilde{C}^p_\bullet = C^p_\bullet(U \cup V, \{U, V\})$, $\widetilde{C}_\bullet = C_\bullet(U \cup V, \{U, V\})$ and $\widetilde{J}_\bullet = \widetilde{C}^p_\bullet / \widetilde{C}_\bullet$. Let $\widehat{C}^p_\bullet = C^p_\bullet / \widetilde{C}^p_s pot$, $\widehat{C}_\bullet = C_\bullet / \widetilde{C}_s pot$, and $\widehat{J}_\bullet = J_\bullet / \widetilde{J}_s pot$. The 3×3 lemma (2.3.12)applied to

$$\begin{array}{ccccccccc}
 & & 0 & & 0 & & 0 & & \\
 & & \downarrow & & \downarrow & & \downarrow & & \\
0 & \longrightarrow & \widetilde{C}^p_\bullet & \longrightarrow & C^p_\bullet & \longrightarrow & \widehat{C}^p_\bullet & \longrightarrow & 0 \\
 & & \downarrow & & \downarrow & & \downarrow & & \\
0 & \longrightarrow & \widetilde{C}_\bullet & \longrightarrow & C_\bullet & \longrightarrow & \widehat{C}_\bullet & \longrightarrow & 0 \\
 & & \downarrow & & \downarrow & & \downarrow & & \\
0 & \longrightarrow & \widetilde{J}_\bullet & \longrightarrow & J_\bullet & \longrightarrow & \widehat{J}_\bullet & \longrightarrow & 0 \\
 & & \downarrow & & \downarrow & & \downarrow & & \\
 & & 0 & & 0 & & 0 & &
\end{array}$$

implies that $\widehat{J}_\bullet$ is acyclic so that the inclusion $\widetilde{J}_\bullet \longrightarrow J_\bullet$ is a homology, hence homotopy equivalence.

The 3×3 lemma applied to

$$\begin{array}{ccccccccc}
 & & 0 & & 0 & & 0 & & \\
 & & \downarrow & & \downarrow & & \downarrow & & \\
0 & \longrightarrow & C^p_\bullet(U \cap V) & \longrightarrow & C_\bullet(U) \oplus C_\bullet(V) & \longrightarrow & \widetilde{C}_\bullet & \longrightarrow & 0 \\
 & & \downarrow & & \downarrow & & \downarrow & & \\
0 & \longrightarrow & C_\bullet(U \cap V) & \longrightarrow & C_\bullet(U) \oplus C_\bullet(V) & \longrightarrow & \widetilde{C}_\bullet & \longrightarrow & 0 \\
 & & \downarrow & & \downarrow & & \downarrow & & \\
0 & \longrightarrow & J_\bullet(U \cap V) & \longrightarrow & J_\bullet(U) \oplus J_\bullet(V) & \longrightarrow & \widetilde{J}_\bullet & \longrightarrow & 0 \\
 & & \downarrow & & \downarrow & & \downarrow & & \\
 & & 0 & & 0 & & 0 & &
\end{array}$$

implies that the bottom row is exact. In conjunction with the previous diagram, this implies the Mayer-Vietoris theorem for J. □

6.5. PROPOSITION. *Suppose $\mathscr{U}$ is a cover of the manifold M. Every simplex and hence every chain of $C_\bullet(M)$ lies in a finite union of elements of $\mathscr{U}$.*

PROOF. This is an immediate consequence of the compactness of simplexes. □

6.6. COROLLARY. *The groups $J_\bullet(M)$ are the union, taken over all finite subsets $\{U_1, \ldots, U_n\} \subseteq \mathscr{U}$ of $J_\bullet(U_1 \cup \cdots \cup U_n)$.*

PROOF. This is true of both $C_\bullet(M)$ and $C^p_\bullet(M)$ and is easily seen to be true of the quotient. □

6.7. COROLLARY. *If $J_\bullet(U_1 \cup \cdots \cup U_n) = 0$ for every finite subset $\{U_1, \ldots, U_n\} \subseteq \mathscr{U}$, then $J_\bullet(M) = 0$.*

PROOF. For any cycle $c \in J_m(M)$, there is a finite subset $\{U_1, \ldots, U_n\} \subseteq \mathscr{U}$ such that c is a cycle in $J_m(\{U_1 \cup \cdots \cup U_n)$. If c is not a boundary in $J_m(M)$, then it is certainly not a boundary in $J_m(\{U_1 \cup \cdots \cup U_n)$. □

6.8. PROPOSITION. *Suppose that U and V are open subsets of M such that $J_\bullet$ is acyclic at U, V and $U \cap V$. Then it is acyclic at $U \cup V$.*

PROOF. This is an immediate consequence of the Mayer-Vietoris theorem. □

6.9. PROPOSITION. *Suppose $U_1, \ldots, U_n$ are open subsets of M such that $J_\bullet$ is acyclic at the intersection of every non-empty finite subset of them. Then $J_\bullet$ is acyclic at their union.*

PROOF. Assume this is true for all sets of $n-1$ open sets. It follows that $J_\bullet$ is acyclic at $U_1 \cup \cdots \cup U_{n-1}$, at U_n and also at

$$(U_1 \cup \cdots \cup U_{n-1}) \cap U_n = (U_1 \cap U_n) \cup \cdots \cup (U_{n-1} \cap U_n)$$

and hence, by the preceding result, at $U_1 \cup \cdots \cup U_{n-1} \cup U_n$. □

6.10. PROPOSITION. *Suppose that U is an open subset of euclidean space. Then $J_\bullet(U)$ is acyclic.*

PROOF. On a convex subset of euclidean space, the homology of both $C^p_\bullet$ and $C_\bullet$ reduce to $\mathbf{Z}$ concentrated in degree 0 and the induced map in degree 0 is the identity, since all 0-simplexes are smooth. Hence $J_\bullet$ is exact on a convex subset. A spherical neighborhood is convex and any intersection of convex sets is convex, so that the cover by spherical neighborhoods has the property that any finite intersection is convex and hence $J_\bullet$ is acyclic on any finite union and hence on all of U. □

6.11. THEOREM. *The chain functor $J_\bullet$ is acyclic, and hence contractible, on any manifold M. It follows that the inclusion $C^p_\bullet(M) \longrightarrow C_\bullet(M)$ is a homotopy equivalence.*

PROOF. A manifold has an open cover by euclidean spaces. Any finite intersection of these is an open subset of a euclidean space and hence $J_\bullet$ is acyclic on all those finite intersections and hence on every finite union of them and therefore on all of M. □

6.12. De Rham's Theorem. One of the reasons that homology theory was captured the interest of mathematicians in the early part of the century was the connection between integrability of forms and the topological properties of the set on which they were defined. A closed (or exact) form on a contractible space was integrable, but on a non-contractible space need not be. This comes to fruition in de Rham's theorem which connects these facts by an equivalence of cohomology theories.

We will not here develop the theory of de Rham cohomology, which is far from the purposes of this book. Nor will we include a proof of the Poincaré Lemma on which all proofs I am aware of depend on. What we will do is show how the methods developed here can help organize the argument. See, for example, [Bredon, 1993] or [Spivak, 1965] for excellent treatments of the de Rham theory.

Although previous results did not require paracompactness, this one apparently does. At any rate, the proof given here uses it. A paracompact manifold has a partition of unity [Bredon, 1993, pages 35–37]. For our purposes, the partition of unity is more basic than the atlas. To explain, if X is a topological space and $\{\alpha_i \mid i \in I\}$ is a partition of unity such that for each $i \in I$, the **support** of α_i, denoted $\operatorname{supp} \alpha_i$, is homeomorphic to an open subset of $\mathscr{R}^n$, then X is obviously a manifold since each point will have a neighborhood that is homeomorphic to an open ball in $\mathscr{R}^n$ and hence to $\mathscr{R}^n$. If on $\operatorname{supp} \alpha_i \cap \operatorname{supp} \alpha_j$, the transition map from the α_i coordinates to the α_j coordinates are smooth, then we have a smooth manifold. Conversely, if we begin with a smooth manifold, we can begin by choosing a smooth partition of unity [Bredon, 1993, pages 89–90] and then we will have a smooth manifold in the sense just described.

Thus motivated, we can define a special category for the proof. An object of the category is a pair $(X, \{\alpha_i \mid i \in I\})$ such that X is a topological space and $\{\alpha_i \mid i \in I\})$ is a partition of unity that gives X a smooth structure. A mapping $(X, \{\alpha_i \mid i \in I\}) \longrightarrow (Y, \{\beta_j \mid j \in J\})$ is a pair (f, ϕ) where $f: X \longrightarrow Y$ is a smooth map and $\phi: I \longrightarrow J$ is a function that induces, for each $x \in X$ a bijection between the $\{i \in I \mid \alpha_i X \neq 0\}$ and $\{j \in J \mid \beta_j f x \neq 0\}$ such that the diagram

$$\begin{array}{ccc} X & \xrightarrow{\quad f \quad} & Y \\ & {\scriptstyle \alpha_i} \searrow \quad \swarrow {\scriptstyle \beta_{\phi i}} & \\ & \mathscr{R} & \end{array}$$

commutes. It is not required, and not generally true, that ϕ be a bijection, but it must be locally so, in the sense described. We will call this category the category of **partitioned spaces** and denote it Part.

Let $G(X, \{\alpha_i\}) = \bigcup_{i \in I} (\operatorname{supp} \alpha_i, \{\alpha_{i'} | \operatorname{supp} \alpha_i\})$. That is, we take the disjoint union of the supports and on each one, restrict the partition of unity to that support. Of course, these will mostly vanish on this support, but the way things are set up, this causes no harm. The index set of the partition of unity in $G(X, \{\alpha_i\})$ is then $I \times I$ with the function $\alpha_{i'} | \operatorname{supp} \alpha_i$ being the function indexed by the pair (i, i'). If $(f, \phi): (X, \{\alpha_i\}) \longrightarrow (Y, \{\beta_j\})$ is an arrow of Part, we have a commutative square, for each $i \in I$,

$$\begin{array}{ccc} \operatorname{supp}\alpha_i & \longrightarrow & X \\ \big\downarrow & & \big\downarrow{\scriptstyle f} \\ \operatorname{supp}\beta_{\phi i} & \longrightarrow & Y \end{array}$$

in which the vertical maps are inclusions and the left hand arrow the unique one making the square commute. This defines the arrow $G(f, \phi)$. The map $\epsilon\colon G(X, \{\alpha_i\})$ is the unique arrow whose restriction to the component indexed by i is $(\subseteq, p_2)$, the latter being the second coordinate projection. This is clearly an arrow of Part and the component at $(X, \{\alpha_i\})$ of a natural transformation. We do not need it, but a natural transformation δ making (G, ϵ, δ) into a cotriple can also be defined. There is a contravariant functor Ω^m that associates to each smooth manifold the group of differential m forms on that manifold. This functor is clearly defined on Part, ignoring the partition. I claim there is a natural transformation $t\colon \Omega G \longrightarrow \Omega$ for which $t \circ \Omega\epsilon = \text{id}$. In fact, a differential m form on $G(X, \{\alpha_i\})$ consists of an I indexed family $\{\omega_i\}$ of m forms ω_i on $\operatorname{supp}\alpha_i$. Then we let $t(\{\alpha_i\}) = \sum_{i\in I} \alpha_i\omega_i$. We see that if $\{\omega_i\} = \{\omega| \operatorname{supp}\alpha_i\}$, then

$$t(\{\alpha_i\}) = \sum_{i\in I} \alpha_i\omega_i = \sum_{i\in I} \alpha_i\omega = \omega$$

which shows that $t \circ \Omega\epsilon = \text{id}$. We still have to show that t is natural on Part. This means showing that if $(f, \phi)\colon (X, \{\alpha_i\}) \longrightarrow (Y, \{\beta_j\})$ is an arrow of Part, that the square

$$\begin{array}{ccc} \Omega^m(GY) & \xrightarrow{tY} & \Omega^m(Y) \\ {\scriptstyle \Omega^m(Gf)}\big\downarrow & & \big\downarrow{\scriptstyle \Omega^m(f)} \\ \Omega^m(GX) & \xrightarrow[tX]{} & \Omega^m(X) \end{array}$$

commutes. We have that

$$\Omega(Gf)\, \{\omega_j \mid j \in J\} = \{f^*(\omega_\phi i) \mid i \in I\}$$

which comes from the commutative square

$$\begin{array}{ccc} \operatorname{supp}\alpha_i & \longrightarrow & X \\ \big\downarrow & & \big\downarrow{\scriptstyle f} \\ \operatorname{supp}\beta_{\phi i} & \longrightarrow & Y \end{array}$$

that tells how G is a functor. Then

$$tX \circ \Omega(Gf)\{\omega_j\} = tX\{f^*\omega_{\phi i}\} = \sum_{i\in I} \alpha_i f^*(\omega_{\phi i})$$

Going the other way, we have

$$\Omega(f) \circ tY\{\omega_j\} = \Omega(f) \sum_{j\in J} \omega_j = \sum_{j\in J} \beta_j f^*(\omega_j)$$

Now when this is applied at an element $x \in X$, we can write each $j \in J$ for which $\beta_j(fx) \neq 0$ as ϕi for a unique $i \in I$ for which $\alpha_i(x) \neq 0$. Thus this last sum becomes

$$\sum_{i \in I} \beta_{\phi i} f^*(\omega_{\phi i}) = \sum_{i \in I} \alpha_i f^*(\omega_{\phi i})\circ$$

What this argument shows is that the de Rham cochain complex is **G** acyclic with respect to natural homotopy equivalence in Part. The Poincaré Lemma shows that the de Rham complex is contractible on convex subspaces of $\mathscr{R}^n$. The same argument as used previously shows that the canonical map from the de Rham complex to the smooth cochain complex is a homotopy equivalence on open subspaces of euclidean spaces and then on all paracompact manifolds.

Bibliography

M. André (1967), *Méthode simpliciale en algèbre homologique et algèbre commutative*, Lecture Notes in Math., vol. 43, Springer-Verlag, Berlin, Heidelberg, New York.

——— (1974), *Homologie des algèbres commutatives*, Grundlehren Math. Wiss., vol. 206, Springer-Verlag, Berlin, Heidelberg, New York.

H. Appelgate (1965), *Categories with models*, Dissertation, Columbia University.

M. Barr (1962), *On the cohomology of commutative algebras*, Dissertation, University of Pennsylvania.

——— (1965a), *A cohomology theory for commutative algebras.* I, Proc. Amer. Math. Soc. **16**, 1379–1384.

——— (1965b), *A cohomology theory for commutative algebras.* II, Proc. Amer. Math. Soc. **16**, 1385–1391.

——— (1966), *Cohomology in tensored categories*, Proceedings of the Conference on Categorical Algebra, Springer-Verlag, Berlin, Heidelberg, New York, pp. 344–354.

——— (1967), *Shukla cohomology and triples*, J. Algebra **5**, 222–231.

——— (1968a), *A note on commutative algebra cohomology*, Bull. Amer. Math. Soc. **74**, 310–313.

——— (1968b), *Harrison homology, Hochschild homology and triples*, J. Algebra **8**, 314–323.

——— (1969b), *Composite cotriples and derived functors*, Seminar on Triples and Categorical Homology Theory (B. Eckmann, ed.), Lecture Notes in Math., vol. 80, Springer-Verlag, Berlin, Heidelberg, New York, pp. 336–356.

——— (1969c), *Cohomology and obstructions: commutative algebras*, Seminar on Triples and Categorical Homology Theory (B. Eckmann, ed.), Lecture Notes in Math., vol. 80, Springer-Verlag, Berlin, Heidelberg, New York, pp. 357–375.

——— (1971), *Exact categories*, Exact Categories and Categories of Sheaves, Lecture Notes in Math., vol. 236, Springer-Verlag, Berlin, Heidelberg, New York, pp. 2–121.

——— (1996a), *Cartan-Eilenberg cohomology and triples*, J. Pure Appl. Algebra **112**, 219–238.

——— (1996b), *Acyclic models*, Canad. J. Math. **48**, 258-273.

——— (forthcoming), *Resolutions and derived functors*, ftp.math.mcgill.ca/pub/barr/derfun.{ps,pdf}

M. Barr and J. Beck (1966), *Acyclic models and triples*, Proceedings of the Conference on Categorical Algebra, Springer, New York, pp. 336–343.

——— (1969), *Homology and standard constructions*, Seminar on Triples and Categorical Homology Theory (B. Eckmann, ed.), Lecture Notes in Math., vol. 80, Springer-Verlag, Berlin, Heidelberg, New York, pp. 245–335.

——— (1969), *Homology and standard constructons*, Seminar on Triples and Categorical Homology Theory (M. Tierney and A. Heller, eds.), Lecture Notes in Math., vol. 80, Springer-Verlag, Berlin, Heidelberg, New York, pp. 245–535.

M. Barr and M.-A. Knus (1971), *Extensions of derivations*, Proc. Amer. Math. Soc. **28**, 313–314.

M. Barr, and G. S. Rinehart (1966), *Cohomology as the derived functor of derivations*, Trans. Amer. Math. Soc. **122**, 416–426.

M. Barr and C. F. Wells (1984), *Toposes, triples and theories*, Grundlehren Math. Wiss., vol. 278, Springer-Verlag, Berlin, Heidelberg, New York.

D. Bayer and P. Diaconis (1992), *Trailing the dovetail shuffle to its lair*, Ann. Appl. Probab. **2**, 294–313

J. M. Beck (1967), *Triples, algebras and cohomology*, Dissertation, Columbia University.

G. Bredon (1993), *Topology and geometry*, Grad. Texts in Math., vol. 139, Springer-Verlag, Berlin, Heidelberg, New York.

H. Cartan and S. Eilenberg (1956), *Homological algebra*, Princeton University Press, Princeton, NJ.

C. Chevalley and S. Eilenberg (1948), *Cohomology theory of Lie groups and Lie algebras*, Trans. Amer. Math. Soc., **63**, 85–124.

A. Dold and D. Puppe (1961), *Homologie nicht-additiver Funktoren. Anwendungen.* Ann. Inst. Fourier Grenoble **11**, 201–312.

S. Eilenberg and S. Mac Lane (1947), *Cohomology theory in abstract groups.* I, Ann. of Math. **48**, 51–78.

——— (1953), *Acyclic models*, Amer. J. Math. **75**, 189–199.

S. Eilenberg and J. Moore (1965), *Adjoint functors and triples.* Ill, J. Math. **58**, 381–398.

S. Eilenberg and J. Zilber (1953), *On products of complexes*, Amer. J. Math. **75**, 200–204.

P.J. Freyd (1964), *Abelian categories*, Harper's Series in Modern Mathematics, Harper and Row, New York.

P. Gabriel and M. Zisman (1967), *Calculus of fractions and homotopy theory*, Ergeb. Math. Grenzgeb., vol. 35, Springer-Verlag, Berlin, Heidelberg, New York.

G. Hochschild (1945), *On the cohomology groups of an associative algebra*, Ann. of Math. (2), **46**, 58–67.

——— (1946), *On the cohomology theory for associative algebras*, Ann. of Math. (2) **47**, 568–579.

D. K. Harrison (1962), *Commutative algebras and cohomology*, Trans. Amer. Math. Soc. **104**, 191–204.

D. Kan (1958), *Adjoint functors*, Trans. Amer. Math. Soc. **87**, 294–329.

H. Kleisli (1965), *Every standard construction is induced by a pair of adjoint functors*, Proc. Amer. Math. Soc. **16**, 544–546.

——— (1974), *On the construction of standard complexes*, J. Pure Appl. Algebra **4**, 243–260.

S. Mac Lane (1963), *Homology*, Grundlehren Math. Wiss., vol. 114, Springer-Verlag, Berlin, Heidelberg, New York.

——— (1971), *Categories for the working mathematician*, Grad. Texts in Math., vol. 5, Springer-Verlag, Berlin, Heidelberg, New York.

M. Makkai and R. Paré (1990), *Accessible categories*, Contemp. Math., vol. 104, Amer. Math. Soc., Providence, RI.

B. Mitchell (1964), *The full imbedding theorem*, Amer. J. Math. **86**, 619–637.

——— (1965) *Theory of categories*, Pure Appl. Math., vol. 17, Academic Press, New York-London.

G. Orzech (1972a,b) *Obstruction theory in algebraic categories.* I, J. Pure Appl. Algebra, **2**, 287–314; II, 315–340.

N. Popescu (1973), *Abelian categories with applications to rings and modules*, London Math. Soc. Monogr., vol. 3, Academic Press, London-New York.

D. Quillen (1970), *On the homology of commutative rings*, Applications of Categorical Algebra, Proc. Sympos. Pure Math., vol. 17, Amer. Math. Soc., Providence, RI, pp. 65–87.

U. Shukla (1961), *Cohomologie des algébres associatives*, Ann. Sci. École Norm. Sup. (3) **78**, 163–209.

M. Spivak (1965), *Calculus on manifolds. A modern approach to classical theorems of advanced calculus*, W. A. Benjamin, Inc., New York-Amsterdam.

N. Yoneda (1954), On the homology theory of modules. J. Fac. Sci. Univ. Tokyo Sect. I **7**, 193–227.

——— (1960), *On* Ext *and exact sequences*, J. Fac. Sci. Univ. Tokyo Sect. I **8**, 507–526.

Index

Titles in This Series

17 **Michael Barr,** Acyclic models, 2002

16 **Joel Feldman, Horst Knörrer, and Eugene Trubowitz,** Fermionic functional integrals and the renormalization group, 2002

15 **José I. Burgos Gil,** The regulators of Beilinson and Borel, 2002

14 **Eyal Z. Goren,** Lectures on Hilbert modular varieties and modular forms, 2002

13 **Michael Baake and Robert V. Moody, Editors,** Directions in mathematical quasicrystals, 2000

12 **Masayoshi Miyanishi,** Open algebraic surfaces, 2001

11 **Spencer J. Bloch,** Higher regulators, algebraic K-theory, and zeta functions of elliptic curves, 2000

10 **James D. Lewis,** A survey of the Hodge conjecture, Second Edition, 1999

9 **Yves Meyer,** Wavelets, vibrations and scaling, 1998

8 **Ioannis Karatzas,** Lectures on the mathematics of finance, 1996

7 **John Milton,** Dynamics of small neural populations, 1996

6 **Eugene B. Dynkin,** An introduction to branching measure-valued processes, 1994

5 **Andrew Bruckner,** Differentiation of real functions, 1994

4 **David Ruelle,** Dynamical zeta functions for piecewise monotone maps of the interval, 1994

3 **V. Kumar Murty,** Introduction to Abelian varieties, 1993

2 **M. Ya. Antimirov, A. A. Kolyshkin, and Rémi Vaillancourt,** Applied integral transforms, 1993

1 **D. V. Voiculescu, K. J. Dykema, and A. Nica,** Free random variables, 1992